TELEMEDICINE

TELEMEDICINE
Theory and Practice

Edited by

RASHID L. BASHSHUR, PH.D.
JAY H. SANDERS, M.D.
GARY W. SHANNON, PH.D.

Supported by The BellSouth Foundation

CHARLES C THOMAS · PUBLISHER, LTD.
Springfield · Illinois · U.S.A.

Published and Distributed Throughout the World by

CHARLES C THOMAS • PUBLISHER, LTD.
2600 South First Street
Springfield, Illinois 62794-9265

© *1997 by* CHARLES C THOMAS • PUBLISHER, LTD.
ISBN 0-398-06731-7 (cloth)
ISBN 0-398-06732-5 (paper)

Library of Congress Catalog Card Number: 96-36746

With THOMAS BOOKS *careful attention is given to all details of manufacturing
and design. It is the Publisher's desire to present books that are satisfactory as to
their physical qualities and artistic possibilities and appropriate for their particular
use.* THOMAS BOOKS *will be true to those laws of quality that assure a good
name and good will.*

Printed in the United States of America
SC-R-3

Library of Congress Cataloging-in-Publication Data

Telemedicine : theory and practice / edited by Rashid L. Bashshur, Jay
H. Sanders, Gary W. Shannon.
    p.   cm.
    "Supported by the BellSouth Foundation."
    Includes bibliographical references and index.
    ISBN 0-398-06731-7 (cloth). — ISBN 0-398-06732-5 (paper)
    1. Telecommunication in medicine.   2. Telecommunication in
medicine—United States.   I. Bashshur, Rashid, 1933–   .
II. Sanders, Jay H.   III. Shannon, Gary William.   IV. BellSouth
Foundation.
    [DNLM: 1. Telemedicine—United States.   2. Delivery of Health
Care—methods—United States.   W 84AA1 T26 1997]
R119.9.T455   1997
362.1'028—dc21
DNLM/DLC
for Library of Congress   96-36746
                                                             CIP

CONTRIBUTORS

Ace Allen, M.D., is Associate Professor in the Department of Family Practice, Division of Oncology, University of Kansas Medical Center. He first became interested in telemedicine when he learned of the Kansas telemedicine pilot project in 1991. He was awarded a National Cancer Institute Career Development Award to set up a series of tele-oncology clinics between the University of Kansas Medical School in Kansas City, and rural Hays, Kansas. He is involved in the research of patient and physician satisfaction, efficacy, and cost-benefit considerations of the Kansas Telemedicine Project, including tele-home health care as well as tele-oncology. Dr. Allen is also active in several other telemedicine project evaluations. He is a founding member of the American Telemedicine Association, as well as of the Telemedicine Research Center, a nonprofit corporation based in Portland, Oregon, that is conducting and promoting telemedicine research throughout the United States. Dr. Allen is also the editor and publisher of *Telemedicine Today,* a bimonthly magazine published since 1992.

Lee Baer, Ph.D., is Director of the Harvard Telepsychiatry Project and Associate Professor of Psychology in the Department of Psychiatry at Harvard Medical School. He is on the clinical staff of the Department of Psychiatry at Massachusetts General Hospital, Boston, where he is also Associate Chief of Psychology, Director of Research in the Obsessive Compulsive Disorder Unit, and Director of the Biostatistics and Computer Unit. Dr. Baer is the author of more than 100 publications, primarily in psychiatry and in obsessive-compulsive disorder, his clinical speciality. He is the co-editor of a textbook on obsessive-compulsive disorder, and the author of a popular book describing the behavioral treatment of this disorder. Dr. Baer was awarded the 1994 Harvard Medical School "Responsible Conduct of Research Award." He serves on the editorial boards of several journals in psychiatry and telemedicine.

Rashid L. Bashshur, Ph.D., is Professor of Health Management and Policy at the University of Michigan School of Public Health. He has been a catalyst for the development and evaluation of telemedicine systems in the U.S. since the early 1970s. While at the

National Academy of Sciences (Institute of Medicine) from 1970 to 1972, he was consultant to the Office of Economic Opportunity on the use of telecommunications in supporting rural health programs. From 1973 to 1976, he was a member of the study section on health care applications at the RANN Program (Research Applied to National Need) of the National Science Foundation (NSF). He was funded by the NSF to evaluate a telemedicine program in rural Maine and to assess the status of telemedicine, the results of which were published in the book *Telemedicine: Explorations in the Use of Telecommunications in Health Care* (1975). Dr. Bashshur has published two books, several book chapters, and many journal articles on telemedicine. He has organized the National Consortium for Telemedicine Evaluation to conduct multistate, multisite research on telemedicine as it relates to issues of access to care, cost, and quality of care, and he is editor-in-chief of the *Telemedicine Journal.*

Achyut Bhattacharyya, M.D., is Associate Chairman of Pathology at the University of Arizona College of Medicine and Chairman of Pathology at Kino Hospital in Tucson. He co-founded the Arizona-International Telemedicine Network and has introduced and analyzed a novel case triage practice model for telepathology services. He subspecializes in gastrointestinal and transplantation pathology.

Brian Berman, M.D., Ph.D., is Co-Director of the joint Jackson Memorial Hospital/University of Miami Telemedicine Project, where he is testing telemedicine systems and equipment for accuracy in dermatologic diagnosis. As Director of Dermatological Services for the University of Miami managed health care system, he has initiated teledermatology consultations to render care to patients of primary care providers of health maintenance and managed care organizations. His clinical research focuses on cytokine regulation of fibroblast proliferation and biosynthetic activity, and seeks to develop novel therapies for conditions of excessive fibrosis such as scleroderma and keloid formation. He is a member of the Clinical Committee of the American Telemedicine Association, the American Association of Dermatology Telemedicine Interest Group, and the Medical Advisory Board of VTEL Corporation.

James E. Brick, M.D., is Professor of Medicine in the section of Rheumatology at the Robert C. Byrd Health Sciences Center of West Virginia University. He is also Chairman of the Board of Directors of the West Virginia Medical Corporation, and the author of numerous publications on laboratory and clinical aspects of rheumatic diseases. Dr. Brick is a fellow of the American College of Physicians and the American College of Rheumatology, and a member of the AOA. He is also a member of the Board of Directors of the American Telemedicine Association and of the Editorial

Board of the Telemedicine Journal, as well as co-founder and Medical Director of one of the oldest active telemedicine programs in the country.

Anne E. Burdick, M.D., M.P.H., has been Chief of the Department of Dermatology, Kaiser Permanente Medical Center in Martinez, California, since 1990. In 1993, she became Chief of the Jackson Memorial Hospital Dermatology Clinics. She is also the Medical Director of the University of Miami/Jackson Memorial Hospital Hansen's Disease Clinic, and Co-Director of the JMH/UM Telemedicine Program at the University of Miami School of Medicine in Miami, Florida. Dr. Burdick is a member of the Board of Directors of the American Telemedicine Association and the Board of Trustees for the Sulzberger Institute of the American Academy of Dermatology. Dr. Burdick is the recipient of the Dermatology Foundation Clinical Career Development Award, which is allowing her to study the accuracy of dermatologic diagnosis by telemedicine.

Joseph T. Coyle, M.D., is the Eben S. Draper Professor of Psychiatry and Neuroscience and Chairman of the Consolidated Department of Psychiatry at Harvard Medical School. Earlier, he served as Director of the Division of Child and Adolescent Psychiatry, Johns Hopkins School of Medicine, and was named the Distinguished Service Professor in 1985. Dr. Coyle's research interests include developmental neurobiology, mechanisms of neuronal vulnerability, and psychopharmacology. He has published over 400 scientific articles. He received the John Jacob Axel Award from the American Society for Pharmacology and Experimental Therapeutics, The Gold Medal Award from the Society for Biological Psychiatry, the Foundation Fund Research Award from the American Psychiatric Association, and the McAlpin Award from the National Mental Health Association. Dr. Coyle is a member of the Institute of Medicine of the National Academy of Sciences, a member of the American Academy of Arts and Sciences, a fellow of the American Psychiatric Association, a fellow of the American College of Psychiatry, served on the National Advisory Mental Health Council for the National Institute of Mental Health, and is a Past-President of the Society for Neuroscience.

Peter Cukor, Ph.D., recently retired from GTE Laboratories where he held the position of Director, Research Administration and Support. In this capacity, his primary responsibility involved the management of GTE's interactions with universities. He gained his doctorate in chemistry and at mid-career, retrained himself in the field of telecommunications in response to GTE's increasing interest in this area. Dr. Cukor serves as a consultant to the Harvard Telepsychiatry Project. His academic appointments include Lecturer at the Harvard Consolidated Department of Psychiatry, Adjunct Associate Professor at Worcester Polytechnic Institute, where he teaches on telecommunications policy, and Adjunct Associate Professor at Tufts University, where he teaches courses in telecommunications and media technologies. Dr. Cukor holds six U.S. patents and has over sixty publications, the most recent being in telemedicine and video conferencing related technology and human factors.

John R. Davis, M.D., is Professor of Pathology at the University of Arizona College of Medicine. He is Director of Surgical Pathology at University Medical Center in Tucson, and has served on the Quality Assurance Review Panel of the Arizona-International Telemedicine Network. He specializes in gynecological pathology and cytopathology.

Gary Doolittle, M.D., is Assistant Professor of Medicine, Division of Clinical Oncology at the University of Kansas Medical Center (KUMC). In 1995, he joined the Kansas Telemedicine Project and began providing tele-oncology consultative services for rural Kansas communities. In addition, a formalized, regularly scheduled tele-oncology practice was established between KUMC and Hays Medical Center, Kansas. To date, over 100 consults have been rendered for patients who suffer from hematologic disorders and a variety of malignancies. Dr. Doolittle is an active participant in studies addressing medical efficacy, cost analysis, and patient/physician satisfaction of telemedicine.

Jesse C. Edwards, Jr., M.S., is an Information Services Senior Consultant for the Sisters of Mercy Health System based in St. Louis, Missouri. He previously served as the Deputy Program Manager for the Medical Advanced Technology Management Office, the office responsible for day-to-day operations of the Department of Defense Telemedicine Testbed. He is also a Major in the U.S. Army Reserve. He has given numerous telemedicine related presentations at conferences, both in the U.S. and overseas.

Bruce A. Friedman, M.D., is Professor of Pathology and Director of Pathology Data Systems and of Ancillary Information Systems at the University of Michigan Medical School. His research interests in informatics involve the study of information systems in hospitals, with a focus on access and control of clinical databases, and on strategic planning. Recently, he has focused on the information management and networking requirements of integrated delivery systems and community health information systems. He has been active in the development of a "virtual" department of pathology informatics that involves both on-site and distance education for pathology house-officers and practicing pathologists. He has authored some 90 scientific articles, book chapters, abstracts, and book reviews.

Joseph N. Gitlin, D.P.H., is Associate Professor in the Radiology Department of the Johns Hopkins Medical Institutions, involved with the development and evaluation of new technology in medical imaging. He is a Fellow of the American College of Radiology and has contributed numerous articles on medical imaging. He is a founder and past Chairperson of the Radiology Information System Consortium (RISC) that developed the specifications and tested the first comprehensive radiology information system. Dr. Gitlin was also a founder of Vortech Data, Inc., now part of Kodak Health Imaging Systems, where he guided the development and testing of a state-of-the-art storage and retrieval system for medical images. During his career with the U.S. Public Health Service, he directed the National X–Ray Exposure Studies, evaluated early teleradiology systems, and designed the Digital Image Network projects.

Anna R. Graham, M.D., is Professor of Pathology at the University of Arizona College of Medicine. She has served as Vice Chief of Staff at University Medical Center and is a member of the Quality Assurance Review Panel of the Arizona-International Telemedicine Network. She specializes in orthopedic surgery and medical education.

Jim Grigsby, Ph.D., is Associate Professor in the Department of Medicine at the University of Colorado Health Sciences Center, and Senior Research Associate at the Center for Health Services and Health Policy Research. Dr. Grigsby has an active research program in cognitive neuroscience, and is a health services researcher, studying the effectiveness of telemedicine, outcomes of stroke rehabilitation, and the use of simulated neural networks and genetic algorithms in risk adjustment and outcome prediction.

James H. Hipkens, M.D., Ph.D., has been the Medical Director for the Georgia Department of Corrections since 1992. Telemedicine has been a favorite project for the State Government of Georgia, and Dr. Hipkens has always supported correctional medicine as an appropriate setting for developing a comprehensive telemedicine system. The Department of Corrections is well on the way to realizing such a system in practice.

Shaun B. Jones, M.D., is a practicing clinical and research otolaryngologist, head and neck surgeon on active duty in the United States Navy, and a staff surgeon at the National Navy Medical Center and the Uniformed Services University of the Health Sciences. Dr. Jones is a Diplomate of the American Board of Otolaryngology, a Fellow of the American Academy of Otorhinolaryngology—Head and Neck Surgery, and an FDA Consultant, Division of Ear, Nose and Throat Device Evaluation. His extensive military operational experience includes undersea medicine, surface warfare and special operations. As a member of the Advanced Biomedical Technology Program at the Advanced Research Projects Agency (ARPA), he is pursuing the application of advanced technologies to medicine and surgery.

Will Mitchell, Ph.D., is Associate Professor of Corporate Strategy and International Business at the University of Michigan Business School, where he has been a faculty member since 1988. His research focuses on business entry and survival in new markets, including business strategy in environments of rapid technological change. He has written extensively about business strategy in several medical sector markets, medical information systems, diagnostic imaging devices, implantable devices, and pharmaceuticals. His current research in the medical sector addresses the role of established firms as the sources of major innovations.

Camille A. Motta is a Ph.D. candidate at George Washington University, School of Business and Public Management, in the field of Information and Decision Systems. She also holds an M.S. in Library and Information Science and has been an information professional for twenty years. Her dissertation work focuses on an evaluation of an army/navy telemedicine project involving a link between a neonatal intensive care unit located in a suburban navy hospital (Bethesda, MD) and the pediatric specialists located in an urban army hospital (D.C.).

Jay H. Sanders, M.D., F.A.C.P., is Professor of Medicine and Surgery, Eminent Scholar in Telemedicine at the Medical College of Georgia, and Senior Research Scientist at the Georgia Institute of Technology, as well as Senior Advisor to NASA on Telemedicine. He earned his medical degree from Harvard Medical School and did his residency training at the Massachusetts General Hospital in Boston. He was Professor of Medicine and Chief of Medicine at Jackson Memorial Hospital at the University of Miami School of Medicine. He has spent the majority of his professional career in teaching, health care research, and for nearly three decades, in the development of telemedicine. Dr. Sanders designed the telemedicine system at the Medical College of Georgia and is overseeing the implementation of a statewide telemedicine system that includes rural hospitals, public health facilities, correctional institutions, and ambulatory health centers. He is the president of the American Telemedicine Association and Senior Editor of the *Telemedicine Journal.*

Richard M. Satava, M.D., FACS, is Clinical Associate Professor of Surgery, Uniformed Services University of the Health Sciences, and a Colonel in the U.S. Army Medical Corps. He works as a general surgeon at the Walter Reed Army Medical Center and is engaged in research in surgery as a Special Assistant in Advanced Technologies at the US Army Medical Research and Materiel Command, and in advanced technology at the Advanced Research Projects Agency (ARPA). He serves on the White House Office of Science and Technology Policy (OSTP) Committee on Health, Food and Safety, is a member of the Emerging Technologies, Resident Education, and Information committees of the American College of Surgeons (ACS), is President of the Society of American Gastrointestinal Endoscopic Surgeons (SAGES), and is on the editorial board of a number of surgical and scientific journals. He has contributed over 100 publications in diverse areas of surgical education, research, and advanced technology, including telemedicine, telepresence surgery, video and 3-D imaging, and virtual reality surgical simulation.

John R. Searle holds a Ph.D. in biomedical engineering. He joined the faculty of the Medical College of Georgia in 1977 in the department of Biomedical Engineering and has served as its director since 1986. He has developed unique research instrumentation for use in the biomedical sciences, dentistry, and medicine. In 1993 he assumed responsibilities as the Technical Director of the Telemedicine Center. In collaboration with the Georgia Department of Administrative Services, he developed the technical specifications for the telemedicine portion of the Georgia Statewide Academic and Medical System, including a set of optical adaptors to permit coupling a miniature video camera to an ophthalmoscope and an otoscope and other devices. He is currently collaborating with researchers at the Georgia Institute of Technology on the design and development of personal computer-based telemedicine systems suitable for deployment in homes and offices.

Gary W. Shannon, Ph.D., is a professor of medical geography in the Department of Geography at the University of Kentucky. He has been researching and writing extensively on accessibility to medical care over the past 25 years. He has co-authored books such as *Health Care Delivery: Spatial Perspectives, The Geography of AIDS* and *Disease and Medical Care in the United States: A Medical Atlas of the 20th Century.* He is currently involved in the evaluation of several state-wide telemedicine systems.

Ronald S. Weinstein, M.D., is Professor and Head of the Department of Pathology at the University of Arizona College of Medicine. A pioneer in the field of telepathology, he designed and tested the first dynamic-robotic telepathology system while working in Chicago. His clinical research team also carried out the first detailed human performance studies that validated the use of video microscopy for rendering pathology diagnoses. His concepts of dynamic-robotic microscopy are being implemented in many countries. After moving to Tucson in 1990, Dr. Weinstein co-founded the Arizona-International Telemedicine Network in 1993 and was appointed Director of the Arizona Telemedicine Program and Arizona Rural Telemedicine Network. He has published over 20 articles on various aspects of telepathology and has been granted U.S. patents for telepathology diagnostic networks. He is Past-President of the United States and Canadian Academy of Pathology. He is an experimental pathologist with an interest in cancer biology and urinary bladder cancer.

To
Kenneth T. Bird
True telemedicine pioneer

PREFACE

The first edited volume on telemedicine was published in 1975 when the promise of telemedicine loomed just over the horizon. But the promise dimmed a short time afterwards, and the field nearly faded from memory. Telemedicine was a novel idea then, and few people took notice. Many viewed it with skepticism. Some dismissed it as a mismatch of a high technology solution to simple problems in primary health care. Others were concerned about excessive intrusion of technology into personal relationships, particularly the doctor-patient relationship. Two decades later, we are witnessing the strong revival of telemedicine. Now, the field is widely embraced with high expectations in both public and private sectors. Telemedicine is viewed not simply as a technological augmentation to health care delivery, but more importantly as an innovative system that provides an effective and versatile solution to many of the intransigent problems in health care delivery in this country and elsewhere. Much of this excitement is generated by the enormous expansion of information technology into everyday life, in education, business, commerce, industry, communication, and entertainment. Indeed, health care is a small bit player in the broader development of information technology and its ubiquitous use. Nonetheless, the excitement about telemedicine specifically derives from intuitive assumptions and reasoned speculation about its laudatory effects on access, quality, and cost of health care.

This book was borne out of a desire to present a comprehensive assessment and a scholarly perspective on telemedicine in the 1990s. Like its predecessor, it is intended to represent the state of the art in telemedicine as viewed by clinicians providing telemedicine services, by biomedical and health services researchers engaged in its evaluation, and by visionary scientists charting its future.

As before, we solicited chapters from leading experts whose work should be recognized and whose knowledge should be shared. We hope we have succeeded in recruiting the leading talent in the U.S. in this particular field.

We undertook the daunting task of producing the book as if it were written by one person. This required homogenizing not only writing styles but also basic approach and content, while respecting the intellectual autonomy of each author and maintaining the integrity of each piece. Our colleagues who wrote the individual chapters were very tolerant of our meddling and changes to their writing style. Without exception, they accepted graciously what seemed to be endless and excessive demands for clarification, editing, and corrections. Our burden as editors was eased considerably by their helpful understanding and cheerful support. When the manuscript was completed, we imposed yet another quality filter by asking our colleague Rudolph Schmerl to help with the copy editing. Dr. Schmerl demonstrated superior skill as a copy editor, and he has earned our endless gratitude for lending his talent, working under pressure, and refining the entire manuscript.

We wish to acknowledge the support and assistance of organizations and individuals who were instrumental in the completion of this book. The BellSouth Foundation served as a catalyst for the project. Their initial funding and ongoing support have been invaluable. Wendy Best, grants manager of the BellSouth Foundation, encouraged the effort as an effective educational tool in telemedicine. The research staff at the National Consortium for Telemedicine Evaluation, University of Michigan, provided valuable assistance throughout the project. Suzanne Worsham worked hard and long preparing numerous drafts of the manuscript, communicating with the authors, and keeping the project on track. Andrzej Kulczycki and Ellen Johnson reviewed and provided valuable editorial changes. Andrew Cameron, Jennifer Schoff, and Rae Reynolds helped with the references. Maryellen Kouba helped with correspondence and clerical chores. While we acknowledge all these people for their valuable assistance and painstaking efforts, together with our colleagues who wrote the individual chapters, we claim full responsibility for any imperfections left in the book.

Finally, we wish to express our heartfelt gratitude to our wives, Naziha, Risa, and Susan, for their tolerance and understanding of our preoccupation with telemedicine, and their helpful support during this labor of love.

CONTENTS

TELEMEDICINE

SECTION I
THE CONTEXT OF TELEMEDICINE

Chapter 1

TELEMEDICINE AND THE HEALTH CARE SYSTEM

RASHID L. BASHSHUR

Introduction

The history of the health care system in the United States over the last four or five decades reveals a preoccupation with three tenacious problems despite various attempts to resolve them:

(1) uneven geographic distribution of health care resources throughout the country, including health care facilities and health manpower;

(2) inadequate access to health care on the part of certain segments of the population, including the under-privileged, isolated, and confined;

(3) unabating rise in the cost of care, including the costs borne by both public and private payers.

All three problems have limited the potential for improving the health status and quality of life for significant segments of American society whether because of their economic status, geographic location, or both. Moreover, and perhaps adding to their significance, these problems are not unique to American society. In varying degrees and intensity, they represent nearly universal phenomena in highly developed as well as lesser developed countries.

Telemedicine has been proposed as a multifaceted response to address all three problems simultaneously through innovative information technologies that expand the productive capability and extend the distributive efficiency of the health care system. The innovations include various combinations of telecommunications, telemetry, and computer technology; integrated organizational structures; health-manpower mixes; as well as a range of diagnostic, clinical, and educational applications.

This chapter assesses the potential role of telemedicine in the health care system once it reaches a reasonable level of maturity and a steady

5

state of operation. The assessment will incorporate an identification of the role of telemedicine in the health care system and descriptions of benefits to be accrued as well as problems to be encountered. The chapter therefore discusses telemedicine in the broader context of (1) previous and current attempts to cope with price inflation in health care; (2) the defining dimensions of quality; and (3) the geographic "maldistribution" of health care resources which forms the basis of inequities in accessibility to health care. The potential role of telemedicine in ameliorating problems pertaining to cost, quality, and accessibility is explained in the light of previous structural and procedural attempts to deal with them. Without understanding the historical and current context within which telemedicine is developing, its promise and limitations may not be fully understood, and more importantly, its practical contributions may not be realized. These considerations are all the more significant because telemedicine technologies and their applications represent only an early, still rudimentary, phase in a fast-moving field. The capability of the information technology supporting telemedicine is expanding rapidly while the price of many of its components is declining. The combination of improving quality and decreasing price is unprecedented and unparalleled in health care technology, thereby assuring continued interest in the field.

An Overview of Telemedicine

Despite agreement on basic principles, what is called telemedicine today is not a monolithic entity or a single unified and well-defined system. Whereas telemedical practice is proliferating in the United States and world-wide, a definition of the concept has yet to be universally accepted, and clear distinctions have yet to be made between categorical clinical or diagnostic services rendered remotely through telecommunications and comprehensive telemedicine or telehealth systems. Because of its newness and the need to identify its parameters, several authors (Bird, 1971; Bashshur, 1975; more recently, Perednia and Allen, 1994) have offered definitions of the concept of telemedicine. Invariably, all definitions repeat the theme of reliance on telecommunications in the delivery of health care, but, with notable exceptions, pay relatively little or no attention to its systemic qualities, or view telemedicine as a new integrated system of care (Bashshur, 1995). Ironically, only when viewed as a complete and integrated network will telemedicine's unique distributive capabilities and integrative functions be maximized. The distributive capabilities derive from the use of telecommunications and computer technology as an effective substitute for face-to-face encounters; and its integrative functions are obtained from its efficient networking capabilities and information exchange. Hence, optimally, telemedicine may be viewed as an innovative **system of care** that can provide a variety of health and educational services to its clients unhindered by space and time. As a system of care, it entails a new organizational form, enhancing our rationality in triaging patients to appropriate sources of care and in efficient clinical decision making. Indeed, in view of its complexity, telemedicine should be viewed as an "innovation bundle" (Rogers and Shoemaker, 1971; Koontz, 1976). As is true of other innovation bundles, it consists of several components, not all new. But this innovation bundle promises to serve the threefold goal of improving access to care while limiting the rate of increases in cost and maintaining quality (Bashshur, 1995).

Ironically, this generation of telemedicine faces the same dilemma confronting its predecessor some two decades ago that led, in large part, to its earlier demise. Both policymakers and the health care community generally expect to see hard evidence of clinical effectiveness, distributive efficiency, and cost savings long before telemedicine systems are designed optimally and used sufficiently, and hence, long before scientifically valid evaluations can be completed. The field has yet to reach a sufficient stage of maturity and stable operation to demonstrate its value and to provide reliable and valid data for unabridged meaningful analysis. Indeed, the lack of a clear definition of the field and the contrived planning and incomplete implementation of some projects threaten to compromise both our ability to deploy optimal systems and our determination of telemedicine's potential role in the health care system (more specifically, to ascertain its ability to achieve its purported benefits). Substantial amounts of public funds were invested in building the technological infrastructure for telemedicine, intensifying the search for quick fixes in the current financially constrained and politically charged environment. Critical decisions about the future of telemedicine may thus be made hastily with incomplete and inadequate information. The result may be a premature political decision to abandon the enterprise, unless it manages to build a viable market on its own without interim subsidies during the developmental stage.

To date, all attempts to deal with the problems of access, cost, and quality in the health care system have been either limited endeavors targeted at one or the other, or large-scale and sweeping health care reforms directed at all three simultaneously. None has proven to be fully successful, as will be explained later. The interrelatedness of these issues and the complexity of trying to deal with them simultaneously was vividly illustrated in 1994, in the unsuccessful efforts by the Clinton administration to implement health care reform on a large scale. The complexity of the task at hand was mired in the political process, and the result was inaction, even though the basic problems that prompted the attempts at reform remained unsolved. But, because of the substantial amount of public responsibility for health care (about 40 percent of all health care expenditures are assumed by the public sector), the issue will remain on the public agenda for a long time, and attempts to control health care expenditures will be repeated. Also, typically, attempts to contain costs result in increases in the uninsured and underinsured, thereby further restricting access to health care for more people.

What might telemedicine do to ameliorate these problems? What does telemedicine "bring to the table" to complement, supplement, or replace previous and current strategies in the design of optimal systems of care? More specifically, is there a proper niche for telemedicine in the health care system or is the whole future of health care, as some have suggested, in telemedicine? If it has a proper "niche," is it, for instance, as much of the current discourse suggests, appropriate only for serving remote, isolated, or confined populations, or is it equally applicable to all service populations in all kinds of areas? If, on the other hand, telemedicine is the new frontier in the information age for all of medicine, then how is it going to reshape the delivery of health care? And ultimately, will it transform the provision of health services and become an indistinguishable part of it in this country and elsewhere?

The capabilities of the technologic infrastructure serving telemedicine are expanding very rapidly. Its widespread application in entertainment, industry, commerce, education, and personal communication will spur further advancement in technology, making it more ubiquitous and less avoidable in normal everyday life. Indeed, the movement toward faster, more efficient, and less costly communication is irreversible in all walks of life. Hence, the future of telemedicine, at least insofar as the development and diffusion of information technology are concerned, is likely to be propelled by forces mostly outside the health sector. While this would help to assure the use of information technology in health care, it might also create financial and logistical pressures to adapt and conform to standards outside of health care. Telemedicine would then lose its independence and, perhaps, the ability to design systems optimally suitable for clinical or health practice.

The evolution and current status of telemedicine are matters beyond the scope of this chapter, but are discussed collectively in this volume and elsewhere (Grigsby et al., 1995; Bashshur, 1976). Pertinent here is that a large number of public and private telemedicine projects are in place throughout most of the United States and the rest of the world. Several more are in the planning stages. Over the next few years, while their scope and scale will vary, there will be at least one operating telemedicine network or system in every state in the United States, all of Western Europe, most of the Near and Far East, and in several African countries and Australia. The ultimate viability of these systems will derive largely from their contributions to an improved health care system through greater distributive efficiency and integrative capacity despite increasing economic constraints. Such systems have to be acceptable to their users, including both providers and clients. Most of the evidence to date, from published and unpublished sources, suggests that clients have readily accepted the use of telemedicine for their care. The primary concern of clients of health care is to get well. If telemedicine can help, they will embrace it. Providers, on the other hand, have demonstrated interest, but remain somewhat skeptical and still concerned about its logistical, financial, and ethical ramifications for their practice (Sanders and Bashshur, 1995; Bashshur, Puskin, and Silva, 1995).

Definitions

*Broadly, **telemedicine** involves the use of modern information technology, especially two-way interactive audio/video telecommunications, computers, and telemetry, to deliver health services to remote patients and to facilitate information exchange between primary care physicians and specialists at some distances from each other. A **telemedicine system** is an integrated, typically regional, health care network offering comprehensive health services to a defined population through the use of telecommunications and computer technology.* The difference between the two concepts (telemedicine and telemedicine system) is important and points to the distributive and integrative capability of information technology. This capability may be fully realized only within the context of an organized network.

Since the health care system is moving toward managed care organizations and corporate structures, it is only a matter of time before these organizations recognize the massive capability of information technology in establishing integrated systems of care and deriving attendant benefits from such integration.

Based on a broad definition of the concept, one might log cally argue that telemedicine began with the first exchange of information between physicians, or the first advice g ven to a patient by a physician, over the wired telephone or wireless radio. In reality, however, the practice of telemedicine, as currently conceived, began in the late 1960s with a two-way, closed-circuit, microwave television psychiatric consultation between Nebraska Psychiatric Institute and a state mental health hospital some distance away (Wittson and Benschoter, 1972). This was followed shortly by a demonstration project linking a poly-clinic at Logan International Airport in Boston and Massachusetts General Hospital via real time interactive audio/video system (Bird, 1971). In the 1960s, as well, the National Aeronautics and Space Administration (NASA) became directly involved in telemedicine as a result of its concern with the effects of zero gravity on the health and physical functioning of astronauts and the need to monitor their vital signs during manned space flight. Subsequently, NASA supported the deployment of a full-service, comprehensive, terrestrial testbed telemedicine system on the then Papago, now Tohono Odham, reservation in Arizona, known as STARPAHC (Space Technology Applied to Rural Papago Health Care) (Bashshur, 1975)[1].

In hindsight, many consider these early attempts, or the "first generation" of telemedicine, unsuccessful because the effort was not sustained (although experimentation continued in few locations), and the innovation failed to diffuse and become an accepted mode of health care delivery. In view of the conditions prevailing at the time and the short-sightedness of the funding agencies, this judgment is partial and does not tell the full story (Bashshur, 1983). First, the information highway was in a nascent stage of construction, both at the state and national levels. Significantly, information transmission capability was, by today's standards, relatively primitive and only a prescient few could predict the

---

[1]Parenthetically, the native residents of the Sonora Desert in Arizona (formerly called Papago, or the beans people) have reclaimed their traditional name, Tohono Odham, or the desert people. In the native culture, desert denotes openness, brightness, and light.

extent and capacity of future developments, especially in digital compression and computer miniaturization and portability (Vivian, 1975; Bashshur and Lovett, 1977). For instance, even dedicated telephone lines could transmit only slow-scan analog images, and the quality of audio transmission was poor with coaxial cables. Although data compression was developed earlier, it was not incorporated into the first generation of telemedicine, and it was too late to redeem the nascent programs at the time. Now, large segments of the nation are connected through fiber optic or coaxial cable, and digital compression expanded their capacity. Moreover, wireless transmission will soon provide comprehensive grids to cover most of the Western Hemisphere.

Second, the experimenters, especially the medical providers, were necessarily inexperienced in the efficient use of the technology, and they were generally faced with perplexing questions regarding the capability and problems presented by the new technology. In fact, as with most innovations, there was a general reluctance to embrace telemedicine because it was mostly unknown (Fuchs, 1979; Grigsby, 1995). Each step forward was a step into an uncertain future with a technology limited in range and sophistication.

Third, the funding agencies decided to "pull the plug" prematurely, before any of the demonstration projects could reach maturity or provide a valid empirical basis for a rational policy in this area. Nonetheless, the first generation of telemedicine provided ample evidence of the feasibility of remote consultation, the clinical effectiveness of several clinical functions, training, and education. These findings were ignored at the time because of a prevailing fear that sophisticated technology would only increase the cost of care. Everywhere else, technology has been a major culprit in the escalation of health care costs. In balance, the first generation of telemedicine projects (and their untimely demise) demonstrated the need for more consideration and rigorous evaluation of the multitude of technological, social, cultural, and organizational dimensions accompanying the introduction of telemedicine.

Today, technologies applied to the delivery of health services and exchange of health information exist over a wide range. Some emphasize telephone-related technologies individually and in various combinations, including the facsimile machine (fax), speakerphone, computer-based image archiving and transmission, audio-teleconferencing, and slow-scan video. Other technologies emphasize electronic communication for such services as mail and conferences; bulletin boards and medical news services; and on-line library catalogs and ordering. Finally, there are projects using broad-band communications systems or combinations of technologies with interactive video and medical telemetry via a land-based T1 link; satellite-based one-way video and two-way audio; interactive video via an optical fiber network and satellite; T1 compressed interactive video; and microwave transmitted video (see Chapter 4, "The Technology of Telemedicine," by Searle).

Despite the over 30-year history of telemedicine, its progress and diffusion may be characterized as slow and limited. The constellation of projects beginning in the late 1980s and representing the "second generation" of telemedicine must still be considered as an early stage of development. This is because the technology is changing very rapidly. Moreover, its contributions to containing costs, improving quality of care, and increasing accessibility to care are yet to be substantiated with empirical data (Bashshur, 1995). It is, therefore, pertinent to consider the barriers to the diffusion of telemedicine and ways of dealing with them.

The Diffusion of Telemedicine

Many technological innovations are either not accepted or not successfully implemented because of resistance resulting from incompatibilities, whether real or perceived, between the nature of the innovation and extant interests, perspectives, and resources. When technological innovations are not accepted or implemented properly, generally the failure may be traced to a poor fit between the nature of the innovation and the vested interests, resources, and expectations of its major gatekeepers. Depending on the source of the problem, and the inherent desirability of the innovation, such resistance may be decreased by one of three strategies: (1) a **marketing** strategy aimed at presenting the innovation in more acceptable form, or a repackaging of the innovation for increased receptivity; (2) an **educational** strategy aimed at reshaping of attitudes and perceptions of major gatekeepers to increase their awareness, knowledge, and appreciation of the innovation; or (3) a **structural** strategy involving the realignment of extant interests and resources to create a better fit between the two (Kimberly, 1981). Also, as Scott (1990) suggested, innovation diffusion within an organization may be viewed either as a process or a discrete event, and it can be investigated on the basis of either the inherent or attributed characteristics of the innovation or the environmental context in which it is introduced. When viewed as a process, though more realistic, the diffusion of telemedicine becomes much more complex than when it is considered as a single event. Subsequently, the different stages of the adoption process have to be explained sequentially.

Based on an exhaustive review of the literature on the adoption of innovations by organizations, Scott identified several factors that influence the diffusion process:

- Initial or continuing cost of the innovation
- Returns to investment
- Improved efficiency
- Payoff (effect on quality of services/products)
- Risk or uncertainty
- Testability (capability of demonstration in a pilot project)
- Social approval (amount of increased public recognition and acceptance)
- Comparability with existing procedures, consistency with prevailing values
- Pervasiveness (extent to which innovation requires other changes in the organization)
- Divisibility (feasibility of implementing part of the program on a trial basis)
- Reversibility (potential for change if indicated)
- Communicability (ease of explaining the changes)
- Complexity (the extent to which new and difficult skills are required to use the innovation)
- Clarity of results
- Social desirability
- Point of origin (internal or external to the organization)

Kaluzny and Veney (1973) found three variables to be critical determinants in the adoption of complex innovations in health care organizations: (1) the expected payoff from the innovation; (2) the rate of recovery of investment; and (3) social approval of the innovation. In brief, the successful diffusion of telemedicine as a system of care will depend upon the ability of planners, developers, and policymakers to identify and decrease the sources of resistance to a socially desirable innovation, and subsequently create a good fit between it and the external environment. The rate at which the diffusion will occur will depend on the expected payoff, the rate at which the investment will be covered, and the expected social benefits to be derived from it. The potential benefits of telemedicine may be grouped into three sets: those pertaining to accessibility to health care, the cost of care, and the quality of care.

The Accessibility Issue

Before describing the nature of the expected benefits of telemedicine in access to care, a discussion of the basic concept of accessibility and its manifestations in the U.S. health care system is in order. Accessibility has geographic, financial, social, and psychological components. Hence, it is a multifaceted concept, and is difficult to define precisely and operationally because it does not simply reflect, for example, the characteristics of health resources such as their location and their price but also people's ability to use them. Despite this complexity, most government health programs have had as at least one of their goals improving accessibility to care for underprivileged groups, thereby assuring equity of access to health care for all Americans. This policy is based on the assumption that health care is necessary to protect health, and hence no American should be denied access to it.

In essence, accessibility of health services reflects the "fit" between health care resources (including hospitals, clinics, doctors' offices) and the health care needs of the people they serve (Donabedian, 1973). Thus, accessibility is high in a given area or population when the needs of the people who live there are well met through the available resources, and low when needs are poorly met. Moreover, the concept incorporates elements of **availability** or the supply and distribution of resources; **affordability** or the ability of clients to pay for the care they need; and **acceptability** or the extent to which the users find the care they need suitable (Penchansky and Thomas, 1981). Of these three elements, the strongest case for telemedicine is its contribution to the availability of health resources, more specifically the virtual redistribution of these resources in time and place.

One of the most persistent and seemingly insoluble problems facing the delivery of quality medical care in the twentieth century is the so-called "maldistribution" of physicians, hospitals, and other health care resources in relation to distribution of need in the population. More specifically, there has been a poor fit between the distribution of need in the population and the distribution of health care resources (Wennberg,

1995). Specialists tend to locate nearer to other specialists rather than closer to their patients' residences. Interestingly, this is not a recent problem. For instance, a survey conducted in 1848 reported that the "less educated" as well as the "uneducated" physicians were practicing in the more remote sections of each state (Johnson, 1848). However, ironically, at the same time rural physicians were in abundance, and there was concern about a "general overcrowding" of physicians in rural areas. Even the smallest towns were oversupplied with them. As a result of the "continuing overproduction of physicians" which was based, in part, on the absence of educational standards, it was speculated that physicians might be forced into every "hamlet of five and twenty souls."

By the mid-twentieth century, however, the increasing number of hospitals in cities; the establishment of educational, certification, and practice requirements and standards; the increase of physician specialization; and the growth of hospital-based practice led to a severe geographic imbalance in the availability of medical care. Rural residents found themselves increasingly removed from tertiary as well as primary and secondary medical care sources. Later, notwithstanding federal programs such as the "Hill-Burton" Hospital Survey and Construction Act of 1946 (P. L. 79-725), and, later, the Emergency Health Personnel Act of 1970 (P. L. 91-623), better known as the National Health Service Corps, residents of rural areas continue to be at a substantial disadvantage in the availability of a comprehensive range of medical services close to home.

Although the number of physicians practicing in rural areas is increasing overall, residents of rural areas are estimated to be more than twice as likely than the national average to face shortages of primary care physicians (National Rural Electric Cooperative, 1994). Of course, the situation is far more critical when specialty care is considered. Since medical specialists and advanced diagnostic services are particularly scarce in rural areas, residents in need of these services often must travel long distances, perhaps missing time from work and disrupting family life, to seek care from specialists located in distant medical centers. As a result, frequently, many may delay seeking care or may do without the care they need.

Rural residents are not the only populations encountering greater than average difficulty in obtaining a comprehensive range of medical services. Other populations isolated from comprehensive medical care include a large and growing prison population which now numbers over one million persons, the elderly in long-term care facilities, and the urban poor and the homeless who are increasingly concentrated in many large central cities. The health care needs of prisoners cannot be ignored because inmates, uniquely, have a constitutional right to health care (Brecht, 1996). The potential of telemedicine to address the needs of these populations is considerable. Practitioners in inner cities, prisons, and other institutions can now conduct consultations in "real time" with specialists in distant medical centers using interactive audio/video links. This results in a marked improvement in access to medical care for people who have been at a severe disadvantage by virtue of their economic or social status or their geographic location.

In view of its distributive capability, now referred to as "telepresence," telemedicine has the potential to resolve important questions about access, questions relevant to clients and providers of care in various contexts. Its impact can be both direct and indirect. The most obvious and direct impact of telemedicine on accessibility is on the remote, isolated, and confined clients. When needing care, these individuals can remain in their local community, home, or institution and, through telemedicine, obtain consultations with specialists who may be located hundreds of miles away—or in the next block. In addition, the sophisticated technology available at tertiary care centers enables clinical evaluations "at a distance" often in the presence of the primary care provider. And, optimally, travel to distant specialty clinics would occur only when the care is not available locally or immediately, or when the specialist requires evaluation not possible via telemedicine.

Indirectly, an improved level of accessibility derived from the establishment of telemedicine can contribute to the recruitment and retention of physicians in rural areas. Whereas many variables contribute to a physician's decision to enter practice or to remain in practice in any location, and their interaction is complex, one of the most common factors in the decision to abandon clinical practice in rural areas is lack of professional interaction with colleagues or the absence of a professional support structure. With telemedicine, professional isolation can be reduced substantially since primary care physicians in remote areas will communicate and interact directly with specialists within their network to the extent they deem appropriate.

By virtue of telemedicine, a larger percentage of remote patients will be able to obtain care in their local community. When patients receive their care locally, local primary physicians or other health providers will retain responsibility for continuity of care for a larger number of their patients. Not unimportantly, this will contribute to the income of local providers, and it should enhance the economic viability of rural practice for current and prospective physicians.

In brief, the reduced professional isolation of primary physicians, the ready availability of educational opportunities, the increased levels of responsibility for a continuing care process, and the increased income resulting from telemedicine should make remote locations more desirable places to practice. Both the recruitment and retention of primary care physicians in remote areas should improve. Similar benefits should accrue to local hospitals through higher occupancy and indirectly to the local economies through a larger financial base.

At this general level of consideration, the potential contributions of telemedicine to improving the geographic availability of both specialty and primary care appear substantial. There is little doubt that, by "bridging" the distance barrier, telemedicine has an important role to play in the future of health care delivery wherever physician and patient are not in close physical proximity. But it is important to consider several other perspectives and dimensions of accessibility.

Most telemedicine projects to date have been aimed at improving the patient's access to sources of care. It is important to remember, however, that the improvement in accessibility brought about by telemedicine is, at least partially, "virtual" rather than wholly actual. Questions may be raised pertaining to the possible creation of another tier, or level, of privilege in the delivery system with the implementation of telemedicine. One level is that in which patients have direct and "real" access to physicians and diagnostic services for the purposes of clinical evaluation and treatment. Emphasizing the use of telemedicine creates a level at which patients must be satisfied with indirect and "virtual" access to specialists via the information highway. However, actual experience with these systems to date does not permit a firm assessment of the validity of this concern. If anything, telemedicine has been valued for enhancing rather than compromising the capability of rural practices.

For practical purposes of clinical evaluation, telemedicine actually may reduce the specialists' "accessibility" to patients who would otherwise be seen in person in the examination room. As explained in ensuing chapters dealing with the clinical applications of telemedicine, there are important clinical issues and technical requirements in each application that must be met for acceptable clinical operations in telemedicine. The most significant of these issues is the type and level of technology necessary for examining, diagnosing, and treating patients via telemedicine. In balance, the question becomes one of whether the improvement in geographic and temporal accessibility for both client and provider, albeit virtual, is achieved at the expense of reduced quality, and if so, whether the "trade-off" is clinically and socially acceptable. Of equal significance is the potential trade-off between cost and quality, but this issue will be discussed later in this chapter.

When fully implemented as a **complete system** of care, telemedicine entails a restructuring of health care. As with all structural changes, this may bring about both intended and unintended, as well as desirable and undesirable, consequences that call for new adjustments. What must be guarded against, therefore, is the potential conflict the value of the new system engenders. For instance, will remote practitioners increase demand for remote diagnostic services beyond those clinically necessary? If so, what will their incentives be? Will they have a lesser or greater incentive to develop their own skills and services in order to match them to the needs of the local population? Will physicians feel threatened by the potential loss of control or income from active participation in a larger network? On the other hand, it is also possible that telemedicine may bestow on the practitioner an added advantage over other practitioners without access to such systems so that those not practicing in telemedicine will be driven out of underserved areas.

Questions of this nature go to the core of the telemedicine movement and the evolution of its supportive technology. To the extent possible, they must be answered with rigorous scientific study. Reflecting the use of off-the-shelf equipment and certain low-cost technologies, some of the authors of subsequent chapters devoted to clinical applications of telemedicine indicate problems in the use of telemedicine, some of crucial significance to the technical quality of care. Thus, at this level, questions pertaining to a possible trade-off between gains in accessibility and losses in quality must be resolved. This requires complex accessibility equations which take into consideration the types of accessibility affected, specific technological configurations utilized, and clinical applications. Such equations should also include the perspectives of the primary care provider, the patient, and the specialist. Telemedicine may be introducing yet another dimension to the concept of accessibility, namely, virtual accessibility. Therefore, if, in fact, the medical care process is compromised via virtual rather than real accessibility, serious questions of equity, let alone quality of care, must be addressed.

Another important consideration is the accessibility of the primary care provider and specialist to telemedicine. To date, cost considerations have prohibited the diffusion of full-scale technology into the office of every primary, secondary, and tertiary care provider. Thus, telemedicine service has yet to become an ubiquitous resource, equally accessible to all providers. Generally, the remote teleconsultation site is located at a local hospital which has been selected in some manner, and is to serve a larger region comprising the service areas of several hospitals. In these situations, the hospital location of the telemedicine consultation site necessitates travel by the remote provider who may be a primary care physician, a nurse practitioner, or a physician assistant. Insofar as the physicians or other providers located at varying distances from the hospital, it follows that they will have varying degrees of access to telemedicine. Thus, travel distances may inhibit the use of telemedicine by some providers. Perhaps to a lesser degree, the same rationale applies to patients. At the same time, current developments in distributive technology will obviate the problem in the foreseeable future.

Acceptability was described earlier as one dimension of accessibility. Regardless of the geographic proximity created between patients and remote providers or between one provider and the other, unless and until telemedicine is accepted generally by providers and clients, there is little likelihood that the full potential and promise of telemedicine in addressing the problem of accessibility will be realized. From the perspective of the providers, both primary care and specialist, this involves, in part, "getting close to the machine." Thus, the increasing level of sophistication and complexity or the "functionality" of the technology must be balanced by bringing it closer to the provider in terms of its "user friendliness." The human-machine-system interface must be accommodated. Unless it is, it will be necessary to add still another person to the clinical encounter, namely, the telemedicine technician or onsite coordinator. Furthermore, providers and clients will be interacting more or less directly with a machine and only indirectly with the humans at various locations along the information highway. To improve acceptability, the engineer's assumptions and presumptions must be replaced by, or at least accommodate, the realities of the provider-patient interaction with the machines.

In addition to the human/machine adaptation, still other significant unresolved impediments exist to adoption of telemedicine by the provider. These include reimbursement, legal liability and litigation, interstate licensing and intrastate credentialing, turf protection, and financial bene-fit (for a detailed discussion of these issues, see Sanders and Bashshur, 1995). The dissemination of the technology may not be supportable on the basis of research limited to dealing with only one or another dimen-sion of accessibility affected by telemedicine or even a sole concern with accessibility without regard to issues of cost and quality.

In the long run, the most appropriate approach to the assessment of telemedicine must be based on its conception as an innovation bundle, involving various mixes of technology, organization, manpower, and clinical services. This bundle, which is not uniform in structure or form, is being introduced into, and is expected to bring change to a complex clinical, physical, cultural, social, and psychological landscape of medi-cal care. Whereas the promise and potential of telemedicine to alter the complex web of accessibility are considerable, determining the specific nature, direction, and extent of the alterations is another matter. The impact is not simple, uniform, or unidimensional.

Quality of Care

Quality is an abstract concept, more so than accessibility. Generically, it pertains to the nature, kind, or character of someone or something and the degree or grade of excellence possessed by the person or thing. Because of the difficulty of direct measurement, quality of health care is often assessed in terms of its effects on health. If these effects are deemed desirable, then quality is judged to be good; if undesirable, bad. In the health care field, Donabedian (1973) offered an all-inclusive framework for explaining and measuring quality on the basis of the **structural** attributes of care, which include qualifications and characteristics of personnel and institutions; **process** variables, which consist of diagnostic and therapeutic steps and activities involved in the delivery of health care; and **outcome** variables, which refer primarily to health status, as well as other kinds of effects on clients. Thus, while quality refers to both tangible and intangible aspects of health care delivery, its effects are both real and measurable. Moreover, quality of care may be measured with respect to individual medical services or sets of services rendered to an individual or group of patients, individual or group of providers, and health programs and facilities.

The dimensions of structure, process, and outcome are quite suitable for assessing the quality of care in telemedicine systems. Structural variables include characteristics of delivery systems and technical configuration and qualifications of providers and other personnel. Process variables include diagnostic and therapeutic processes, the way services are ordered and patients served, volume and intensity of care, as well as timeliness and continuity of care. The quality of process variables can be determined on the basis of their conformity with established norms and standards. Quality outcomes can be determined on the basis of the results of treatment on the patient's health status, quality of life, and satisfaction. While operationalization and subsequent measurement of structural factors are relatively straightforward, process is more complex, and outcome still more so. Outcome assessment poses serious questions of attribution since, typically, health outcomes are slow to emerge, and rarely are they solely attributable to medical interventions.

Further, with regard to the organization of quality assessment, it has been suggested that the quality of telemedicine can be evaluated either on a biomedical/bioengineering or health services basis (Bashshur, 1994). Biomedical evaluation focuses on clinical performance and tries to determine clinical efficacy, effectiveness, and safety of specific technologies. Concerns to be addressed include accuracy, precision, reliability and sensitivity, as well as potential risks and harmful side effects of specific technologies. Biomedical evaluation in telemedicine answers questions regarding the appropriateness of specific configurations and components in technology for case finding, diagnosis, treatment, follow-up, and prevention of disease.

Health services evaluation focuses on the effects of telemedicine, as a defined system of care, on cost, quality, and accessibility as well as its acceptance by the intended users, both providers and clients. Hence, health-services research picks up where biomedical research ends, and it answers questions pertinent to policy and decision making regarding the merit of telemedicine and justification for its support. Also, because of the pluralistic nature of the health care system, all these effects, including quality, must be considered from the perspectives of patients, providers, and society at large.

Quality of care assessment implies the use of criteria and standards, while specific variables can be measured quantitatively and/or qualitatively. As it pertains to telemedicine, quality assessment necessitates reliance on criteria—namely what aspects are indicative of quality and the standards of how these aspects are judged on a uniform universalistic basis. Further, quality can be assessed in terms of **technical** and **interpersonal** dimensions (Donabedian, 1973). The **technical** dimension of quality is concerned with all aspects of the clinical process (including case finding, diagnosis, treatment, and follow-up) and its outcome (including changes in health status and quality of life). The **interpersonal** dimension of quality is concerned with the human and interpersonal aspects of treatment, including the doctor-patient relationship, satisfaction, user acceptance, and confidentiality.

Perhaps the most direct contribution that information technology generally and telemedicine specifically can make in improving the quality of health care is to provide the clinician with better and timely information about the patient, his or her health problems, and the latest knowledge for diagnosis and treatment, typically at the point of care. Within this context, the core activities are the **transfer** of information, including patient records, diagnostic data, and clinical expertise, and the **development** of information through patient evaluation and examination. In the former instance, electronic patient records, diagnostic and other data accessed at various points within the network via portable computers can facilitate faster and more accurate entry and retrieval of information about the patient and his/her problem. In addition, on-line inquiries of remote or local computerized data bases containing relevant information on the patient's health problem can be made readily. In turn, the quality of information exchanged can be assessed by developing standards for the acquisition, storage, transmission, and retrieval. Standards can also be developed for the content and format of electronic patient records, specifying necessary data elements, structure of the record, etc., as well as content and quality of imaging, including resolution, precision, and scale, for diagnostic information. Finally, conformity with practice protocols and standards for telemedicine practice can be utilized as structural indicators of quality.

Telemedicine and related information technology have the potential to improve quality in several different contexts (Grigsby et al., 1994). In emergency care, for example, telemedicine can improve health outcomes in urgent situations by providing immediate access to providers at a central facility for the initial evaluation of patients and trauma cases, triage decisions, stabilizing critically ill patients, and pretransfer arrangements. In chronic disease management, telemedicine can contribute specialist care in those communities where one is not available locally. It also provides a regular source of care for the home-bound chronically ill. When a physician is not available to provide primary care, telemedicine can provide supervision and consultation for primary care encounters rendered by nonphysician providers. Also, it can facilitate routine consultations and second opinion for primary care physicians confronted with conditions requiring such services. Finally, telemedicine can contribute positively to the quality of life and health status of individuals through public health, preventive medicine, and individual patient education. It can do so through the provision of interactive user-friendly educational materials that can be accessed from people's homes and workplaces. Indeed, ways have to be devised to involve patients more directly in the choice of their care, such as rendered in Shared Decision Making (SDM) programs that utilize interactive video technology (Kasper, Mulley, and Wennberg, 1992; Barry et al., 1995). The SDM program provides state-of-the-art information, in interactive format, about various treatment modalities and prognoses for certain health problems (such as benign prostatic hyperplasia, breast cancer, and ischemic heart disease), and patients participate actively in the choice of treatment. The program supports informed consent and informed decision making on the part of the patient, and it is expected to improve quality outcomes while controlling cost. The assumption is that the central objective of the health care system is to produce health with full informed consent of patients. The production of health consists of disease prevention through a variety of interventions aimed at reducing risk for disease and appropriate patient care for the sick and infirm, as well as health promotion through encouraging a healthy life style. The role of the informed patient is critical to success. All things considered, therefore, the potential contribution of telemedicine to an improved quality is considerable.

In current practice, telemedicine is utilized for two-way educational experience for primary care physicians in rural areas and consulting physicians in academic centers. Rural physicians gain from interacting directly with specialists on a case-by-case basis, as well as by having greater access to and participating in formal continuing education courses. Academic physicians, in turn, learn about the types of health problems facing rural practitioners. Moreover, the potential of this technology to improve the efficiency and proficiency of the educational process in a variety of settings is widely recognized. A simple example is the National Library of Medicine's creation of digital image data of a complete human male and female as an educational tool (see NLM, The Visible Human Project: http://www.nlm.nih/extramural research.dir/getting data.htm). More relevant to the topic at hand, the new technology is expected to produce qualitative changes in the educational process. This is described by Koop (1995) as a breakthrough: "Through telemedicine, we are going to be able to enjoy some things that we were never able to do before. If this system eventually works in the way we envision, every encounter you have with the medical system will not only inform the next one; it will also inform all the encounters other people have with the medical system." Koop expects telemedicine's expansion of the educational process to create more homogeneity in clinical practice. "Our real effort is to change the cognitive ability of physicians who are isolated from normative medicine." Indeed, if this is accomplished, some of the observed variability in practice patterns between areas should diminish.

Research pertaining to the evaluation of the quality of the telemedicine experience is rudimentary and so is our insight into these issues. In sharp contrast to continuous and rapid improvement in the quality of the technology serving telemedicine and the few clinical studies of the efficacy and effectiveness of specific clinical procedures as reported elsewhere in this book, evaluation of the effects of telemedicine systems on the care process and health outcome has only been superficial. Better methods are needed to evaluate the impact of particular technologies on the health care system and clearer improvements in standards of care need to be demonstrated (McLaren and Ball, 1995).

The Issue of Cost

The national preoccupation with the price inflation in health care has prompted a variety of methods to control the rate of increase in cost. The public (both civilian and military) and private sectors (both small businesses and large industrial firms) have actively searched for effective means to limit their liability for health care expenditures. Until recently, it appeared that a free market and the attendant free competition have not led to significant cost containment. Over the past several years, and in the shadow of proposed major health care reform, there is evidence of change as the medical care Consumer Price Index (CPI) has risen rather slowly (Metropolitan Life Insurance Company, 1994). For instance, spending for hospital care, physician services, and purchases of prescription drugs grew nationally at 10.5 percent annually from 1980 to 1991 (Levit et al., 1993). The same CPI rose 6 percent in 1993 and 5.7 percent in 1994. Nonetheless, more than 13 percent of the gross domestic product is derived from health care, and it is widely believed that this amount of health care expenditure is excessive and must be controlled (Sulmasy, 1995).

The continuing escalation of medical care costs has only exacerbated the increasing disparity in access to care between various segments of American society, especially among those who are and are not geographically, economically, or socially challenged. Affordability of health services is often considered the prominent dimension of access to care. Affordability is necessary, though not sufficient, for assuring access to care, and it is a function of people's ability to pay and the price of care. Other impediments to care include a variety of geographic, social, cultural, and psychological factors. Nonetheless, because of its importance, major health care reforms in the latter half of the twentieth century have focused on affordability.

The national concern with costs is intensified and complicated by the continuing debate about whether health care is a right or a privilege. Also contributing to the debate is the steady emergence of sophisticated and costly diagnostic and therapeutic procedures and technology, and the uncertain evidence of their positive cost-benefit ratios. Moreover, the real benefits of clinical medicine relative to a healthy environment, a communal responsibility, and a healthy life style, an individual responsibility, are often called into question. Indeed, there is strong evidence that substantial investments in medical technology have not been matched with commensurate improvement in the health status of the population as measured by life expectancy, infant mortality, or similar indicators.

The search for cost-containment strategies has shied away from interventions that challenge the prevailing modes of producing health services, including the supply and distribution of health manpower and resources in this country. Hence, existing strategies have heavily emphasized various modalities of limiting demand for care, especially expensive care, by imposing higher and higher amounts of co-payment by clients and now more directly through managed care. Relatively recently, as a proposed means of controlling costs, we have witnessed the apparent beginning of a major reconfiguration of payment methods for medical care in the United States. In the public sector, for example, Diagnostic Related Group (DRG) reimbursement schedules were established and implemented in Medicare in order to control the rapidly rising cost of that program. In the private sector, the beginning of a general restructuring of medical care in the United States is reflected in the growth of Health Mainte-

nance Organizations and other forms of managed care since the early
1970s.

One of the most significant developments in the 1980s was the growth
of selective contracting. The message of selective contracting was clear —
payers and insurers would do business only with providers (hospitals
and physicians) who kept costs down (Bodenheimer and Grumbach,
1995). In the 1990s, increasingly, employers are directing their employees
into one or a few managed care plans with which they have contracts.
Outside of rigidly defined emergency situations, noncontracted physicians and hospitals are not reimbursed for providing care to members of
such plans unless referral to them is approved by the plan.

There are several proposals to reform the health care system, none
dealing with modes of production of health or health services, or the
supply of manpower. There is only limited attention to restricting the
supply of physicians. Within the current political climate, it appears that
some form of "managed competition" will play an important role in
government and employer efforts to control health care costs. What has
been envisioned under managed competition is a three-tier system. The
government or public/private entities manage HMOs, HMOs manage
physicians, and physicians manage patients. At each level, the gatekeepers will maintain incentives to limit spending on health care.

Under this system, the key to economic success is the concept of
"managed care" (Iglehart, 1992). In managed care, patients are assigned
primary care physicians who act as "gate-keepers" and determine how
much and what type of care is necessary and appropriate for the patient.
The assumption is that patients drive up the cost of care by demanding
too much care, over and beyond what is necessary for them to protect
their health. Managed care takes the decisions about the use of service —
when, where, and how much care is used — away from the individual
patient. A primary care provider assumes the gatekeeper function for
such decisions, and the gatekeeper is vested with a strong economic
incentive to minimize expenditures. Hence, "keeping patients from con-
suming health care dollars by limiting the use of expensive tests,
treatments, and specialty care" is the solution (Bodenheimer and Grumbach, 1995). While patients normally make the initial decision to initiate
care, once care is initiated, the physician will take over and determine
the amount and nature of service rendered to the patient. In order to
dampen the initiation of care, co-payment is imposed in amounts considered effective to rule out frivolous use. At the same time, the physicians

are expected to maintain quality of care because they assume responsibility for all subsequent expenditures. If appropriate care is foregone, more expensive service may ensue later. Also, in managed care settings, primary care providers deliver the majority of services themselves, and they minimize the use of complex and expensive procedures by restricting patient access to specialists. Primary care physicians who prove to be ineffective "gatekeepers" are penalized through financial disincentives. Under this type of incentive, it is believed that primary physicians increasingly will be encouraged to provide a more comprehensive range of services and emphasize prevention. The challenge of managed care, therefore, is to design a system in which rules and incentives together create an effective system of control over demand without compromising quality of care. To accomplish this goal, payment needs to be linked both to quality and productivity and appropriately risk-adjusted (Hillman, 1995).

The growth of managed care is changing the practice of medicine, including the quality of care. In light of these changes, some are calling for a reformulation of medical education and training for primary care physicians to reflect the changes taking place in the organization and delivery of medicine in the United States (Koop, 1995; Hoffman and Johnson, 1995). Koop has suggested that the primary role of telemedicine is to support the decision-making process for physicians, particularly those physicians isolated from normative medicine.

The current generation of telemedicine is being introduced into a highly charged and complex playing field of medical care delivery. Fee-for-service, private and public insurance programs, and managed care programs exist side by side and, sometimes, co-exist within the same delivery system. Into this uncertain environment of continually increasing costs and increasingly complex interplay of attempts to control them, telemedicine must find its niche. In terms of the "niche theory" of ecology, the unique role for telemedicine derives from its unique attributes and capabilities. This niche represents the optimal competitive position for telemedicine as compared to all alternative systems or arrangements.

The practice of telemedicine will combine some characteristics of conventional medicine with some derived from the new technology. The result may be a hybrid, sufficiently distinct from any practice previously seen to warrant, at a minimum, a reconfiguration of cost modeling. The cost of telemedicine needs to be considered in relation to how it contributes, through access to information and communication, to improving the health of the population by preventing disease, treating illness, and ameliorating pain and suffering; how it permits new mixes of providers, with and without an M.D.; and how it changes the types of interactions and relationships between providers among themselves and between providers and clients (Bashshur, 1994).

Again, views of the issue of cost savings and cost containment with regard to telemedicine depend upon the observer. For example, there is little doubt that the opportunity costs in seeking care will be reduced for patients. By reducing travel time to specialists for evaluation, treatment, and follow-up, telemedicine can significantly reduce the opportunity costs of obtaining medical care. Savings from less travel and reduction of lost wages due to travel for care can be substantial. Alternatively, itinerant specialists in multisite rural delivery systems will also reduce their opportunity costs since they no longer need to travel to clinics held in remote regions. Costs to patients hospitalized in their own community hospitals rather than remote tertiary hospitals will also be substantially reduced. The cost of a hospital bed is lower in a local community hospital than in one at a distant university medical center.

Structural costs may also be reduced through development of an alternative manpower mix appropriate to the new technology. For instance, it may be possible to staff follow-up clinics with allied health professionals such as nurse practitioners and physicians' assistants. In other situations, telemedicine may be used to deliver primary medical care services to remote areas without a resident physician. There are potential cost savings associated with developing different manpower mixes where appropriate.

The inseparability of cost from accessibility and quality is reflected in the possibility of increased medical expenditures resulting from improvements in access and quality (Bashshur, 1995). For example, if distance barriers to health care are reduced via interactive video and audio communications, *ceteris paribus,* the demand for care should increase from two sources: the patient and the primary physician. For patients, the pent-up demand will become manifest when people who have postponed or foregone needed health services are presented, by virtue of telemedicine, with improved access to these services. The more health services they utilize, the greater the cost because the total cost of care is the product of quantity of use and unit price. Therefore, cost savings which might be realized by unit price decreases may be offset by an increase in volume. Additionally, in many instances, higher quality care (which may include more tests and application of higher levels of technology) may be associated with higher costs per episode of patient care. Remote primary physicians, previously hesitant to refer patients hundreds of miles to a specialist, may now be willing to prescribe more tests with ready access to a specialist and diagnostic facilities via telemedicine. Alternatively, under certain conditions, health outcomes enhanced by telemedicine may ultimately reduce the need for care, subsequent utilization, and costs of care.

Economic analysis of telemedicine necessarily entails a comparison of specified sets of inputs and outputs under telemedicine systems versus alternative arrangements in the provision of personal health services. Economic analysis, therefore, has to limit its focus on assessing the effects of known quantities of personal health services (such as physician visits, episodes of care, hospital stays, prescriptions, and consultations), under two arrangements—with and without telemedicine—on the health status outcomes of individuals and populations. Both inputs and outputs are then translated into costs (including time costs). These services consist of a specified set of inputs involving a level of medical expertise, medical technology, facilities, service personnel, and the clients' preferences and characteristics. Moreover, cost-benefit or cost-effectiveness analysis of telemedicine requires a precise definition of "fixed" system characteristics (i.e., quantitative and qualitative capabilities and expected outputs or objectives). Both types of analysis involve the development of associations between individual costs, production process, and output. Cost-effectiveness analysis will determine the least costly system (telemedicine or traditional practice) that is capable of delivering a specified set of objectives. Cost-benefit analysis involves a complete listing of all desired objectives, rank-ordered by the least costly means for their attainment, in order to maximize the benefits for given levels of expenditures. As a result of these analyses, cost containment may be achieved by substituting lower-cost for higher-cost providers and facilities, reducing need for care, increasing benefits, and streamlining the care process. However, to date, these expectations are based upon assumptions yet to be verified in a mature program that has reached a steady state of operation in an actual field setting.

Conclusion

The resurgence of telemedicine for the second time in so many decades reflects both the unfulfilled promise and the intractable problems of the health care system. The dilemma of the 1990s is the same as that of the 1970s. If properly implemented, telemedicine has the potential to reshape the structure and process of health care delivery in ways that maximize its efficiency and minimize the disparities that exist between various communities in this country.

Optimal systems of patient care should encourage the appropriate use of service, in both quantity and intensity. Optimal telemedicine systems introduce a realistic opportunity to create more rationality in the health care process. Hence, decisions about the initiation of care, the type and site of provider sought, and the amount and intensity of care utilized are determined on the basis of explicit and logical criteria. They should facilitate the use of the appropriate providers commensurate with the patient's needs and choices; and these services should be provided at the appropriate site, taking into account the patient's condition and the availability of needed services nearest to the patient's residence. Also, when clinically indicated and medically necessary, patient referrals (or transfers) should be facilitated and encouraged. Fully integrated telemedicine networks have the potential of matching the patient's clinical needs and personal choices with the appropriate level of care with the appropriate provider at the appropriate site.

At the same time, many of telemedicine's potential benefits may be lost with suboptimal system design, myopic vision, inadequate preparation, and incomplete planning. Ironically, the rush to install these complex systems without adequate preparation and planning is likely to produce mixed, and possibly inaccurate, results. Moreover, because of contraction in public spending and the heavy emphasis on cost-containment strategies aimed at diminishing demand for care, public support for the development of telemedicine may be short-lived. The intuitive appeal of telemedicine, though much touted, will not be sufficient for the long-term sustainability of operating systems, unless, of course, these systems become self-supporting.

Much of the current discussion has focused on the remote, confined, and other geographically disadvantaged populations whose assistance has been the subject of many projects. But ultimately, the true merit of telemedicine systems, as determined by their distributive efficiency, clinical effectiveness, productive capacity, and cost savings, will have to be demonstrated on the basis of serving mainstream normal populations under normal market conditions. Indeed, as the information technology permeates the health care system and become an integral part of it, telemedicine will come of age and occupy its proper niche in the health care system.

Still, telemedicine may have to demonstrate its merit to various decision-makers at the national, state, and institutional levels long before optimal telemedicine systems are designed and implemented and before their performance can be investigated meaningfully. Only optimal systems will produce optimal results, and a consensus about the essential requirements and characteristics of such systems is yet to be achieved. The opportunity, through telemedicine, to improve both the distributive efficiency of health care resources and to provide a more rational basis for the use of these resources to promote health and reduce disease, disability and dysfunction may be squandered by the rush to install these complex systems without adequate design, preparation, and planning. Ideally, programs should be designed to meet explicit and demonstrable needs in the community being served, and also to satisfy clinical requirements necessary for accurate diagnosis and proper treatment at minimal cost.

Need assessment, or demand analysis, will determine the size and characteristics of the service population or market for telemedicine services in a given area or region. Biomedical research will determine the technological configurations necessary for specific diagnostic and treatment processes.

Evaluation of the complex impact of telemedicine on the interrelated aspects of health care cost, quality, and accessibility requires innovative approaches to evaluation and an experiential data base sufficient to sustain meaningful research. Without meaningful evaluative research which demonstrates the potential benefits as well as limitations of the current generation of telemedicine technology and application, current systems cannot be redesigned for maximal efficiency, and new systems may repeat errors of previous telemedicine efforts. Much of the evaluative work now underway suffers from limited attention to critical prerequisites for valid evaluation of telemedicine as a system of care rather than the mere use of information technology as an adjunct to patient care.

In the best-case scenario, the beneficial effects of telemedicine systems on medical care delivery may be substantial. *Improved access* may be achieved directly through the reduced need for travel and reductions in opportunity cost. *Improved quality* may be achieved through the adherence to clinical treatment protocols and the ready availability of consultations and referrals. *Cost containment* may be achieved by substituting lower-cost for higher-cost providers and facilities, reducing need for care, increasing benefits, and streamlining the care process. However, to date, these expectations have been based on assumptions yet to be verified in a field setting. It is time for telemedicine to move beyond the preoccupation with the latest technology and there are indications, given the several evaluative studies being conducted, that this is occurring. This is especially the case as more expensive technologies are developed which may drive services toward high-cost diagnostic practice. Research is needed on how existing technology can be integrated into health care delivery systems so as to improve the effectiveness and efficiency of those systems, and methods are needed to evaluate the impact of communications media on those systems that have been developed. Unless and until the positive contribution of telemedicine to improving access to care, maintaining and increasing quality of care, while containing or even reducing costs of medical care, is demonstrated, as one observer stated, the history of telemedicine may continue to be one of "false dawns."

We are currently at a critical threshold in the development of telemedicine. We must determine the sum total of the alterations (both benefits and disbenefits) as well as costs produced or saved by telemedicine and whether its salutary effects are sufficiently positive to warrant support for further implementation. Thus, the challenge facing the development of telemedicine is to demonstrate how it can contribute significantly to improving the quality of health care delivery and the health status of the population, and to do so in an affordable fashion.

REFERENCES

Allen, A. 1994. *The Telemedicine Newsletter,* 1:1–3.

Barry, M. J., F. J. Fowler, A. G. Mulley, et al. 1995. "Patient Reactions to a Program to Facilitate Patient Participation in Treatment Decisions for Benign Prostatic Hyperplasia," *Medical Care,* 33:771–82.

Bashshur, R. L. 1995. "On the Definition and Evaluation of Telemedicine," *Telemedicine Journal,* 1:19–30.

Bashshur, R. L. 1995. "Telemedicine Effects: Cost, Quality, and Access," *Journal of Medical Systems,* 19:81–91.

Bashshur, R. L., D. Puskin, and J. Silva, eds. 1995. "Telemedicine and the National

Information Infrastructure: Second Invitational Consensus Conference on Telemedicine and the National Informational Infrastructure," *Telemedicine Journal,* 1:321–75.

Bashshur, R. L. 1994. "Base Evaluation on Access and Quality as well as Cost," *Telemedicine,* 2:8ff.

Bashshur, R. L. 1983. "Telemedicine and Health Policy," *Proceedings from the Tenth Annual Telecommunications Policy Research Conference.* Annapolis, MD: Ablex Publishing Corporation, 348–60.

Bashshur, R. L., and J. E. Lovett. 1977. "Assessment of Telemedicine: Results of the Initial Experience," *Aviation Space Environmental Medicine,* 48:65–70.

Bashshur, R. L. 1976. "Telemedicine's History," *IEEE Spectrum,* 33.

Bashshur, R. L., P. A. Armstrong, and Z. I. Youssef, eds. 1975. *Telemedicine: Explorations in the Use of Telecommunications in Health Care.* Springfield, IL: Charles C Thomas Publisher.

Bird, K. T. 1971. *Teleconsultation: A New Health Information Exchange System,* Third Annual Report to the Veterans Administration. Boston, MA: Massachusetts General Hospital.

Bodenheimer, T., and K. Grumbach. 1995. "The Reconfiguration of US Medicine," *Journal of the American Medical Association,* 274:85–90.

Brecht, R. M., C. L. Gray, C. Peterson, and B. Youngblood. 1996. "The University of Texas Medical Branch—Texas Department of Criminal Justice Telemedicine Project: Findings from the First Year of Operation," *Telemedicine Journal,* 2(1): 25–35.

Donabedian, A. 1973. *Aspects of Medical Care Administration.* Cambridge, MA: Harvard University Press.

Fuchs, M. 1979. "Provider Attitudes Toward STARPAHC: A Telemedicine Project on the Papago Reservation," *Medical Care,* 17:59–68.

Grigsby, J., R. Schlenker, M. Kaehny, et al. 1994. *Analysis of Expansion of Access to Care through the Use of Telemedicine, Report 4: Study Summary and Recommendations for Further Research.* Denver, CO: Center for Health Policy Research.

Grigsby, J., M. Kaehny, E. J. Sandberg, R. E. Schlenker, and P. W. Shaughnessy. 1995. "Effects and Effectiveness of Telemedicine," *Health Care Financing Review,* 17: 115–31.

Grigsby, R. K. 1995. "Telemedicine," *Journal of the American Medical Association,* 274 (6): 461–62.

Hillman, A. 1995. "The Impact of Physician Financial Incentives on High-Risk Populations in Managed Care," *Journal of Acquired Immune Deficiency Syndrome and Human Retrovirology,* 8 (Suppl 1): S23–30.

Hoffman, E., and K. Johnson. 1995. "Women's Health and Managed Care: Implications for the Training of Primary Care Physicians," *Journal of the American Medical Women's Association,* 50: 17–19.

Iglehart, J. K. 1995. "Health Policy Report: Medicaid and Managed Care," *New England Journal of Medicine,* 332:1727–31.

Iglehart, J. K. 1992. "Managed Care," *New England Journal of Medicine,* 327:742–47.

Johnson, C. P. 1848. "Report on the Number of Practitioners of Medicine in Virginia," *Transactions of the American Medical Association,* 1:359–64.

Kaluzny, A. D., and J. E. Veney. 1973. "Attributes of Health Services as Factors in Program Implementation," *Journal of Health and Social Behavior,* 14:124–33.

Kasper, J., A. Mulley, and J. Wennberg. 1992. "Developing Shared Decision-Making Programs to Improve the Quality of Care," *Quality Review Bulletin,* 18:183–90.

Kimberly, J. R., and M. J. Evanisko. 1981. "Organizational Innovation: The Influence of Individual, Organizational and Contextual Factors on Hospital Adoption of Technological and Administrative Innovations," *Academy of Management Journal,* 24:689–713.

Koontz, V. L. 1976. *Determinants of Individuals' Level of Knowledge, Attitudes Towards and Decisions Regarding a Health Innovation in Maine,* doctoral dissertation. Ann Arbor: The University of Michigan.

Koop, C. E. 1995. "The Interface of Health Care Reform and Telemedicine," *Bulletin of the American Academy of Arts and Sciences,* 49:36–51.

Levit, K., H. Lazenby, C. Cowan, and S. Letsch. 1993. "Health Spending by State: New Estimates for Policy Making," *Health Affairs,* 12:7–26.

McLaren, P., and C. Ball. 1995. "Telemedicine: Lessons Remain Unheeded." *British Medical Journal,* 310: 1390–91.

National Library of Medicine. Internet: http://www.nlm.nih.gov/research/visible/ visible-human.html Bethesda, MD: U.S. National Library of Medicine, National Institute of Health, Department of Health and Human Services.

National Rural Electric Cooperative. 1994. *Health Care Needs, Resources, and Access in Rural America.* Alexandria, VA: Analytical Services, p. 14.

Penchansky, R., and W. J. Thomas. 1981. "The Concept of Access: Definitions and Relationship to Consumer Satisfaction," *Medical Care,* 19: 127–140.

Rogers, E. M., and F. F. Shoemaker. 1971. *Communication of Innovations: A Cross-Cultural Approach,* Second Edition. New York: Free Press.

Sanders, J. H., and R. L. Bashshur. 1995. "Perspective: Challenges to the Implementation of Telemedicine," *Telemedicine Journal,* 1: 115–23.

Scott, W. R. 1990. "Innovation in Medical Care Organizations: A Synthetic Review," *Medical Care Review,* 72(2): 165–92.

Sulmasy, D. 1995. "Managed Care and Managed Death," *Archives of Internal Medicine,* 155: 133–36.

Metropolitan Life Insurance Company. 1994. "Developments in Health Care costs—An Update," *Statistical Bulletin of the Metropolitan Insurance Co.* 75: 30–35.

Wennberg, J. E., et al. 1995. *Dartmouth Atlas of Health Care in the United States.* Chicago: American Hospital Publishing.

Witherspoon, J. P., S. M. Johnstone, and C. J. Wasem. 1993. *Rural TeleHealth: Telemedicine, Distance Education and Informatics for Rural Health Care.* Boulder, CO: WICHE Publications.

Wittson, C. L., and R. Benschoter. 1972. "Two-Way Television: Helping the Medical Center Reach Out," *American Journal of Psychiatry,* 129: 136–39.

Vivian, W. 1975. "Status of Video Communication Technology for Medical Care," in: Bashshur, R. L., P. A. Armstrong, and Y. I. Youssef. *Telemedicine: Explorations in the Use of Telecommunications in Health Care.* Springfield, IL: Charles C Thomas Publisher, pp. 41–54.

Chapter 2

TELEMEDICINE:
RESTRUCTURING RURAL MEDICAL CARE
IN SPACE AND TIME

GARY W. SHANNON

In fact space and time always go together and we might as well, the sooner the better, try to get accustomed to seeing space and time as united into one compact four-dimensional entity.

—Hagerstrand, 1970

Introduction

Almost every chapter in this volume expresses the need for evaluation of telemedicine and of its role in specific clinical and institutional settings. This chapter also deals with the question of evaluation of telemedicine, but at a more general level. The purpose here is to discuss the fundamental characteristics of telemedicine as they contribute to the restructuring of medical care of isolated populations. Proponents of telemedicine systems should understand their complexities and their impacts on the configuration of medical care and related social behavior. The approach taken here to the basic development and impact of telemedicine combines consideration of time and space in a variety of settings, focusing on telemedicine and the reconfiguration of medical care in rural areas. First, however, it is necessary to review briefly: (a) selected conceptual and contextual notions related to the approach used here; and, (b) major efforts employed in the past to improve rural residents' access to medical care.

Activities in Time-Space

In discussing the distribution of medical care and related behavior of physicians and patients, it is important to consider each medical care opportunity as occurring at a specific place and time. In other words, medical care opportunities are arranged not only geographically in offices, clinics, and hospitals with specific addresses, but also have temporal addresses or locations (i.e., appointments and schedules). For patients to obtain medical care and for others to provide general or specialty medical care, each must arrive at a particular place at a particular time. Therefore, patients and physicians must adapt their activities to this temporal-spatial distribution of opportunities in what can be termed a *relatively* "fixed-featured time-space" (Zerubabel, 1981). This simply means that the combined temporal-spatial addresses of opportunities for patients to interact, to transact the delivery and receipt of medical care, are relatively fixed in time and space. Therefore, we need to consider not only *where* an opportunity for medical care is located but *when* it is available. Knowledge of only one or the other is insufficient to make informed decisions about and evaluations of the provision and use of medical care.

Admittedly, the fixed-featured time-space of medical care opportunities reflects to some extent the general distribution of need within a general population. However, few if any patients, physicians, and other medical care providers can arrange the time-space distribution of medical care for personal convenience. Patients do not have direct control over the location of a general or specialist physician's office. Hospitals are not located to suit the personal convenience of potential patients or physicians. Nor can we determine when opportunities for medical care in these facilities will be available for our use. This is determined by appointment, queue, or chance. Both patients and physicians must adapt their behavior to the relatively fixed-featured medical care time-space.

The distribution of medical care opportunities in time-space is determined by a complex interaction of social values and economic forces. And the temporal spatial distribution of these opportunities may reflect some underlying theoretically optimal pattern or, at least, movement toward one. For example, according to economic theory, in a free-market system, the optimal distribution of medical care reflects a relative competitive position. At the same time, however, while perhaps economically optimal, variations in the fixed-feature distribution[1] of medical

---

[1]Since change is constant, it may be misleading to speak of a truly "fixed-feature" time-space opportunity. Work places are opened and closed, grocery stores and gas stations come and go. The same is true of physicians and hospitals. The landscape of opportunities in time-space is continuously changing. For most of us, however, the specific locations we visit for the necessities of life and the times they are available remain fixed for considerable periods. Only occasionally do changes in our opportunity landscape force us

care opportunities are sometimes identified as socially inappropriate or inequitable.

Periodically, identification of large-scale "inequitable" distributions of medical care has been followed by social engineering and/or legislative attempts to reconfigure the spatial or temporal structure of the medical care environment. Rehabilitative strategies are developed and programs offered to modify the environment suitably. Unfortunately, in providing medical care to isolated populations, especially to those located in rural areas, success has been modest at best.

Visiting and Revisiting the Rural Medical Care Problem

The rural-urban imbalance in medical care was recognized in the early twentieth century. But it would be 1946 before major federal legislation—the Hill-Burton Hospital Construction Act (P.L. 79-725) addressed the problem by financing state-wide hospital inventories and, subsequently, hospital construction in those areas determined to be underserved. Urban areas were eligible for consideration, but priority was given to rural areas, where the shortage of beds was determined to be acute. Community hospitals were built in these areas in hopes of attracting physicians to the countryside (Lave and Lave, 1974).

Though several hundred thousand hospital beds were developed under the Act, the differential between rural and urban areas remained. Some years later, recognizing that the problem had not been resolved, the federal government developed the National Health Service Corps program (NHSC 1976). The Corps was established under the Emergency Health Personnel Act of 1970 (P.L. 91-623). This program funded the education of physicians who contracted to provide two years of service in medically underserved areas, mostly rural.

On the supply side, efforts were also made to increase the total number of physicians in hopes that some would "trickle down" from larger cities to the towns in rural areas (Hynes and Givner, 1983).

Regardless of the source or type of effort, public or private, supply or demand, previous major efforts to improve rural medical care have focused on physical restructuring through building hospitals and relocating

---

(¹Continued) out of our usual pattern of behavior. Therefore, viewed from an individual perspective, it is appropriate to speak of a fixed-feature landscape of opportunities in time-space.

health providers. None has resolved the problem. Today, we are faced with an epidemic of closings of community hospitals, many in small rural towns. While the increased numbers of general practitioners in rural areas appear to be increasing, primary care remains in short supply, and the level of available clinical expertise is relatively low in many communities. Therefore, in many rural areas, medical care is at best unstable and in some situations critical. Attempts to alter rural medical care physically by building hospitals and redistributing physicians, particularly specialists, have not been successful. Now a new technology and concept of medical care, embodied in telemedicine, are being promoted to improve the environment via a virtual rather than physical restructuring.

Restructuring via Telemedicine

As described elsewhere in this volume, telemedicine was first introduced in the late 1960s and, shortly thereafter, was almost totally abandoned. In the 1990s, it is being offered again as a technological solution to the limited availability and instability of medical care resources in rural areas. The content of telemedicine varies considerably from place to place. In the more basic formulations, telemedicine may consist only of store-and-forward "picture-" or Picasso telephones. More sophisticated systems include cameras with powerful zoom lenses; electronic stethoscopes equipped with real-time transmitters for EKGs or echocardiograms; camera adapters to provide ophthalmic, endoscopic, laporoscopic, and cytoscopic examinations; telemicroscopic adapters to permit remote pathology; and equipment to transmit x-rays, CAT- and ultrasound scans.

In these systems, physicians and patients view one another on color monitors, and omnidirectional microphones pick up audio from both participants. Graphics' tablets provide the system interface. They control local and remote video selection and a multitude of camera capabilities. At the heart of the system is a computer-controlled switching matrix which facilitates the creation of networks between a medical hub and a single or multiple satellite locations.

The goal of telemedicine is to transform medical care electronically. It virtually transports the expertise of specialists at medical centers to remote consult sites, sometimes hundreds of miles away. Or, viewed alternatively, the system transports the patient and attending primary care provider from remote or otherwise isolated hospitals and health centers to major medical centers. In either case, at least for the duration of the consultation, the medical care process is altered dramatically. Some proponents of telemedicine suggest that it will eliminate problems related to the geographic distribution of medical care. As illustrated below in the time-space approach, this is premature, at best. Rather than eliminating the geography of medical care, telemedicine introduces a new geography.

Time-Space

In the time-space approach to the assessment of human behavior within any environment, both animate and inanimate objects follow time-space paths (Hagerstrand, 1970; Thrift, 1983; Parkes and Thrift, 1980). The major difference between the two objects is that, generally, inanimate objects are stationary in geographic space while animate objects change location. In any event, the distribution of opportunities for obtaining goods and/or services to satisfy our everyday needs, as well as the distribution of medical care opportunities, can be characterized as a distribution of paths in time-space (Figure 2-1). As illustrated, each opportunity has a geographic and temporal location. For example, our home (A) has a fixed address and the path in space is geographically fixed. However, it also has a specific temporal location (i.e., when it is available to us). The duration of the opportunity is indicated in Figure 2-1 by a cylinder. Since our homes are available to us 24 hours a day, the cylinder (or window of opportunity) extends throughout the day.

Other places, however, such as grocery stores (B) and schools (C) with fixed locations have limited temporal availability. This is illustrated in Figure 2-1 by the cylinders of varying sizes superimposed on the time-space paths. Medical care opportunities can be characterized similarly as a distribution of opportunities in time-space. For example, a physician's office (D) has a fixed address, but the opportunity to visit a physician is limited to predetermined office hours. Thus, the geographic location of the physician's office taken together with the temporal location or duration of the office hours comprise the time-space location of this opportunity for medical care. Similarly, a hospital (E) has a fixed address, but departments and clinics within the hospital may have different temporal locations. The Emergency Department (I) is available 24 hours a day. However, other services, such as outpatient clinics (II), operate on a fixed and limited temporal schedule. Thus, we can speak of a time-space path for the location of each medical care opportunity.

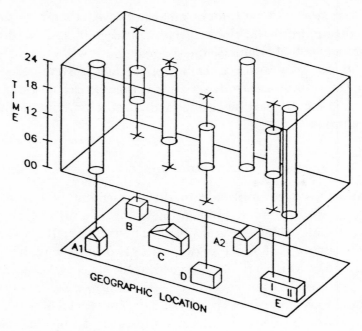

A1, A2 = HOMES
B = SCHOOL
C = GROCERY
D = PHYSICIAN'S OFFICE
E= HOSPITAL
   I = CLINIC
   II = EMERGENCY ROOM

Figure 2-1. Time-space landscape.

Movement in Time-Space

We must travel to a selected number of locations in this time-space to complete projects and take advantage of opportunities which provide the necessities and pleasures of life. We must alter or adjust our time-space path to join temporarily with other locations in time-space to obtain the desired and necessary goods and services. In the parlance of time-geography, we must "couple" our path with others, and this coupling forms a "bundle" of paths. Once this bundle is formed and the exchange is completed, we can move on to the next opportunity. Our daily experience, therefore, can be conceptualized as a sequence of projects (a combination of goals and tasks) represented by the creations of bundles in time-space linked by travel from one project to another (Hagerstrand,

1982). This daily journey for everyday activities and medical care is illustrated in Figure 2-2. Our regularly needed and scheduled activities such as grocery shopping and school are viewed as time-space paths. The occasional need for medical care requires us to alter our time-space trajectory away from everyday projects to couple with a medical opportunity (physician, office, clinic, hospital), situated at a specified location in time-space.

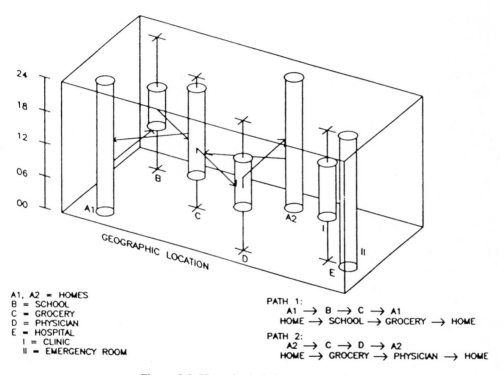

A1, A2 = HOMES
B = SCHOOL
C = GROCERY
D = PHYSICIAN
E = HOSPITAL
I = CLINIC
II = EMERGENCY ROOM

PATH 1:
A1 → B → C → A1
HOME → SCHOOL → GROCERY → HOME

PATH 2:
A2 → C → D → A2
HOME → GROCERY → PHYSICIAN → HOME

Figure 2-2. Hypothetical time-space paths.

Within this framework, the lack of specialized medical care opportunities in isolated communities can be recast in a time-space configuration. Within the time-space landscape, the distribution of opportunities for medical care is such that they cannot be reached with a reasonable alteration of an individual's projects and time-space path. In other words, the individuals in these situations cannot couple with distant medical care opportunities. The medical care opportunities are said to lie outside the maximum time-space (travel) volume of the individuals.

While this situation applies to primary care in many isolated communities, it is most pertinent when accessibility to specialized medical care is considered. The remote location of secondary and tertiary care necessitates at best a difficult and often impossible alteration of individual time-space paths (Figure 2-3). From the perspective of the prospective patient, the decision to use a remote medical care opportunity requires consideration of often excessive investments in travel time and associated costs. These must be weighed against: (a) the perceived threat to health of a particular symptom or set of symptoms; and (b) the time-space commitments and requirements of other projects such as work, school, child-care, and the like. Often, therefore, use of distant medical care opportunities is rendered difficult or impossible. The goal of telemedicine is to restructure the time-space configuration of medical care opportunities and bring them within reasonable range of the isolated individual's time-space path. This is accomplished by electronically transporting the specialist to a more accessible location via telecommunication, as illustrated in Figure 2-3.

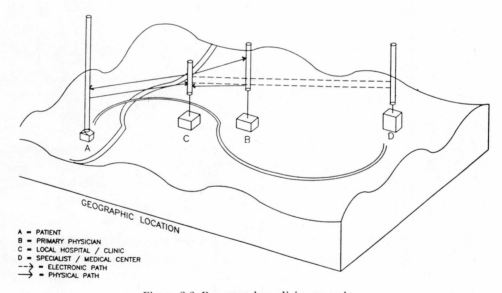

A = PATIENT
B = PRIMARY PHYSICIAN
C = LOCAL HOSPITAL / CLINIC
D = SPECIALIST / MEDICAL CENTER
--→ = ELECTRONIC PATH
—→ = PHYSICAL PATH

Figure 2-3. Remote telemedicine consult.

Obviously, this significantly reduces travel distance and time for the patient. Telemedicine reconfigures the time-space medical care landscape in such a way that the opportunity is brought closer to the patient. This new configuration generates a much greater probability that the patient will take advantage of the medical care opportunity. In most telemedicine systems established to date, however, the patient still must travel to a clinic or hospital where the consult station is located. Nevertheless, the advantage to the patients is obvious. They can be seen locally by a specialist physically at a distant medical center, because the medical care opportunity can now be reached with reasonable adjustments of their time-space paths. This virtual reconfiguration of the medical care landscape does not, however, eliminate the problems associated with the geographical distribution of medical care. Rather, a new and more complex geography is created.

Complications of Telemedicine in Time-Space

While proponents of telemedicine envisage the day when each physician will be able to interconnect with a telemedicine network from the convenience of his office, the current telemedicine scenario requires both the primary provider and specialist to restructure their time-space paths to accommodate the telemedicine consult. As illustrated (Figure 2-3), a primary physician must leave the office and travel to the consult site, usually in a community hospital or clinic. Travel by both the primary and specialist physicians results in lost time for other, on-site patients.

In addition, the complexity of the bundle (Figure 2-4) necessary to complete the project or consult increases consumption of the physicians' time and, perhaps more importantly, the probability of project failure. The primary physician or other local provider must couple with the path of the telemedicine equipment; with the patient, with (usually) a local facilitator in the form of a nurse or technician; and with the remote specialist and facilitator who have been electronically transported to the local hospital or clinic. At the medical center, the specialist and facilitator must travel to the telemedicine station in the hospital. Once the project or consult has been completed, the physicians must return to their offices or clinics.

The delay of any participant, or failure to arrive at the specified consult location at the designated time, or equipment problems and failure, can significantly extend the coupling time or even prevent the coupling from taking place. In the latter instance, the bundle must be rescheduled or is lost altogether. The complexity and number of paths which must come together, the extended duration of the consult bundle, and the increased potential for lost time is particularly important to

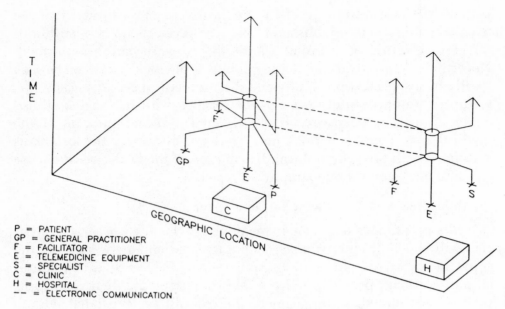

Figure 2-4. Telemedicine consult bundle.

physicians and patients. Not unexpectedly, resulting delays and missed opportunities reduce the enthusiasm of all participants for telemedicine. Reportedly, newer systems are being designed to facilitate "desk-top" computer linkages to the telemedicine networks. Even if successfully developed and generally distributed, however, it is unlikely that these linkages will provide the full range of capabilities available at the main telemedicine consult station.

For physicians, the time-cost in travel to the telemedicine consult site and the duration of a successful consultation is significant for their other activities. It has been observed that activities of a specified or undetermined duration carried out at one locale have a direct bearing on those activities subsequently transacted in a different setting, which, in turn, have a direct bearing on the activities in another locale (Cohen, 1989). For physicians, given such a web of causation, the time taken to participate in a telemedicine consult must be evaluated in the reduction in the number of patients seen in the office. Each telemedicine consultation which requires physician travel (and there is anecdotal evidence suggesting that the telemedicine consult lasts longer than an in-person consult) reduces the number of patients who can be seen in the office or clinic. And patient appointments must be rescheduled or lost. For the physician, this means lost opportunities to "dispense" medical care and loss of income. For the nontelemedicine patients, this means an increasingly complex and restricted opportunity for care. The reduced time available for office appointments places additional coupling constraints upon potential patients as they must adjust and readjust their daily activities to include medical care which has become more elusive. This increases the probability of delayed medical care or neglect on the part of the patients.

Without systems which connect physicians in their offices to the telemedicine networks, it is very likely that a strong distance-decay function will be a factor in determining which physicians in remote areas will use the centralized telemedicine consultation sites. In other words, one would expect that the greater the distance and travel time of physicians' offices from the consultation site, the fewer physicians will be able or willing to take advantage of telemedicine. This proposition is supported by anecdotal evidence[2] indicating that the rural primary physicians who use telemedicine are those whose practices are located in, or near, the hospital or clinic which houses the telemedicine consult site. Rather than the patient, now it may be the physicians who are located too far from participation in a medical care opportunity.

Thus, as the vast majority of telemedicine networks are now configured, telemedicine may not reach the more remote physicians and, hence, the more isolated populations dependent on them. In addition, the more remote physicians are likelier to lose patients to physicians with practices closer to, and who use, the centralized telemedicine facilities. Reminiscent of the early twentieth century and the introduction of the telephone and automobile, telemedicine may bring rural physicians into greater competition with one another. If medical care available via telemedicine is perceived as necessary and superior by prospective patients, they may "shop around" and patronize those physicians using the technology, thereby creating a competitive advantage for those physicians with reasonable access to telemedicine and those who use it.

Telemedicine and the Time-Space of the Specialist

At what might be termed the microlevel of time-space analysis, the distribution of specialists relative to the telemedicine consult site within the hospital may be assessed. Here, it is not a question, as it was for the

---

[2]Information from interviews conducted among physicians and nurses in central Georgia.

primary physician, of traveling from an office practice to a hospital with the telemedicine equipment. Rather, it is a matter of traveling from a clinic within a major research hospital to the consult room. Again, there is anecdotal evidence from specialists who perceive a "serious" misplacement of the consult rooms within hospitals. In many instances, the consult stations are located near the emergency suite which is generally some distance from the outpatient clinics and physicians' offices.

The location of the telemedicine consult room within the teaching-research hospital may be as important to acceptance by specialists as the location among a group of rural counties is to acceptance by primary physicians. Time-consuming travel for specialists through hospital corridors and up and down elevators to different floors to reach a "remote" section of a hospital may be just as important to the specialist as, to the primary care provider, is his travel from his office practice to the community hospital or clinic. And, analogous to the experience of primary providers, specialists traveling to telemedicine consult sites are losing valuable office/clinic time which could be spent seeing additional patients. Some may argue that a one-way five- or ten-minute intrahospital trip is inconsequential. However, even one trip might occupy the time for one or two additional clinic consults. The "travel" problem necessarily increases with the volume of nonsequential telemedicine consults. The constraints imposed by travel time and the resulting loss of even one or two patient consultations in a clinic might contribute to a loss of enthusiasm for the technology.

"Time-costs" associated with physicians' use of telemedicine are important factors associated with the ultimate acceptance and success of telemedicine. Efforts must be made to minimize these costs. Some programs are already developing specialty clinics and attempting to schedule requested consults from remote sites to have the consulting specialist available during such clinics. In this context, several patients from more than one remote location can be examined. To date the relatively low volume of and sporadic demand for the wide range of specialty consultations obviate this type of reorganization on a large scale. Nevertheless, a regularly scheduled time for multiple consults may well contribute to greater acceptance and increased use of telemedicine by physicians of all types and in more widely distributed locations.

Telemedicine and Rural Hospitals

Not only rural physicians, but rural hospitals may be placed in a new situation resulting from the introduction of telemedicine. In the current generation of statewide telemedicine systems, only few community hospitals receive a telemedicine consult station. Most networks link selected rural hospitals to the major medical center as secondary hubs. The selection process varies from place to place. In some instances the process involves examination of hospital-area market shares and directed negotiations between competing hospitals. In other cases, the selection may be determined by political considerations. In any event, the rural community hospital ultimately selected as a secondary hub may have a competitive advantage over its neighbors. Already precarious, the fiscal situation of those rural hospitals not included in a telemedicine network may be further destabilized. Patients as well as physicians may "shop" for hospitals which have telemedicine capability.

SUMMARY AND CONCLUSION

Time-space analysis can be used to understand better the phenomenon of telemedicine and its impact on the rural medical care landscape. Time-space concepts allow us to illustrate fundamental principles of telemedicine as well as to identify and illustrate current benefits and potential problems which may act to restrict the use and diffusion of the technology.

At what might be termed the macro level, telemedicine is restructuring the fixed-feature time-space of medical care by "transporting" specialists from teaching-research hospitals to remote community hospitals and clinics often many miles away. As illustrated here, the benefits which accrue to the rural patient are obvious significant savings in travel time and real costs as well as in increased quality of care. The benefits to the isolated physician include increased probability for continuity of care, enhanced income, and more effective medical education. When viewed simply at this level, it is easy to understand proponents' enthusiasm for telemedicine and its transformation of medical care.

Things are rarely as simple as they appear, however, and a new geography is being developed with a new set of costs which should be considered. Within the current configuration of most of telemedicine systems, the burden of excessive travel has been shifted to rural primary care providers, especially those with offices located at some distance from the centralized telemedicine consultation sites. Valuable office time is lost in pursuit of a project which, to be successful, requires a complex coupling of technology, medical and technical personnel, and patients. This increased complexity leads to a greater probability of delay, duration, or failure which may result in rejection of the innovation.

While "desk-top" systems may be in the offing, it is likely that, in the foreseeable future, primary providers with practices located at greater distances from the telemedicine sites may be effectively prohibited from using telemedicine in its present time-space structure. In turn, the availability and use of telemedicine by selected physicians may lead patients to reconfigure their medical care time-space path to utilize those physicians with access to telemedicine. In the minds of some patients, the trade-off may be between using a local physician without access to specialists via telemedicine, and travel to a more distant physician with such access. To the extent this becomes pervasive among isolated residents, it is conceivable that telemedicine systems not designed to alleviate or to offset these potential negative factors may actually destabilize the medical care environment. This would be especially true for those physicians who, were it not for time-space considerations, might otherwise use telemedicine. Thus, certain telemedicine networks may improve access to more specialized medical care among physicians (and their patients) located near the telemedicine consult site at the expense of those providers and their patients located some distance away.

What of the rural community hospitals and clinics which fail to become telemedicine consultation sites? It is quite possible that the presence of telemedicine in one rural hospital may give it a competitive advantage in developing its market among both physicians and patients over a neighboring hospital unsuccessful in obtaining a telemedicine linkage to a major medical center. This is a particularly important question for those community hospitals in precarious financial situations. Some people might suggest, ceteris paribus, that this increased pressure to close these hospitals as a result of the absence of telemedicine may not be bad. But not all things are equal. For example, as discussed in other chapters in this volume dealing with the clinical applications of telemedicine, important questions persist about to the quality of care delivered via telemedicine versus that delivered "in person." In many clinical situations, telemedicine has not been demonstrated to be equal, let alone superior to, in-person care. Unless and until the questions of equal or superior quality are addressed and answered satisfactorily by careful and critical research, the rush to develop telemedicine may be premature.

At a minimum, the current generation of telemedicine is being promoted as a technological solution to one of the more intransigent medical care problems of the twentieth century, namely, the relatively limited availability of medical care in rural and other isolated communities. Regardless of the level of medical care delivery proposed, it is important to assess thoroughly the impact of telemedicine on medical care and associated behavior. Some proponents suggest that telemedicine will eliminate distance and time barriers to care. In reality, the restructuring of medical care associated with the introduction of telemedicine introduces a new set of time-space dynamics and related issues which should be carefully assessed and addressed.

REFERENCES

Cohen, I. 1989. *Structuration Theory.* New York: St. Martin's Press.

Hagerstrand, T. 1970. "What About People in Regional Science?" *Papers,* Regional Science Association, 24:1–21.

Hagerstrand, T. 1982. "Diorama, Path, and Project," Tijdschrift voor Econ. en Soc. Geografie, 73:323–39.

Hynes, K., and N. Givner. 1983. "Physician Distribution in a Predominantly Rural State," *Inquiry,* 20:185–90.

Lave, J., and L. Lave. 1974. *The Hospital Construction Act.* Washington, D.C.: American Enterprise Institute.

NHSC. 1976. "National Health Service Corps," in *Discursive Dictionary of Health Care.* Washington, D.C.: U.S. Government Printing Office.

P. L. 79-725. 1976. "Hill Burton," in A *Discursive Dictionary of Health Care.* Washington, D.C.: U.S. Government Printing Office.

Parkes, D., and N. Thrift. 1980. "Time-Geography: The Lund Approach," in *Times, Spaces, and Places: A Chronogeographic Perspective.* Chichester, NY: John Wiley & Sons.

Thrift, N. 1983. "An Introduction to Time Geography: Concepts and Techniques in Modern Geography," No. 13, Study Group in Quantitative Methods. London: Institute of British Geographers.

Zerubabel, E. 1981. *Hidden Rhythms: Schedules and Calendars in Social Life.* Chicago: University of Chicago Press.

Chapter 3

COMMUNITY HEALTH INFORMATION NETWORKS (CHINS) AND THEIR RELATIONSHIP TO TELEMEDICINE

Bruce A. Friedman and Will Mitchell

Introduction

This chapter will describe the relationship between telemedicine and community health information networks (CHINs). Hence, the first task is to define a CHIN.

> A community health information network is an organizational and technical entity designed and operated to facilitate the electronic data interchange and integration of various types of health care information for the benefit of those organizations and health care professionals that participate in the network.

The term "electronic data interchange (EDI)" refers to electronic exchange of information among organizations. The term was coined outside the health care system. For example, a clothing manufacturer might create electronic links with its upstream cloth suppliers and its downstream retailers. If inventory of a clothing item falls below a specified level on the shelves of the retailer, an electronic message would be sent to the clothing manufacturer who, in turn, would automatically initiate an order to an upstream cloth supplier to ship more cloth to the clothing manufacturer. Electronic data interchange in health care can be taken to mean the creation of electronic links between payers and providers of services to enhance the quality and efficiency of health care services. EDI has been described as potentially one of the most significant cost-saving measures in health care, particularly relevant to health care transactions, such as: (1) materials management, (2) coordination of benefits, (3) referrals and authorizations, (4) prescription ordering, (5) test ordering/result communication, and (6) appointment scheduling (Schaich, 1993). Participants in a CHIN may include health care providers such as hospitals and private physicians, suppliers such as pharmacies and reference laboratories, payers, employers, regulators, inspection and accreditation bodies, and other health care entities.

CHINs are recent developments, and there is thus little literature about them so far. In fact, the primary discussions about CHINs are taking place in the business press (Zinn, 1994). A valuable source of current information about CHINs is the *Journal of the Community Medical Network Society* (ComNet), based in Atlanta, Georgia. The most up-to-date profile about the current status of CHIN development was recently published in this journal (Furukawa and Peake, 1995). Included was a list of what are described as "CHINs in progress," 224 CHINs either operating or being formed in nine geographic regions as of August, 1995. The Community Medical Network Society also publishes the "CHIN 100 Market Directory," which profiles over 500 "CHIN–Ready" health information networks that might form the basis of future CHINs. The number of CHINs will undoubtedly grow every month for some time to come.

The notion of a CHIN as both an organizational and technical entity can be confusing. The distinction between the two will be discussed in greater detail in ensuing sections. Confusion and ambiguity also arise when one attempts to list the existing CHINs in the U.S., such as in the article just cited. Generally speaking, the CHINs recognized in such lists are groups that publicize themselves as CHINs, even though their ability to share information on a regional basis is often only partly developed. Conversely, a for-profit hospital chain may have deployed highly sophisticated systems for sharing information across hospitals or routing financial information to a centralized operations center for the hospital chain, but may make no claim to having achieved the status of a CHIN. Most CHINs are in embryonic stages of development at present, and the CHIN mantle is largely self-bestowed.

Two of the most extensive existing CHINs are based in Wisconsin and Colorado. The Wisconsin Health Information Network (WHIN) was established by the Ameritech Corporation and Aurora Health Care of Milwaukee during 1992 and operates in metropolitan Milwaukee. In mid-1995, WHIN connected about 13 hospitals, 1,700 doctors, 5 payers, 8 clinics, and 12 ancillary service providers. The WHIN system processes patient claims and insurance eligibility information and also provides some access to clinical information such as laboratory results, drug prescriptions, and patient records. Integrated Medical Systems, Inc. (IMS), Golden, Colorado, has operated the Colorado Medical Information Network (MEDACOM Colorado) since about 1990. In 1993, the Colorado CHIN included 27 hospitals, 800 physicians, five pharmacies, a laboratory, an insurance company, a diagnostic center, and an imaging center. Physicians can use the network to access laboratory reports, send prescription refill orders to pharmacies on the network, and communicate with other physicians through electronic mail. About 45 IMS MEDACOM physician networks are in operation in the U.S. Although other CHINs plan to achieve similar or greater levels of functionality, most networks now emphasize transfer of current information between hospitals and physicians with little or no central storage and retrieval of integrated patient records.

Few CHINs offer extensive functionality as yet. The first simple CHINs were organized in the United States during the mid-1980s. Although several hundred CHINs existed or were being formed by mid-1995, many were in planning stages and most others offered rather limited functionality in the information that they allowed participants to exchange. The term *functionality* is computer system jargon for which there is no adequate substitute, and refers to the capabilities of an information system to perform various tasks or applications adequately. The reader should also be aware that "vaporware" abounds in the CHIN sector as it does in all information-system development fields. Vaporware refers to a computer software product that is announced and developed in a prototype form, but is not yet available in the commercial market with all the features cited in brochures or advertisements.

The rapid and fluid development of CHINs should create several opportunities for health care organizations that offer telemedicine services or are contemplating doing so. Telemedicine is health care delivery at a distance, while information technology overcomes time and distance barriers between health care providers and clients. Telemedicine and CHINs complement each other because a CHIN provides the basic information technology and network infrastructure to permit telemedicine initiatives. As discussed later, managed care and capitated payment also provide potent economic incentives for health care CEOs to invest in telemedicine. Therefore, synergies between CHINs and telemedicine should be made more explicit in health care organizations engaged in CHIN planning. Such organizations should explicitly identify deployment of telemedicine services as one of the primary goals for CHIN development projects. Stated another way, CHINs now provide useful platforms for providing health services at a distance, but the telemedicine opportunity may not be obvious, or a high priority, for CHIN planners at the highest level. It is helpful, therefore, for those most knowledgeable about telemedicine to highlight the clinical, as opposed to financial and administrative, potential of electronic exchange of information.

Types of Information Exchanged by CHINs

Typically, the delivery of health care services requires the generation, storage, and communication of three types of information: **clinical, patient management,** and **financial.** A better understanding of CHIN requires further description of these three types of information. **Clinical information** consists of diagnostic and therapeutic data concerning patients. In telemedicine, much of the diagnostic information is available in digital form. The chances for success of telemedicine initiatives under the organizational umbrella depend on the extent to which a CHIN facilitates the exchange of *all* digitized information throughout an enterprise or region. **Patient management information** refers to demographic information, admission-transfer-discharge information for hospital inpatients, and scheduling information for ambulatory care patients. **Financial information** consists of billing and other financial data. Under the previous fee-for-service system, the major goal of financial information has been to generate charges for services delivered to patients. Financial management has been changed under managed care. Patients no longer receive a bill for the services delivered to them. Consequently, financial systems for generating bills are now rapidly evolving into systems for tracking and monitoring the allocation of resources.

CHINs as Organizational Entities

As described earlier, a CHIN is both an organizational and technical entity. Typically, discussions about CHINs focus on their hardware, software, and architectural elements. The architectural aspect refers to the information-technology infrastructure of the health care network. This primary focus on CHIN technology is unfortunate, because the organizational underpinnings greatly affect the goals and objectives of the network. The type of organization that owns and operates such a network will control the degree to which it emphasizes the exchange of clinical information. For example, a network owned or heavily influenced by one or more health care provider organizations is more likely to have an interest in the exchange of clinical information than one owned by a health insurer. The physician acting as an advocate for the integration of telemedicine objectives into the strategic plans of a CHIN therefore needs to take into account the goals and interests of the CHIN organization with which he or she is interacting.

No single organizational model dominates this early stage of CHIN development. Indeed, organizational variety is likely to continue in the future because the goals and objectives of any particular CHIN need to be shaped to the strategies of its organizers and to suit heterogeneous health care environments. CHINs will also undoubtedly change as the structure of the health care environment itself changes. Emerging federal and state health care legislation may shape future evolution. For example, legislation might require electronic data interchange between health care providers and the federal and state governments when they act as third-party payers. If so, public agencies might become important technical and organizational participants in CHINs.

CHINs as Technical Entities

Three key technical aspects of a CHIN are connectivity, interoperability, and the data repository. Connectivity refers to the physical linkage (e.g., telephone lines) between host systems. Interoperability refers to the ability of two host systems to exchange information in a meaningful way. A data repository, simply stated, is an information storage depot. Information can be copied to the repository from multiple host systems, thus integrating common information such as historical and clinical data for a single patient in a manner not possible on any one of the host systems that copy information to the data repository.

Given the extant technology, connectivity as it relates to CHINs is relatively simple. Systems consist of a local area network or wide area network that links the various host computers comprising the CHIN. Connectivity to physicians' offices in the community is usually achieved with modems and standard telephone lines. To emphasize the previous point, the various host computers may provide patient management, financial or clinical information. Authorized users with PCs can access the CHIN network via direct network connections or via a modem connection. Connectivity is sometimes referred to as "bandwidth," referring to the ability to transfer information from one place to another.

Because of improvements in network technology, bandwidth will soon cease to be a constraining factor in CHIN development.

Unlike connectivity, interoperability raises many technical problems for the near future. Interoperability refers to the ability of distributed host computers with different hardware platforms and running under different operating systems to communicate with each other. The presence of so-called legacy systems within nearly all health care organizations complicates interoperability. Legacy systems consist of older installed dedicated systems that usually perform specialized tasks adequately. Hospital personnel are often reluctant to eliminate legacy systems because of the previous investment in them and also because the systems frequently are meeting the perceived current needs of the organization. The basic problem with legacy systems is that they are often highly customized and inflexible. It is thus often technically challenging to create interface links between customized proprietary systems installed in a single hospital facility. Before the advent of so-called "open computing," hardware and software vendors, often the same companies, purposefully designed systems that could not easily communicate with a system supplied by another vendor. In fact, interoperability was often difficult to achieve even across product lines from the same vendor. This strategy was pursued to achieve "client control" and ensure a continuing income for the vendor. Some firms achieved remarkable success with such closed system approaches until networked architecture began to supplant mainframe-centric architecture and hospital clients began to balk at the purchase of highly proprietary systems. Achieving interoperability with a CHIN across multiple hospitals and connecting multiple legacy systems poses a substantial challenge.

The data repository of a CHIN, sometimes referred to as the data warehouse, is the most technically and strategically challenging aspect of CHIN deployment. It is at the level of the data repository for a CHIN that much of the information copied to it from various production systems is to be integrated. Queries supporting outcomes research will also be directed primarily to data warehouses. Such integration provides the patient-view of information that clinicians find highly desirable, as opposed to the departmental view of information provided by specialized departmental systems such as the laboratory information system. Most data repositories being developed today use relational database technology that allows sophisticated queries and *ad hoc* report development that cross individual patient records and individual patient encounters.

Relational database technology presents a tradeoff between the ability to retrieve data in a structured way across patients and the rapidity with which the system can respond to a request for such support. Clinicians appropriately demand sub-second system response time when retrieving clinical information necessary to treat an individual patient. System response time is the time required by a computer to respond to a user's command. Although response time depends on the complexity of the task that the computer is being asked to perform, response time is generally understood to refer to the "speed" of the computer.

Strategic issues concerning data sharing present what may be even a greater challenge than technical issues concerning data repositories. Creating a data warehouse to which multiple CHIN partners have unlimited access raises vexing issues for organizations that compete in a particular health care market. We expect an inverse relationship between the breadth of participation in a CHIN within a city or region and the level of detail in the information stored in the CHIN's data repository. CHINs that provide the most open access to health care-organizations and professionals will tend to emphasize short-term information exchange between pairs of organizations rather than long-term data storage and retrieval available to most participants in a city or region. CHINs that provide relatively open access to information can thus be distinguished from those that offer more restricted access. The tensions that exist between open versus restricted access to information are likely to have a major impact on the technical capabilities developed by particular CHINs.

Open Versus Restricted Access to Information

Some CHINs function as open systems, where most or all health care-related organizations and professionals in a region are eligible to join and are encouraged to do so. For instance, a network established in a city or region might accept as clients all hospitals, pharmacies, private practitioners, and payers willing to accept its fee structure. The Wisconsin Health Information Network exemplifies such an open CHIN. As an incentive for participation in the network, the CHIN fees for physicians are often underwritten by hospital sponsors.

Open systems provide opportunities for maximum exchange and integration of health care information. Open systems also allow smaller health care organizations to gain access to more sophisticated systems than they themselves are able to develop in-house. However, a CHIN pursuing an open system approach to development will need to address several major issues: maintaining system compatibility, establishing price schedules that allow less affluent health care organizations to participate yet also provide sufficient revenues to maintain and improve the system, and coping with conflicting strategic demands of competing participants in that CHIN. Many open systems will probably maintain a relatively simple level of technical sophistication to be compatible with other systems, keep costs low, and avoid strategic information conflicts. In addition, the difficulty of achieving a technical consensus among multiple participants will inherently limit the sophistication of such open systems.

Operating as a network hub, services offered by open CHINs are likely to include payment processing and relatively simple access to clinical data such as clinical test results. Such a system is analogous to a telephone system that provides basic communication connections among many people and organizations. To the extent that a CHIN maintains patient management, financial, or clinical databases; however, issues of breaches of security, patient confidentiality, and loss of control of proprietary information inevitably will arise. A hospital or other provider in a community might join an open system initially in order not to be excluded and perhaps achieve some of the basic benefits of information transfer. However, such a client might then oppose the creation of a more sophisticated network to protect patient confidentiality or to keep information away from its competitors. A hospital that does not achieve a high level of expertise as a participant in such an open coalition may also have an incentive to withdraw from the coalition and create its own restricted-access CHIN to develop more sophisticated information management capabilities.

In contrast to open systems, some CHINs will operate as proprietary systems in which the CHIN operator explicitly or implicitly restricts access to only some of the health care organizations in a city or region. The terms "restricted" and "proprietary" CHINs are synonyms here. Restrictions might take the form of interconnection requirements that effectively exclude smaller organizations or competitors. Proprietary networks are likely to be more sophisticated than open systems because of the possibility of enforcing stricter interconnectivity standards on all clients. Moreover, the ability to screen out competitors within a proprietary network will increase the use of the system to exchange strategic information. For instance, hospitals increasingly will need to compete to provide contract-based services for payers, employers, or lower-tier hospitals and managed care facilities. In this situation, knowledge of a hospital's cost structure and clinical volume become critical strategic issues. A hospital will be willing to share such information with other CHIN clients only if hospitals with which it is competing for contracts cannot gain access to the information. Similarly, pharmacies increasingly are viewing on-line patient records of prescription history as a strategic advantage and will be reluctant to share their records with competing pharmacies.

In the near term, it is likely that substantial tensions will arise within many regions concerning the extent to which CHINs will take an open versus restricted approach. Vendors that operate CHINs as open systems are likely to face conflicting pressures to maintain broad-based open systems and also to protect the strategic interests of individual organizations. Such vendors will also face issues concerning breaches of security and patient confidentiality, especially when the CHIN maintains extensive patient management, financial, or clinical databases. Some health care providers and payers undoubtedly will feel locked out from proprietary systems and will use whatever financial and political pressure they can to gain access to the CHIN organization and to the information and services provided by that organization. Some attempts to organize proprietary systems are likely to falter due to political pressure or the refusal of key organizations to join. In open access to CHINs, meanwhile, some central players are likely to attempt to limit their competitors' access to the system. Some open systems are likely to fail due to controversies that arise as a result of such attempts.

Because of the tension between open versus restricted access, we expect that CHINs evolving within hospitals or integrated delivery systems (IDSs) will become increasingly important. We also expect that more CHINs will form around a dominant health care provider than around multilateral coalitions of equal-status providers because it is easier to negotiate and develop a network in a less fractious organizational and political setting. Nonetheless, relatively small coalitions of providers are likely to continue to play important roles as organizers of hospital-based networks in some regions. Hospital coalitions that create CHINs will tend to offer access to those organizations and professionals whose interests complement those of the coalition. The complimentarity of interests should increase the incentives to provide detailed access to current information and to data repositories. In addition, the presence of a CHIN established by a local provider or small hospital coalition will create a concentrated enough set of interests to undertake long-term financial and managerial commitment necessary for long-term survival and success.

Organizational Models

Table 3-1 presents four basic organizational models for CHINs: (1) the for-profit vendor model; (2) the health care insurer/payer model; (3) the hospital or integrated health care delivery system (i.e., provider) model; and (4) the community-foundation model. The second column of the table summarizes several organizational features of current information networks representing each model, including ownership and management, expertise, information emphasis, and access.

Table 3-1. Four Basic Organizational Models for CHINs

| BASIC MODEL | FEATURES | | | |
	Ownership & management	*Expertise*	*Information Emphasis*	*Access*
(1) For-profit vendor model (e.g., startup firms, telecommunications companies, computer manufacturers, health care software vendors)	Vendor, often in partnership with local health care providers and payers	Varies with vendor background	Financial and patient management, with some clinical information (e.g., test results, medical records, prescription ordering, image transmission)	Sometimes operates as a common-access regional provider but will often offer primary access to partners
(2) Health care insurer/payer model	Health care insurer, payer, preferred provider organization, sometimes in partnership with local providers	Claims processing	Financial	Relatively open access to providers, but might have limited access to other payers
(3) Hospital or integrated delivery system model	Local hospital or local hospital coalition owns the CHIN; vendors frequently provide capital, develop systems, and manage networks	Clinical and administrative	Financial and patient management, with some attention to clinical information	Primary access to local provider or coalition partners
(4) Community-foundation model	Regionally based community coalitions, often with employer organizations as key members	Varied	Financial and patient management	Open

The For-Profit Vendor Model

The for-profit vendor model includes systems developed by startup firms as well as by companies with other established business lines. Several startup firms, most notably Integrated Medical Systems (IMS), have created community health-information networks in various regions and cities throughout the United States. IMS, which currently operates the largest number of CHINs in the U.S., is active in about 45 markets. However, most CHINs that are established by new firms offer relatively simple functionality to users because of the technical and financial constraints that new companies face. Startup firms will likely continue to play roles in creating new CHINs but will often face limits on new opportunities for expansion. IMS, for instance, recently was acquired by pharmaceutical manufacturer Eli Lilly and Company. Eli Lilly also owns PCS Health Systems, Inc., which manages most of the real-time retail pharmacy prescription transactions in the U.S. for managed care organizations.

Established companies in various industries are also keenly aware of the growth potential of health care and have started information networks. These are often developed in partnership with local health care providers and payers. Examples of for-profit network developers involved in CHINs throughout the U.S. include telecommunications companies such as Ameritech, hospital information-system software vendors such as Shared Medical Systems Corporation (SMS), and computer companies such as the International Business Machines Corporation (IBM).

Most of the CHINs initiated by established companies are in early stages. In addition to the Wisconsin Healthcare Information Network in Milwaukee, Ameritech launched the Regional Health Information Network of Northern Ohio (RHINNO) in Cleveland in late 1994 and is involved in projects in Tennessee and Illinois (1995). SMS, which is based in Malvern, Pennsylvania, is the prime contractor for the Metropolitan Chicago Community Health Information Network. The Chicago CHIN, which awarded its vendor contract in late 1994, is developing a network that eventually will link providers, physicians, payers, employers and others for the electronic exchange of clinical, financial and administrative data. The initial demonstration phase of the project, in fall 1995, allowed 500 physicians within six hospitals to receive laboratory results electronically. In 1995, IBM initiated two pilot-stage CHIN projects in Dayton and Cincinnati, Ohio.

The growth potential of the CHIN market will probably continue to attract many vendors of health care and EDI software as a new market for their products. Such vendors may position themselves as purveyors of CHIN software or they may sell turnkey CHINs (hardware plus software) to hospitals and health care insurers. Several factors underlie the appeal of the evolving CHIN market for software and hardware vendors. First, the health care vertical market is attractive owing to its size. Second, the hospital-based market for information systems is fairly mature, so new sales will be difficult to generate unless they are compatible with a CHIN strategy pursued by a hospital or integrated delivery system. Third, a health care software vendor without CHIN software runs the risk that its installed hospital client base may be eroded by other vendors that are able (or perceived to be able) to offer products that allow information to be shared across heterogeneous organizations.

The relationship between such for-profit software or turnkey system vendors and payers and providers will be analogous to that which currently exists between hospitals and their vendors of hospital information system (HIS) or laboratory information system (LIS) software. That is to say, the vendor licenses the use of the software to the client and also charges monthly software support fees that cover the cost of new software releases as versions. Although the software and hardware supplied by such CHIN software vendors may be critical to the success of the CHIN, the vendor will not be a financial stakeholder in the CHIN organization and will be compensated on the basis of software and hardware purchase price and software licensing fees rather than on the basis of being an equity stakeholder in the CHIN venture.

The experience and expertise of the vendor developing a health network, as well as the demands of payers and providers participating in the organization, will affect the functionality of a vendor-operated network. Although this may seem intuitively obvious, it is important to state it to counteract some of the hyperbole that often accompanies CHIN announcements and press releases. Different vendors bring different expertise to bear on their development efforts. For instance, firms dealing with hospital information systems usually have prior experience with information-system development and deployment in the medical sector. On the other hand, telecommunications companies, especially large ones, have vast experience with data transmission, system connectivity, and interoperability among disparate systems. Computer companies have substantial computer network architecture and operating system experience. By contrast, many vendors, even HIS software vendors, tend to have limited experience in the day-to-day management of clinical databases. To date, the primary emphasis of most vendor-operated CHINs has been on financial and patient management information. Many systems also offer some management of clinical information such as laboratory test results but, for most, large-scale clinical data exchange remains in the future.

The Health Care Insurer/Payer Model

The second category of CHINs is the health care insurer/payer model. Preferred provider organizations (PPOs) are included in this category. Some health-insurance organizations such as the Blue Cross/Blue Shield in Virginia, Minnesota, and Illinois are establishing CHINs to provide two-way electronic financial links with health care providers. Similarly, some PPOs such as the ProviderLink subsidiary of United HealthCare, based in Minneapolis, Minnesota, have also set up such networks. A preferred provider organization is a network of designated providers that accepts discounted fees for services in exchange for an agreement to abide by certain guidelines regarding use of service. Typically, insurers and preferred provider organizations have experience in electronic data interchange of claims processing data. Therefore, most insurer/payer CHINs are heavily oriented toward financial transactions.

Hospital or Integrated Health Care Delivery System Model

The third category listed in Table 3-1 includes a local hospital, hospital coalition, or an integrated health delivery system (IDS). An integrated delivery system is a vertically and horizontally integrated health care provider organization. Nearly all large tertiary care medical centers or large networked hospital systems are now integrated delivery systems. Examples in Michigan would include the University of Michigan Medical Center in Ann Arbor and the Henry Ford Health System in Detroit. An integrated delivery system may also provide an insurer function, which spans the second and third categories in Table 3-1, and can be defined as an accountable health plan. Examples of CHIN ownership by providers include a coalition of hospitals in a single city (e.g., the Metropolitan Chicago Healthcare Coalition), a health system encompassing multiple hospitals under a single corporate umbrella (e.g., Inova Health Systems in Virginia), and a large tertiary care hospital that has purchased smaller hospitals and developed a series of free-standing clinics to create an integrated health care delivery network and a health-maintenance organization (e.g., Henry Ford Health System in Michigan and its Health Alliance Plan). We also refer to integrated delivery system CHINs as provider-based systems.

Provider-based CHINs are organized as organizational units within a hospital or a coalition of hospitals, and tend to offer the broadest range of financial, patient management, and clinical information to users. The scope of their functions reflects the need to increase the efficiency or productivity of physicians and nurses who are the employees of the enterprise. By way of contrast, a CHIN owned by a PPO would have less perceived need to improve the efficiency of physicians who contract to provide services to that organization, since they are contractors rather than employees.

In practice, many CHINs in the first three categories listed in Table 3-1 include joint participation among for-profit CHIN vendors, payers, and providers. MEDACOM Minnesota, for instance, is a joint venture of Minnesota Blue Cross Blue Shield and Integrated Medical Systems. IMS, which operates CHINs in multiple large metropolitan areas throughout the country, also licenses health information network software to payers or providers in smaller markets. Joint participation in CHIN development is often critical because of the complexity of the network design and installation, the technical challenges of creating data warehouses, and the political ramifications of sharing health information on a regional basis.

The political ramifications of making health care information widely available on a regional basis are beyond the scope of this discussion. Some of the political issues involve confidentiality and privacy, antitrust legislation, Food and Drug Administration (FDA) regulatory environment, professional accreditation, reluctance of provider organizations to share what they perceive to be proprietary information with competitors, and, finally, control of the storage of information. The latter issue concerns both CHIN vendors who try to position themselves as regional brokers of health care information, and the providers who may perceive this as a usurpation of their natural roles.

Provider-oriented CHINs frequently involve one or more CHIN vendors that serve as investors, developers, and managers. For instance, a vendor partnership known as ChinAlliance is developing the Metropolitan Chicago CHIN. Shared Medical Systems (SMS) acts as the prime contractor. AT&T Health Care Solutions provides networking and network management, videoconferencing, telemedicine and teleradiology experience. Coopers & Lybrand's health care consulting practice provides systems integration experience. Healthcare Data Exchange offers broad access to payer eligibility files. IMS/Illinois Medical Information Network provides community health care products and services that allow the electronic communication of text, data, and voice messaging among providers. National Electronic Information Corporation (NEIC) provides electronic claims processing experience. PCS Health Systems, Inc., which is also the parent of IMS, offers prescription services. Synaptek provides all-payer claims processing for physicians. The partnership contracted in 1994 to develop and operate the CHIN for a consortium of Chicago-area hospitals. Thus, the same vendors that operate for-profit networks are also actively developing provider-oriented ones.

The Community-Foundation Model

The fourth organizational model for development listed in Table 3-1 is the community-foundation model. An example is the John A. Hartford Foundation of New York, which has provided funds and acted as a catalyst for the preliminary development of CHINs within several communities, including MidSouth Health Care Alliance in Memphis, Tennessee, and the Foundation for Health Care Quality in Seattle, Washington. Ownership of such a network resides in a not-for-profit organization that encompasses health care providers, employer groups, foundations, consumer groups, and public agencies within the community. Private foundations have been willing to provide seed money for the creation of CHINs because the exchange of health care information may provide positive benefits for the community, such as ready access to a patient's medical record, sharing of health records across different providers, and the presumed efficiency gains for the health care industry associated with enhanced EDI referred to earlier. The community foundation category is similar to the hospital or integrated health-system model. The foundation model is not likely to be a major force in CHIN development in most communities, however, owing to the difficulty of organizing broadly based groups with diffuse goals. Nonetheless, the model may succeed in particular locations where coalitions already exist or are easy to form.

Comparison of For-Profit and Other CHINs

The for-profit vendor CHIN variants share several distinct advantages in comparison to insurer-based, hospital-based, and community-based CHINs. Established companies in the telecommunications and computer sectors tend to have substantial capital to invest in CHIN ventures, as well as requisite technical expertise. Additionally, vendor-based networks are often able to respond to technical and market changes more quickly than CHINs controlled by health care organizations, especially when compared to CHINs that involve coalitions of health care providers and other organizations. Further, for-profit health information networks can sometimes act as "disinterested brokers" among health care organizations with competing interests, much as telephone companies and electric utilities provide basic services to competing businesses.

In spite of these advantages, both technical and political problems have slowed the diffusion of for-profit CHINs. On the technical side, some vendors have invested insufficient capital or lack expertise needed to develop sophisticated systems. Perhaps more importantly, vendors generally have underestimated the inherent political difficulties of bringing together disparate health care organizations to create a single information network. The political difficulties become particularly pronounced as the sophistication of systems and sensitivity of information increase. Health care professionals and delivery organizations have become increasingly wary about becoming too dependent on outside for-profit vendors for the day-to-day control of their proprietary information. Health care professionals understand that they run the risk of being unable to deliver effective health care if system problems disrupt information transfer in any way. System failure can also occur if information management is handled totally in-house, of course, but many health care organizations want to play a major role in managing information transactions. Moreover, hospital CEOs have become increasingly aware that information management plays a strategic role in health care delivery, and organizations risk losing competitive advantage if competitors gain access to sensitive information about them. The potential for security breaches exists when providers delegate information management to vendors.

Such issues are likely to have a strong impact on the success of CHINs. For instance, political concerns have slowed the expansion of the Regional Health Information Network of Northern Ohio (RHINNO), which Ameritech established in late 1994. The Health Action Council of Northeast Ohio, which is a business coalition of health care purchasers, contends that the network cannot "serve the complex and divergent needs" of the area health care community with Ameritech running the network (Schrimpf, 1995). The business group and the Cleveland Academy of Medicine have withheld their support from RHINNO, which had only one subscriber by fall, 1995. Similarly, more than half the respondents to a survey by *Modern Healthcare* magazine said that consortia of hospitals, physicians, payers, and business interests should own CHINs, compared with 6 percent who thought vendors should own the networks (Morrissey, 1995).

Revenue issues also may favor the development of provider-based CHINs relative to for-profit CHINs. Revenue sources among for-profit CHINs include combinations of initial connection charges, annual membership charges, and usage fees. One company charges a client $75,000 to $150,000 to connect to its system, plus a monthly charge ranging from about $80 for doctors to as high as $24,000 for high-volume payer organizations. Some for-profit networks assess fees on local hospitals who then provide free or inexpensive connections for local physicians. Some companies operate by charging an initial licensing fee for the software, yearly software support fees, and consulting fees for their roles as a systems integrator, while others charge a small fee per transaction.

No matter what revenue format is selected, however, many for-profit CHINs are likely to reach quickly the limits of what health care organizations and professionals are willing and able to pay for their services. These financial constraints may in turn limit the growth of for-profit CHINs. By contrast, many hospital-based networks will be able to include capital and operating costs within their broader information-systems budgets. To the extent that it contributes to the strategic position of a hospital, the direct financial self-sufficiency of the CHIN will be perceived as less important than it would be in the case of a for-profit CHIN. In the near term, therefore, health care providers, payers, and other health care organizations are likely to play substantial roles in most successful CHINs. Many for-profit vendors will play larger roles as subcontractors for system development and management of provider and insurer-based CHINs than as operators of stand-alone for-profit health networks.

A CHIN Evolution Trajectory

As noted earlier, most existing CHINs offer relatively simple information exchange and relatively limited access within a region. Over time, however, CHINs will probably become more technically sophisticated and serve a broader base of health care professionals and patient populations. Table 3-2 provides a simple taxonomy for a CHIN trajectory of the evolution of CHINs.

Table 3.2. The CHIN development trajectory from an enterprise health network to an integrated health system network to community health information network (CHIN). Under each of the three categories are two bullets listing its major distinguishing features.

Enterprise Health Network	*Integrated Health System Network*	*Community Health Information Network*
• Confined to enterprise	• Links affiliated and contracted entities	• Community-based separate organization
• Minimal integration of information	• Early development of data repository or data warehouse	• Full regionalization of health information

The simplest form of a CHIN in Table 3-2 is an enterprise health network that establishes access via a local area network among various host computers within a single enterprise. We also refer to this model as an enterprise CHIN. Such enterprise networks usually develop within large tertiary care medical centers that include multiple hospitals and geographically distributed free-standing clinics or surgicenters. Users access the various host computers and databases within the enterprise using workstations and PCs. The users of an enterprise network can connect sequentially with the mainframe computer for patient accounting and patient management information, the laboratory information system for laboratory test results, and the pharmacy system to order medications. An enterprise health network provides relatively high volume and highly functional linkages between the main hospital and geographically distributed clinics, surgicenters, and purchased physician practices. Enterprise CHINs serve many needs of complex health care organizations. A large medical center has a compelling need to establish two-way electronic communication with its own medical staff and also with its cadre of referring physicians. In essence, such a complex tertiary care center creates a new goal in its strategic planning when it decides to create a CHIN, that is, regional integration and communication beyond that offered through smaller-scale, facility-wide computer links.

Many CHINs organized to serve a single integrated health care delivery system will be enterprise CHINs, with a complex tertiary care hospital system as the central organizer. An enterprise CHIN encompasses a single provider organization (e.g., the Henry Ford Health System). An enterprise health network will often be easier to plan and deploy than a health network that facilitates communication between independent organizations because competing hospitals are not part of the process. Nonetheless, legacy systems pose the same interconnectivity problems for enterprise CHINs that they do for health networks spanning different organizations. In addition, an enterprise health network is as much a moving target as an interenterprise network because large medical centers are in constant negotiation with other providers and payers and frequently establish new strategic alliances and merged organizations. The new enterprises that emerge from such negotiations will have new and enlarged appetites for information exchange.

The integrated health system network in Table 3-2 is the second stage in the evolution. An integrated health system network is broader than an enterprise health network, providing wider ranging communication links to the various entities that have contractual relationships with the managed care organization that controls the integrated network. Such a network might involve the creation of communication links between a tertiary care hospital that has organized itself as an HMO and an outside group of primary care physicians with which the health maintenance organization has a contract to provide care and gatekeeping functions. Clearly, brief telemedicine consultations between the primary care physicians and the specialists in the tertiary care hospital about the need for patient referrals would be valuable. Similarly, the hospital might create links with specific insurers, employer organizations, pharmacies, reference laboratories, and other organizations with which it has close relationships. Many vendor-operated CHINs that offer open participation also will provide at least part of the functionality associated with an integrated health system network.

The last category in Table 3-2 is the fully formed CHIN. Criteria for such an entity include the seamless transmission of clinical, patient management, and financial information among participants. A fully-formed health network would be a separate business organization and would have open participation to a broad range of health care organizations and participants. A fully-formed CHIN would help achieve regionalization of health information, by which we mean nearly total portability of information across various providers and payers. This would be achieved by the creation of a regional data repository that stores data generated by multiple providers in the region. These criteria for a fully formed health information network are far beyond the technical and organizational reach of any current CHIN. Nevertheless, there are compelling reasons for such a goal. Even partial realization will be a major step forward.

CHINs and Managed Care

The phenomenal growth of managed care organizations has been the major impetus for the growth in interest in CHINs, especially for the increasing interest in creating them as an integral part of integrated health care delivery systems. Several reasons underlie the tight coupling between health network development and managed care. First, the movement toward managed care has stimulated the consolidation of smaller hospitals into larger geographically dispersed organizations. Such consolidated health care corporations require integrated financial and patient management information systems, if for no other reason than to manage the organizations from a business point of view. In this context, hospital CHINs can provide system integration on a regional basis. The pursuit of interoperability and the integration of multiple host systems within a hospital chain present substantial difficulties, given that many or most of the consolidating hospitals will often wish to retain, or cannot afford to eliminate, their legacy financial and patient management systems.

Second, the movement toward managed care stems from the need to reduce costs while simultaneously maintaining quality. The cost of HMO and PPO contracts is now declining after an extended period of yearly increases for health care coverage that substantially exceeded increases in the yearly cost of living. Because health care is information-intensive, many of the anticipated cost reductions under managed care will flow from enhancements in automated information management (Davidson, 1992; Johnson, 1994; Grandia, 1994; Wyatt, 1994). Robert Kennedy, the Chief Information Officer at the Metropolitan Chicago CHIN, estimated at the "Second Annual CHIN Summit" held in June, 1995, that approximately 2–4 percent of total health care dollars could be saved through administrative cost savings related to CHINs, without taking clinical cost savings into consideration. With a $1 trillion health care economy, this would translate into about $20–$40 billion in savings.

CHINs have thus become largely synonymous with the notion of efficient data management and data sharing across regions and complex organizations, providing a kind of mantra for efficient and automated information management in health care organizations. This enthusiasm for integrating systems and information across disparate organizations and across geographic distances may be premature, given the relative lack of success of system integration within many individual hospitals. Nonetheless, while some of the goals and assumptions about CHIN functionality may be overstated, CHINs are likely to continue to evolve as organizational and technical entities.

The movement toward managed care and the need for health care reform have also provided opportunities for health care consultants. Many of the consulting system integrators of former times are now being engaged as CHIN consultants and much of the impetus for projects may be coming from such consultants. Although there is support for the notion of top-down, long-range, strategic CHIN planning, there is also a fear that such plans may be unrealistic and might never be acted upon, in the same way that many prior system integration plans were never implemented at the facility level. Strategic network planning that considers the needs of an organization for data communication and exchange, including the provision of health care delivery at a distance, is likely to be more successful.

Managed Care as a Driver for Telemedicine Within CHINS

Current developments concerning CHINs in managed care systems will affect the potential relationship between CHINs and telemedicine. Some CHINs undoubtedly will evolve as networks that link many independent health care facilities. In the near future, however, most CHINs and similar systems will be controlled by single managed care systems or, at most, by small groups of health care organizations. Managed care systems have the incentives to use more complex forms of information exchange, both to provide health care and to gain competitive advantages, and the resources to develop the information networks. Therefore, many of the opportunities for linking telemedicine initiatives to CHIN development will arise within the managed care systems now emerging, rather than within multiparty networks of traditional health care organizations.

The evolution of CHINs within managed care systems has implications for the development of telemedicine.

Historically, telemedicine programs have been designed primarily to deliver health care services to isolated and underserved patient populations, such as rural populations, military personnel, nursing home residents, and prison populations. The funding for such programs often came from federal and state governments or from foundations with a special interest in the health care needs of such populations. Under managed care, financial incentives are now rapidly evolving that can serve as a stimulus for hospital and health maintenance organization CEOs to invest in telemedicine programs that span shorter distances. Such incentives have been weak in the past.

It is useful to consider how such telemedicine initiatives might arise in a large tertiary care teaching hospital. Under fee-for-service, physician specialists and the hospital itself derived financial benefits when geographically removed primary care clinics referred patients to the specialty care ambulatory clinics or referred patients for inpatient admission to the hospital. Under managed care and capitated systems, by contrast, there are strong disincentives for such referrals to specialists. There is undoubtedly a high percentage of cases in which a primary care physician in a free-standing clinic could manage effectively, if that physician could have a brief real-time consultation with a specialist to buttress his or her proposed treatment plan for the patient.

Of course, such a consultation could and does take place today with little capital investment, via the telephone. This opportunity notwithstanding, real-time spontaneous teleconsultations will be more acceptable to physicians and patients alike with both audio and video connections and with both the referring and consulting physicians simultaneously viewing identical laboratory test results, diagnostic images, and a patient's current list of medications on their PC screens. Such teleconsultations will be common and well accepted in the near future, particularly as patients begin to realize that they are also direct beneficiaries of the efficiencies of the system. The nature of such intra-organizational teleconsultations, in summary, is that they will be brief, the distance spanned will be relatively minor, and the goal of such a consultation will be to avoid a patient referral to the very specialist who may be providing the telereferral. The major driver in the growing incidence of teleconsultations will be that CEOs will invest in and promote such systems because their emergence will have a substantial effect on the cost and efficiency of medical services. In other words, the telemedicine investment "goes right to the bottom line," especially when it can be combined with investment in a CHIN that also provides other clinical, patient management, and financial functionality.

This discussion has significant implications for the future geographic span of telemedicine. Telemedicine is generally thought of as spanning long distances in order to reach remote locations. In addition, though, telemedicine that develops in a CHIN environment and contributes to CHIN development will also commonly span much shorter geographic distances. Distances to be spanned in the delivery of health care will include a few miles between suburban outpatient facilities and central hospitals, or even contact of a few city blocks between adjacent buildings that are part of the same medical center. This variant of telemedicine is referred to here as "enterprise telemedicine" and is likely to become the most common variant of "health care delivery at a distance" in a short time.

REFERENCES

Anonymous. 1995. "Introducing the New Comprehensive Reference Tool that Profiles over 500 Health Information Networks," *Journal of the Community Medical Network Society,* 1: 32–33.

"Careful Aim for Community Connections a Worthy Goal," *Journal of the Community Medical Network,* 1: 52–57.

Davidson, R.J. 1992. "Computerized Patient Records: CPR–The Key to Reform," *Health Systems Review,* 25: 18–19, 22.

Furukawa, M.F., and T. Peake. 1995. "Profiling America's Health Information Networks," *Journal of the Community Medical Network Society,* 1: 25–28, 30–31, 34.

Grandia, L. 1994. "How a Regional Health Provider Uses Information to Meet Its Strategic Objective," *Healthcare Information Management,* 8: 35–43.

Johnson, G. 1994. "Computer-Based Patient Record Systems–A Planned Evolution," *Health Information,* (January): 42ff.

Morrissey, John. 1995. "Structure of Vendor Deal Stalls Cleveland's CHIN," *Modern Healthcare,* (August 21): 64.

Schaich, R.L. 1993. "Making the Case for EDI," *Computers in Healthcare,* (December): 18, 20, 22.

Schrimpf, N. 1995. "Area Health Info Network Seeing a Sluggish Start," *Crain's Cleveland Business,* (October 2): 39.

Wakerly, R.T., ed., in cooperation with the First Consulting Group. 1994. *Community*

Health Information Networks: Creating the Health Care Data Network. Chicago: American Hospital Publishing.

Wyatt, C. 1994. "Automation is at the Heart of This Cost-Control Effort," *Health Data Management,* 2: 60, 62–64.

SECTION II
THE TECHNOLOGY OF TELEMEDICINE

Chapter 4

TELEMEDICINE TECHNOLOGY

JOHN R. SEARLE

INTRODUCTION

Technology has long been associated with the practice of medicine. Typically, those technologies which practitioners and patients evaluated as desirable were perceived as providing diagnostic information or therapeutic effect at a reasonable cost. The wide range of technologies applicable to telemedicine, from the telephone to telerobotic manipulation of surgical instruments, and the diversity of telemedicine applications have broadened and complicated the evaluation process.

The supply of and demand for technologies applicable to telemedicine have played a major role in the development of telemedicine applications. For some applications, appropriate technologies are either lacking or are not widely available. Some applications once tied to a single technology now benefit from multiple competing technologies, especially emerging technologies. The degree to which telemedicine is able to play significant roles in health care must be partly due to the appropriate application of both mature and emerging technologies to well-defined applications. Therefore, this chapter is intended to help the reader understand and critically evaluate the essential technical elements of telemedicine applications, elements often given little attention in the telemedicine literature. To evaluate telemedicine systems properly, one must know not only the medical and operational requirements, but also the costs and benefits of competing technological responses to clinical requirements in the diagnosis and treatment of disease.

Enabling Technologies

One of the difficulties of telemedicine in the 1970s was that the technology of the day could not economically meet what was needed. In some cases, the technical needs were not fully appreciated. Since those first telemedicine experiments, various technologies have emerged from the research and development phase and are now being used in diverse operational settings. The core technologies of telemedicine today are: digital video, video compression, digital imaging, digital communication, and multimedia computing.

Digital Video

Introduction. Many telemedicine applications utilize video and individual still-frame images obtained from that video for interactive visual communication and medical diagnosis. National Television System Committee (NTSC) video, adopted in 1953, is the analog signal format used in the United States for broadcast and cable transmission of television. For point-to-point connections at distances under about a kilometer, it is feasible to transmit two-way NTSC video over permanently installed coaxial cables or over a line-of-sight microwave link. But, beyond that distance, economic considerations usually justify digitizing the signal (representing the analog signal as a stream of digital data), compressing the resulting digital video (removing redundant data), and transmitting the data over digital communication links provided by telephone, cable television, or other telecommunication service provider. Not only is this method less expensive, but more choices exist for switching the signal to one or more distant sites.

The NTSC video format consists of 30 image frames per second; each frame is an interlaced raster scan of 512 horizontal lines (typically 484 lines are displayed) sized to obtain a 4:3 aspect ratio (Conrac Division, 1980). Luminance (intensity) and chrominance (color) information are combined into one composite signal for broadcast over a bandwidth limited channel or coaxial cable. Vertical resolution is limited by the number of scan lines and horizontal resolution is limited by the specified bandwidth of the luminance portion of the signal, 4.2 MHz. An alternative expression of resolution is the term television lines (TVL), the number of alternating light and dark bars an observer can resolve in the vertical dimension or along 75 percent of the horizontal dimension (an equal distance for 4:3 aspect ratio). NTSC luminance resolution is 336 TVL. Hue and saturation data are embedded within the NTSC signal as a modulated subcarrier at lower bandwidth (resolution) than the luminance signal. A complex relationship exists between the hue, intensity, and saturation (HIS) color-coding system used in NTSC and the red, green, and blue (RGB) coding used in computer display systems (Conrac

Division 1980: Appendix III). NTSC is a composite signal including luminance, chrominance, and vertical and horizontal synchronization signals.

In component video formats, unlike composite NTSC, the intensity and chrominance components are separate signals and are not limited to an arbitrary bandwidth because component video is confined to non-broadcast applications. The most common component video formats include luminance/chrominance (also known as Y/C and Super-VHS), red/green/blue (RGB), and luminance/red chrominance/blue chrominance (YCrCb). Component video is superior to composite NTSC for interconnecting the cameras, local display monitor, video tape recorder, and frame grabber components of a telemedicine system because none of the spatial detail captured by the camera is lost.

The Digital Camera. An image-forming lens system, one or more light-sensitive integrated circuits with a matrix of light sensitive elements, and circuits for timing, nonlinear amplification, and encoding color constitute a digital camera. The most common type of sensor is a charge-coupled device (CCD). Incident light exposes an array of discrete light-sensitive regions called pixels which accumulate electric charge in proportion to light intensity. The CCD sensor in a camera with a single sensor (single-chip camera) contains a microscopic array of red, green, and blue filters covering the pixels, while a camera with three CCD sensors (3-chip camera) contains large prisms and filters to form separate red, green, and blue images. After a specific integration time, the accumulated charges in the pixels are sequentially converted into conventional analog (continuously variable) intensity and color signals. The integration time can be varied automatically as an electronic iris to adjust exposure and reduce the blurring of brightly illuminated moving objects. Single CCD cameras offer 450 TVL or higher resolution at moderate cost. High-quality 3-chip CCD cameras offer 700 TVL or higher resolution. It should be noted that high resolution is not only a function of the number of pixels in the sensor(s), but also can be augmented by digital signal processing, which effectively interpolates additional points between pixels and enhances intensity transitions at object borders. Increased resolution from larger numbers of sensor pixels, the technique of choice for still-frame images, is much more costly for video. Larger CCD arrays are exponentially more difficult to manufacture and require more sophisticated and, therefore, more costly electronics to accommodate shorter sample times for each pixel. As the number of pixels increases, the amount of time available to sample each pixel decreases at a fixed frame rate. High-resolution cameras typically do not use standard video format but instead scan at a low rate and offer more scan lines. Noise, dynamic range, and sensitivity are also factors; as the number of pixels per unit area increases, the area per pixel decreases along with electron storage capacity (noise and dynamic range) and the amount of light collected per pixel (sensitivity).

Digitization. Digital video is a coded sequence of data which represents the intensity and color of successive discrete points along the scan lines. It is a sampling process which occurs first in the CCD camera. Another sampling process occurs in the frame grabber or digitizer. Here, the signal is first decomposed into red, green, and blue (RGB) components which are simultaneously sampled at a fixed rate and resolution (typically 640 8-bit samples per color per scan line times 480 scan lines). These components are stored in video random access memory (VRAM). One such frame thus requires 921,600 bytes (7.4 million bits). When the image is displayed, a third sampling process takes place; after the digital RGB data are converted back into analog values, a scanned image is formed of pixels composed of discrete red, green, and blue light emitting elements. The errors and distortions which occur in each of the sampling processes are more significant to the quality of still images than to video.

Video Compression

At the resolution and frame rate associated with NTSC video, digital data are produced at a rate of over 100 Mbps (megabits per second). Communication and storage costs usually dictate that some form of compression be used to reduce this data rate. The codec (coder/decoder), a device responsible for compressing and decompressing the signal, finds and removes both spatial and temporal redundancies and reconstructs the video in such a way that the missing data are not readily perceived. Using a variety of techniques, codecs typically found in telemedicine applications operate at data rates between 336 Kbps (kilobits per second) and 1536 Kbps, although rates as low as 56 Kbps and as high as 45 Mbps are sometimes used (see Table 4-1). In most telemedicine applications, the data rate of the communication channel is static while the information content of the video sent to the codec is dynamic. The greater the amount of movement and detail within the visual field, the less redundant information is present and either the frame rate or spatial resolution of the transmitted signal must be lowered. In commercial systems, control of the coding parameters which determine frame rate and spatial resolution is usually fixed within the design of the codec. The temporal, spatial, intensity, and color resolutions required by a telemedicine application along with available communication services and rates determine the minimum level of codec performance needed. The temporal resolution (frame rate), spatial resolution, and operational cost afforded by a data rate of 384 Kbps might be perfectly adequate for face-to-face discussion of therapy choices, but is inadequate for diagnosis of muscle tremor. A codec operating at 45 Mbps would display pediatric cardiologic ultrasound with excellent fidelity, but the recurring cost of the communications channel would not justify its use primarily for continuing education.

Table 4-1. Video Compression Formats

Name	Spatial Resolution H X V	Maximum Frame Rate	Communication Rate (video + audio)
H.261	352 X 288 (FCIF)	30	P x 64; P=1, 2, 3, 4, 6, 12, 18, 23, 24, 30
	176 X 144 (QCIF)	30	
MPEG-1	352 X 240	30	1.2 - 3 Mbps
CCIR 601	720 X 486	30 Hz	5 - 10 Mbps
CTX-Plus*	256 X 240	30 Hz	384 Kbps - 1.5 Mbps
Blue Chip*	352 X 288	30 Hz	112 Kbps - 1.5 Mbps
Indeo 3.2*	320 x 240	15 Hz	106 Kbps
	160 x 120	30 Hz	
DVI* (PLV)	320 X 240	30	1.2 Mbps

 * Proprietary: CTX-Plus owned by Compression Labs Inc.; Blue Chip owned by VTEL, Inc.;
    Indeo and DVI (Digital Video Interactive) are owned by Intel Inc.

Compression may be lossless or lossy, and is expressed as the ratio of uncompressed data to compressed data. Lossless compression may be indicated for preserving fine detail or rapid motion, but the maximum amount of compression is usually less than 2:1. Lossy compression can achieve large compression ratios, and when applied within specific constraints to medical images, lossy compression can reproduce images indistinguishable from the originals for many types of images. There are two approaches to lossy video compression: interframe coding and intraframe coding. (Compression of still frame images such as radiographs will be discussed in the section on digital imaging.)

Interframe Coding. Motion compensation is the principal technique available to reduce temporal redundancy, that portion of an image which changes very little from one frame to the next. In a sequence of video frames, each frame is defined as an array of blocks, each usually 16 by 16 pixels. A *reference* frame is one in which all the blocks contribute to the coding process. If the size of the blocks is small enough to preserve their spatial content over a short period of time, then motion in near-future frames can be modeled as rearrangements of the same blocks translated by arbitrary distances and directions. The movement of these blocks to new locations in a subsequent *predicted* frame is coded by a motion prediction algorithm. The content of a predicted frame is derived from the content of the reference frame; the data transmitted for a prediction frame are only those needed to describe the motion of the blocks from the previous reference frame. In a typical sequence of nine frames, the first is a reference frame and the fourth and seventh are predicted frames. Frames groups 2 and 3, 5 and 6, and 8 and 9 are bidirectionally *interpolated* frames for which the spatial differences between neighboring frames are coded as residual errors. This process yields smooth transitions between reference and predicted frames, and greatly enhances the amount of temporal compression. The combination of reference, predicted, and interpolated frames can be adjusted statically or dynamically to maintain image quality (resolution and frame rate) at the expense of variations in data rate, or to vary image quality to meet a static data rate constraint.

Intraframe Coding. The data stored in reference frames and predicted frames both contain spatial redundancies. If the blocks which make up the frames are small enough so that the spatial features of the image contained within a block may be assumed to be constant, the pixels within the block may be represented by a series of terms generated by a two-dimensional (horizontal and vertical) mathematical transformation. Although an infinite number of terms may be required to define the image perfectly, the value of most of the coefficients is close to zero, and the image may be represented in its new transformed state without redundancy by only a few coefficients. The most efficient transform, the Karhunen-Loeve transform (KLT), requires the computation of unique mathematical functions for each unique region of the image, an expensive process in time and transmission of the transform functions. The discrete cosine transform (DCT) is less efficient but more suitable: not only does it operate on data organized in blocks of pixels, but the transformed data define spatial frequencies which have a direct psychophysical interpretation, the coefficient terms can be grouped into subbands, efficient and therefore fast algorithms for computing the DCT are well known and can be implemented in software or in hardware, and mathematical functions are the same for each block in the image. The range of subject responses of the human visual system to spatial frequency, contrast, and brightness as well as objective measures of these parameters for various types of medical images have been studied (Chesters, 1982; Ji, Sundareshan, and Roehrig, 1994).

The DCT coefficients are transmitted with less precision than their calculated value in a process called quantization. Rounding the coefficients up or down to the nearest quantum value reduces the number of data needed to represent them. Quantization of the DCT coefficients can be tailored to the type of image to allow for optimum image compression without sacrificing acuity needed for diagnosis (Lee, Kang, and Lee, 1993). The quantization process may be modified from one moment to the next, better to match the coding process to variations in the video content to preserve image quality or to vary the image quality in order to maintain a constant transmission bit rate. This adaptation process can be readily seen when a large spatially detailed moving image comes to rest; the spatial detail of the still image becomes fully evident over a period of about a second as more coefficients representing higher spatial frequencies are transmitted and as their values are more accurately defined. The nonzero quantized coefficients are Huffman-coded by indexing in a data dictionary. Common coefficient values have shorter codes than less frequently encountered values. The coefficient matrix of horizontal and vertical coefficients is scanned diagonally to maximize runs of zeros which are economically coded by counting their number (run length coding). More detail on interframe and intraframe coding can be found in Le Gall (1991).

Digital Imaging

The traditional media used to record static diagnostic images include sheet and roll film (photography and radiography) and that used for recording dynamic images include cine film and video tape (fluoroscopy, ultrasonography). Digital media for recording diagnostic still images include magnetic disk and tape and optical film and disks. The digital recording process is not direct but involves a sequence of processes which can be lumped into the general categories of acquisition and display.

Acquisition. The usual types of diagnostic images used in telemedicine include:

- Images obtained from direct visualization by a video camera and a lens system for direct observation (e.g., a skin lesion) or an optical adaptor to a conventional scope (e.g., laparoscope, microscope, otoscope) that provides magnification or remote access using fiber optics
- Computer-generated images (e.g., ultrasound, CT) available in standard video format (NTSC), computer format (SVGA), or computer-file format (TIF).
- Images stored on traditional film or print media (e.g., x-ray film) and converted into digital format by direct imaging or scanning in a raster sequence under controlled lighting conditions.

Video is captured one frame at a time typically in a 640 by 480 pixel format, with an intensity scale typically consisting of eight bits for monochrome and 24 bits for color (8 bits each for red, green, and blue). Other formats exist for high-resolution cameras; a 1024 by 1024 pixel format at 8, 10, or 12 bits per pixel can be achieved with a capture time which depends in part on the time needed to integrate a sufficient number of photons from dark areas of the image. High-resolution slow-scan cameras and interface boards cost considerably more than conventional video equipment. Sequences of video frames can be captured in real time and stored in memory subject to limitations of speed and size of memory. One-half second of color video (15 frames) requires ten megabytes of memory. By comparison, most personal computers running Windows require a minimum of eight megabytes of memory in addition to the memory needed by active application programs. Longer video clips therefore require real-time data compression and either extensive memory or high-speed data transfers to disk storage.

Modern digital computed tomography (CT) images have about one-tenth of the spatial resolution of conventional radiographs but can provide a series of planar images in any sectional orientation. With data processing for contrast enhancement and three-dimensional reconstruction on video displays, the radiologist can more readily visualize subtle contrasts in tissue and anatomic features. Once CT images are optimally processed, they can be acquired as still-frame images and transmitted by a telemedicine system for display. The resolution of a processed CT image is typically 256 by 256 pixels with a depth (intensity resolution) of 8 bits per pixel (256 gray levels). Single-photon emission computed tomography (SPECT), positron emission tomography (PET), and magnetic resonance imaging (MRI) are other imaging modalities which create digital images. Ultrasonography and Doppler ultrasound are imaging modalities which produce monochrome and color video respectively.

Images may also be copied from x-ray film using a variety of techniques. A typical 11-by-17 inch chest film requires at least 2000 by 2000 pixels and an optical dynamic range of at least 4000 to 1 (12 bits) to represent the image adequately. The relative cost and performance characteristics of a number of film digitization techniques are reviewed in Trueblood, Burch, Kearfott, and Brooks (1993). No one technique emerges as a clear cost/performance choice because of the varying resolution and dynamic range requirements of different imaging applications. The American College of Radiologists standards for teleradiology including spatial resolution and dynamic range required for given applications are discussed in the section on telemedicine standards.

Display. The weakest link in the imaging function of a general-purpose telemedicine system is usually the display. Unless the system includes a monochrome display system designed for radiology, visualization of diagnostic images will likely be limited to 640 by 480 pixels and 24-bit color with limited adjustments for brightness and contrast. Since the contrast ratio of a good color display monitor is about 30 to 1, the available contrast cannot be appreciated over the full range of intensity (256 to 1) without manipulation of the contrast and brightness of either the image or the display. Furthermore, when the brightness and contrast of a color monitor are changed from their calibration values, the color rendition changes. Fortunately, most color images do not span the full range of intensity values, and during acquisition, the exposure can be adjusted to optimize a region of interest. Since a monochrome (black and white) display monitor has considerably greater luminance (brightness) and spatial resolution, high-resolution images are often displayed in monochrome. To standardize the image appearance independent of the display used, one must calibrate for gamma (the nonlinear relationship between equal increments of perceived brightness and the corresponding increments of digital intensity values), maximum luminance, and spatial resolution. Tests and calibration procedures for quality control of display monitors are described by Parsons and Kim (1994).

System gamma is an exponential parameter which relates the luminance of the input to the luminance of the output of an imaging system. To obtain the best perceived quality, the gamma and average response of each light converting element of the system (i.e., camera, film scanner, film printer, or display monitor) must be chosen according to the type of image being presented. These parameters can be measured, stored in lookup tables, and selected at display time. In angiography, the relationship between perceived image quality and gamma has been studied for film and monitor displays (van Overveld, 1994).

Interchange. Unlike processed film images, digital images are not transported in the medium in which they are captured. A digital image format consists not only of the overall image size and sample resolution, but also of the actual codes representing the location, intensity, and color values of the pixels and the sequence in which these and other related data are stored or transmitted.

The topic of compression arises eventually in every discussion of images. Compression is needed to reduce storage costs and to reduce the time to access an image, particularly for low data rate transfers such as obtained when using a telephone modem. For specific diagnostic tasks, compressed (10 to 1 lossy compression) radiologic images viewed on a high-resolution monitor yield no difference in observer performance compared to uncompressed images as measured by the receiver operating characteristic (ROC) method of analysis (Collins, Lane, Frank et al., 1994). Comparable studies of compressed imaging of other modalities have yet to be reported in the literature.

Digital Audio

Realistic-sounding interactive conversation is the most important part of the conferencing component of telemedicine. A room audio system must be able to pick up a wide range of sound intensity within the working area of the room, digitize and delay the audio so that it is synchronous with the video, reproduce the sound in another room with possibly different acoustic characteristics, and suppress pickup of sound from the room speakers by the microphones (echo cancellation).

In addition to conferencing audio, one or more separate audio channels are usually provided for diagnostic instruments such as an electronic stethoscope, Doppler ultrasound, or audiometer. The overall audio characteristics must match the requirements of the particular application. To reproduce heart and lung sounds accurately, an electronic stethoscope must have a uniform frequency response from 20 Hz to 2Khz, while Doppler ultrasound requires a uniform frequency response from 100 Hz to 10 Khz. Audiometry requires an even wider bandwidth of 20 Hz to 20 Khz. It is important to consider the audio characteristics of every component in the system, including the sound reproduction. To reproduce frequencies below about 100 Hz accurately, the listener must wear earphones which completely enclose the ear, forming an air-tight seal, or stethophones that fit within the external ear canal.

Digitization. Audio used for conversation and medical diagnosis in a telemedicine system must be digitized and compressed before it can be combined with digital video and other information. Typical audio compression algorithms, operating at data rates from 16 Kbps to 64 Kbps, differ in the extent to which high frequencies can be reproduced (see Table 4-2). The limit of low-frequency reproduction is not usually a function of the compression algorithm but of the analog acquisition and reproduction components. Medical diagnostic applications which require fidelity at higher audio frequencies will require higher data rates; 128 Kbps is sufficient to reproduce the full auditory frequency spectrum (20 Hz to 20 KHz) over a dynamic range of 90 dB which is adequate to reproduce sound intensities within the normal physiologic hearing range.

Table 4-2. Standard Digital Hierarchies

Asynchronous				Synchronous	
North America		Europe		SDH (SONET)	
Designation	Mbps	Designation	Mbps	Designation	Mbps
				OC-12/STS-12	622.080
				OC-3/STS-3	155.250
				OC-1/STS-1	51.840
DS-3[*]	44.736	DS-3E[*]	34.368		
DS-2[*]	6.312	DS-2E[*]	8.448		
DS-1[*]	1.544	DS-1E[*]	2.048		
DS-0	0.064	DS-0	0.064		
Analog	0.004	Analog	0.004		

[*] Can be aggregated into STS-1.

Echo. In full duplex communication, independent circuits are provided to transmit and receive data. An audio system in which sound reproduced in a speaker system can be picked up, amplified, and reproduced again is susceptible to feedback. If there is little or no delay in the feedback loop, the result is ringing or sustained oscillation depending on how much acoustic energy is picked up and amplified. When an appreciable delay in the feedback path occurs due to the propagation time of sound reflecting from the walls, floor, and ceiling of a room and electronic delay due to communication processing and video synchronization, echoes result. Echo delays of less than about 100 milliseconds (ms) are not noticeable, while delays of more than about 500 ms can completely disrupt normal conversation. Echo suppression can be achieved by opening the feedback path so that only one end at a time is active. This method is based on setting a switching threshold high enough not to be triggered by the far side audio coming from the sound system, but low enough so as not to lose much of the talker's initial sounds. To obtain simultaneous two-way conversation, one must implement echo cancellation. Figure 4-1 outlines the principle of echo cancellation as described by Keiser and Strange (1985). A digital signal processor (DSP) implements a digital filter matched to the echo characteristics of the room and subtracts the predicted echo from the room microphone(s). The matched filter coefficients are obtained during a "room training" session in which an internally generated wideband noise signal is substituted for the audio signal from the far site. The matched filter effectively models the echo and attenuates it by a factor of 100. The separation between the microphones and speaker system must be at least one meter to obtain effective echo cancellation (Shure, 1994).

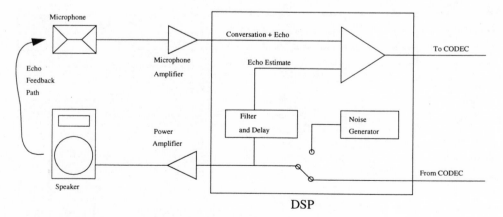

Figure 4-1.

Digital Communication

Telemedicine systems exchange two types of data transfers: short bursts as from images or patient information files, and streams of data as from digitized video and audio. The former are not particularly time-critical and may be transmitted through any of a number of means, including mailing of disks or tapes. A data stream, however, demands a continuous and robust communication link for the duration of the exchange. To appreciate the various communication choices available today, we must first examine the nature of the digital data stream.

Teleconferencing Data. The bits that make up digital video and audio are combined by a communication controller, a subunit of the audio/video codec, or by a multiplexer if audio and video codecs are separate units. Added to the video and audio data are communication and control data which combine into a continuous stream of bits delimited by codes, bit patterns known only to the codecs at each end of the communication circuit. The communication link treats all bits equally without regard to their origin or significance. Each packet contains error-checking and correcting codes, sent along with the data. A typical forward-error correction code can recover from errors which occur at a rate of not more than one in 100,000 or one every tenth of a second at a T-1 transmission rate of 1536 Kbps. Depending on the coding algorithms used for video, uncorrected transmission errors can result in loss of decoding synchronization and a reduction in frame rate and resolution in all or part of the image, a phenomenon known as tiling. Tiles are blocks of data which have failed to be updated or interpolated with other blocks and therefore stand out as visual errors.

Carrier Services. In communication terminology, a carrier is both a physical means to move data from one place to another and, in the marketplace, a business which sells communication services. Some of the types of digital data services available, and of the physical media (wire, radio, microwave, and optical fiber) which carry them, are discussed below.

Using a modem (modulator/demodulator), the analog telephone system (Plain Old Telephone Service—POTS) can accommodate digital signals at data rates up to about 30 Kbps. Depending on the quality of the circuit, the maximum reliable data rate may be less than half of this rate. Analog transmission is adequate for data file transfers of still images; it is even possible to transmit data unobtrusively in the quiet spaces of a telephone conversation. Since 1962, telephone traffic between major switching hubs has been carried in digital format through the digital data system (DDS). Based on this infrastructure, digital data service has been available as a part of the analog telephone system. Today the all-digital integrated services digital network (ISDN) is available in many parts of the world for voice and data. Both the analog and digital modes of communication, however, use the same type of wire in buildings and in neighborhoods to connect subscribers to switching equipment and long-distance carriers.

Switched-56, the most common type of DDS service, provides a data rate of 56 Kbps. Within two miles of a switching office, switched-56 is available on a single pair of wires. Distances beyond that range require two pairs of wires, and service therefore is more expensive. Switched-56 service can be obtained almost anywhere, but in rural locations, existing wiring may need to be upgraded by the insertion of repeaters. Long cable spans attenuate any signal, and repeaters are used to regenerate the signal to its original level. Unlike analog voice repeaters which amplify both voice and noise, digital repeaters regenerate the signal with almost no corruption. To obtain increased data communication capacity, an inverse multiplexer can be installed at a user's site to aggregate a number of switched-56 lines (usually 2 to 7) into a single channel. Six switched-56 lines aggregated to 336 Kbps give realistic replication of detail and motion when used for applications such as video conferencing and distance learning. Switched-56 is a switched service meaning that the customer can call other switched-56 customers using area codes and phone numbers analogous to the analog (voice) telephone network.

The digital data system (AT&T, 1977) was developed as the basis for increasing the capacity of telephone interconnections (trunk lines) between switching offices. The system is based on the conversion of an analog voice signal into a digital equivalent at a data rate of 64 Kbps. This basic channel rate, DS-0 (digital signal, level zero), represents the bandwidth needed to encode telephone quality audio using the pulse code modulation (PCM) encoding method. Channels are aggregated into increasingly larger units by time division multiplexing (TDM) of digital voice data, one from each channel in sequence, plus some additional synchronization data. The most frequently used levels of aggregation are shown in Table 2. If carried on conventional copper wire cables, the digital signal is denoted by T in place of DS. Since the early 1980s, T-1 (and fractional T-1) service has been available for sale to private customers. Service may be dedicated (point-to-point) for which permanent connections are "nailed-up" within the switching office (i.e., directly connected on a patch panel instead of a switch). Or the service may provide access to a digital switch, the benefits of which are covered in the section on data networks. In telemedicine systems that use T-1 or fractional T-1 service, the physical signal path from the "demarc" (point of demarcation between telephone company wire and equipment and the customer supplied wire and equipment) to a switching office is two cable pairs with repeaters spliced in at intervals of about 6,000 feet. At the switching office, switched T-1 circuits are aggregated with other T-1 circuits and DS-1 voice channel banks to yield higher data rates, and the signals are physically carried by less expensive high-capacity media such as microwave, satellite, or optical fiber for bulk transmission across town or around the world. The customer must supply equipment (CSU/DSU) to interface between the T-1 termination equipment provided by the phone company and the customer's data terminal equipment (DTE), the telemedicine codec (Minoli, 1991, pp. 128–34).

Recently cellular radio telephony has become pervasive in metropolitan areas. It offers a means of mobile digital transmission, albeit limited to low data rates and limited specific coverage areas. Using ultrahigh frequencies (UHF) in the 900 and 1,200 MHz bands, the spread spectrum technique developed for military communication makes digital cellular data transmission relatively more immune to interfering radio signals than current fixed frequency analog cellular telephone communication. As digital cellular telephony and data services become more widely available, mobile telemedicine applications such as emergency medical team support will be one of its first applications.

Low-power UHF broadcasting is also used for wireless interconnection of computers to local area networks. Over relatively fixed and limited distances, a wireless local area network can provide a communication capacity greater than that provided by T-1, the gold standard for high-quality compressed video. Such flexibility is ideal for cart-mounted equipment that can be moved to a patient's room or into a clinic as needed.

Nonbroadcast VHF and UHF carriers for digital communication are available through community antenna television (CATV) facilities (i.e., cable TV). The frequency spectra in a CATV system are divided roughly into thirds: lowband, midband, and highband. Signals originating from the "head end" are sent out on midband and/or highband channels. The lowband channels are available for transmission in the reverse direction, toward the head end, from any point in the network, provided each amplifier module in the connecting path has a reverse amplifier installed. Since the signal pathways for high and low frequencies are separated by filters, simultaneous two-way transmission is possible. Some cable companies now offer metropolitan area network (MAN) ethernet services for residential and business customers. Not only can sites within the network exchange data, but the cable company can also offer links to various off-network services such as the internet, airline reservation systems, and commercial information services. Such data transmission can also be used to link telemedicine applications at a data rate comparable to DS-1.

Terrestrial microwave radio transmission is commonly used for private point-to-point communication (e.g., television studio to transmitter link, business television and data links between branch offices, and medium and long-distance telephony). Businesses with extensive networks may have unused capacity appropriate for linking telemedicine applications. Satellite microwave is usually limited to one-way broadcast uses because of the complexity and cost of a transmitting earth station site and the cost of satellite transponder time. When a satellite link is used for video, the propagation delay from the earth to the satellite and back (approximately 240 milliseconds) adds to the existing video processing delay (150 to 350 milliseconds). This makes interactivity cumbersome. Using terrestrial communication for two-way audio out of synchronization with the video can improve interactivity. Satellites in low earth orbit offer significantly less delay, but are not yet available for continuous point-to-point two-way communication.

Three characteristics of optical glass fiber make this technology of great significance to communications:

- Light can travel in silica-based glass fibers up to 60 km before needing a repeater to detect and regenerate the digital signal it carries. New glass formulations promise a 100-fold increase in that distance.
- The theoretical maximum bandwidth of digital information which can be carried by light is over a thousand times the bandwidth of the fastest electronic circuits today
- Light transmission is inherently immune to electrical interference, and thus fiber-optic transmission exhibits very low error rates.

On the negative side, deployment of fiber cable, network interface equipment, digital switches, and multiplexing equipment is expensive.

Compared to microwave and satellite, fiber communication has a higher cost per channel-mile until aggregations exceed 20 to 40 thousand subscribers (Minoli, 1991, pp. 368–69). While network interface standards exist for optical carrier (OC-1, OC-3, etc.), expanded subscriber use of digital services is needed to justify the capital investment to extend fiber beyond the local exchange to the subscriber. As long as the most lucrative accounts for digital services are large metropolitan users, the expansion of fiber-based services into rural areas, where they are most needed for telemedicine applications, will proceed slowly.

Switched Networks. Networks provide flexible connection pathways for communication between two or more users. Three types of switching make this possible: circuit switching, packet switching, and cell switching. In a circuit-switched network such as switched-56 or ISDN, once a circuit is established between two points, the channel bandwidth established by the call is continuously available at a fixed data rate. Circuits may be automatically added or dropped through multiplexing to accommodate variations in data-rate requirements. The multiplexer is part of the telemedicine interface equipment provided by the system integrator. The communication circuit is continuously and exclusively available to the call through fixed routes. A circuit route built up of many spans between switches may be rerouted automatically if trouble occurs, provided the customer has reserved idle links for backup service. Since a call consumes bandwidth whether data are being sent or not, the cost of a call is related to connect time and distance. DDS and ISDN are circuit-switched digital networks, and switched-56, T-1, and ISDN BRI (basic rate interface) are switched digital services. For DDS, dialing and other signaling functions are provided "in band" and consume 8 Kbps of the 64 Kbps total bandwidth leaving 56 Kbps available for the user. The network interface equipment (CSU/DSU/multiplexer) provide the signal formatting functions which keep the user's data separate from the control data while sending them on the same line. The common form of ISDN, BRI, is composed of two 64 Kbps bearer (data) channels and one 16 Kbps data (control) channel (2B+D) multiplexed on two wire pairs. The bearer channels can be aggregated into one 128 Kbps channel by the user's equipment, for example, a codec used for desktop video conferencing. ISDN dialing and other control functions are handled "out of band" in the D channel. The recurring cost of BRI varies widely, but is often not much higher than the recurring cost of switched-56 service. Like switched-56, multiple BRI lines can be aggregated for higher bandwidth. In some locations, ISDN PRI (primary rate interface) is available and provides 23B channels at 1472 Kbps and one D channel at 64 Kbps (Motorola 1992a). PRI service requirements differ from BRI not only in the network interface equipment, but also in the need for repeaters in those portions of the circuit carried by wire. Not all telephone ISDN services are the same, and there can be a large number of signal format options from which to select, depending on the user's equipment and the service provider's brand of ISDN switch.

Another service offered by switched network providers is multipoint conferencing where from three to eight (or sometimes more) users connect to a centrally located multipoint control unit (MCU) to form a conference call. In the MCU, audio and video data from one transmitting site are duplicated and sent out to all the other sites. The site selected to transmit is determined manually by the conference chairman or automatically by protocols within the MCU software, based on which site has the highest audio level for a minimum period of time, typically two seconds.

Packet Networks. A packet data network parcels out data bits in well-defined chunks called packets. Equipment within the network reads the destination address encoded in the header of each data packet and allocates whatever path (if any) is available at the moment to transmit that packet. A packet structure allows for:

- An orderly combination of audio, video, text, and other data streams to be transmitted synchronously and apparently simultaneously
- An orderly recovery from the occasional error in transmission, equipment failure, or loss of a communication link
- The aggregation of multiple sources of packet data into low-cost high-capacity communication channels
- Automatic routing of data through a network of dedicated or switched communication channels.

There are different general designations of packet networks, depending on the physical distances involved. A network confined to a building or a campus of buildings where the end-to-end wiring distance is less than one kilometer is generally referred to as a local area network (LAN). One characteristic of a LAN is that when a device transmits, its data are received by some or all other devices on the network. All devices not addressed ignore the data while the addressed destination stores them in a buffer for use by an application program. Because of the broadcast nature of each data exchange, a LAN cannot provide secure communication. But if the LAN is physically confined to secure locations and authorized users, then controlled access can be implemented so that a connection to a computer or network outside of the LAN will receive only packets specifically addressed to it. If the data network needs to include LANs separated beyond the physical constraint of one kilometer, a carrier service such as ISDN is required and the network becomes a wide area network (WAN). There are three types of carrier services which can be used to implement a WAN:

- Dedicated private circuits such as fractional T-1
- Switched circuits such as switched-56 and ISDN
- Packet network services such as X.25 and Frame Relay.

Permanent leased lines are commonly used to link physically separated local area networks using routers as the interface between the telecommunications link and the LAN. To keep unnecessary packet traffic off the telecommunications link, each router passes only those packets whose destination addresses are outside its LAN. Much of the internet is a WAN composed of both public and private communication circuits. When traffic becomes heavy, congestion occurs because there is a definite limit to the number of packets the network can transfer in a given amount of time. At other times, the communication links may be nearly idle. To achieve a closer match between capacity and demand, a *switched* packet network is designed to dial up and disconnect communication circuits. Often a combination of leased and dial-up circuits is the most economical solution. In the third type of implementation, the customer purchases a packet data service in which the communication link between the customer and the service provider is a local dedicated carrier service like fractional T-1. The service provider then manages the circuits and switches necessary to aggregate packets from many customers on high-capacity (and thus low cost per packet) circuits. A number of telecommunication service providers offer private packet network services based on standard protocols such as X.25 and Frame Relay. Since the network bandwidth may be allocated on demand, the user is charged for use of the network in proportion to the number of packets and the peak rate at which they are transmitted.

Some packet network protocols such as UDP over Ethernet do not guarantee delivery of every packet. Therefore either the customer or the service provider must check the data stream for missing or damaged packets and request retransmission to assure reception of 100 percent error-free data. A series of individual packets originating from one site may each take different routes to a common destination depending on the dynamic allocation of available communication circuits. Hence, each may encounter a different amount of propagation time through the network. At the destination, the packets must be reassembled in the correct order and may be tested by network equipment or by the customer's equipment for transmission errors. As a means of transferring patient data and medical image files over long distances, packet networks can be cost-effective; but their use for video or audio is not appropriate because of inherent transmission delays and uncertainty in packet arrival time (Motorola, 1992b).

Virtual Networks. Broadband ISDN (B–ISDN) is a network service which employs a combination of circuit and packet switching known as cell switching. The asynchronous transfer mode (ATM) protocol used in B–ISDN defines both the switching and transmission of cells, small packets of any sort of data. Like Frame Relay, the ATM protocol assumes error-free transmission channels and thus avoids the overhead of checking for errors and retransmitting cells. When an ATM call is initiated, a series of transactions take place to establish a virtual route through the network switching elements. But unlike frame relay which sets up specific connections for the duration of a call, an ATM connection can reserve a specified channel capacity and modify it dynamically as needed to fulfill variable data-rate requirements. This bandwidth on demand type of service can, for instance, accommodate a codec which adjusts its data rate as needed to assure constant video quality as the information content of its video source varies.

ATM is usually associated with high-capacity communications linking a multitude of widely dispersed sites. A network of ATM switches distributed over distances of 1000 km or more and linked by fiber communication channels such as OC-3 (155 Mbps) or OC-12 (622 Mbps) can serve hundreds of clients with a relatively simple ring structure. Because communication capacity can be negotiated on a per-call basis, and transmission latencies are both small and predictable, ATM is well suited for transmitting digital video and audio. The major obstacle to widespread adoption of ATM for telemedicine is that fiber facilities are concentrated in metropolitan areas, and where connecting digital services are available to rural areas, they are usually low-bandwidth and expensive. A technical description of video applications of B–ISDN and ATM can be found in Ohta (1994).

Multimedia Computing

As audio and video codec hardware and software, removable and inexpensive mass storage, high-speed network interfaces, high-resolution displays, and fast microprocessors become integrated into relatively low-cost workstations, the industry devoted to developing multimedia applications will continue to grow. For example, on a single ISDN BRI line at home or office, one can video conference on a PC, interactively annotate still images, access internet-based information resources, and share database files for an investment of as little as $2,000 (in addition to the PC and line installation). It is becoming common to find MPEG-1 decompression hardware included in PCs configured with a compact disk for multimedia applications such as home entertainment, business training, and educational programming. Multimedia resources enable developers to integrate a high level of interactivity into routine health care activities such as patient information gathering, problem solving, therapy management, and follow-up. The value of multimedia to telemedicine is as a means to merge the audio-visual world with the computer-based information processing world, using tools and computer components developed for mass markets. The new patient-centered focus on health care is creating a wide range of opportunities for home-based multimedia telemedicine applications which enable patients to monitor, treat, and learn more about their specific health problems. The underlying premise is that patients who are informed and participate in their own health care process gain a greater feeling of autonomy over therapy decisions and are more likely to comply with life-style modifications beneficial to their health. Technically unsophisticated patients can successfully operate diagnostic equipment like a pulse oximeter or blood pressure monitor when these instruments are coupled to a multimedia computer. The usual computer interface and instrument control panels, designed for college-educated operators, are hidden from view, and the patient interfaces with a touch screen or similar input device using a simple set of buttons to operate the equipment and seek information. Each action includes automatic instructions and multilevel help using combinations of text, voice, graphic images, and video. The telemedicine unit could record monitored data and transmit them automatically over a communications link to a central monitoring station in a hospital or clinic.

Kleinholz and Ohly (1994) describe a system which acquires, remotely accesses, and manipulates multimedia patient data on desktop computer workstations in diverse locations and linked by primary rate ISDN (2 Mbps). Their goal of "fast and flexible cooperation between physicians" was largely a technical success. The real design challenge they reported was to integrate the system more completely into an existing clinical practice routine.

Patient Data. Medical signs and symptoms are interpretations of data obtained from direct and indirect patient observations. Direct observations include data obtained from the senses (sight, sound, touch, smell) and through mental and physical interaction with the patient, while indirect observations include data obtained from diagnostic instruments (electrocardiograph, MRI, etc.). Both direct and indirect data must be interpreted within the context of relevant observations from the patient's past and present history. Multimedia data (and some of their attributes) include still and sequential images (velocity, absolute and relative location, brightness, contrast, hue, saturation, size, orientation, region of interest), sound (amplitude, frequency, temporal pattern), text (patient history, diagnostic reports), and graphics (interactive visual and verbal annotation). The signs and symptoms derived from these data are used singly and in combination to confirm or reject diagnoses, and to calculate and to adjust therapeutic protocols through follow-up monitoring. In a conventional medical exam, the clinician makes verbal notes and sketches documenting the characteristics observations. Multimedia tools can be used to quantify those observations as measurements of area, optical density, and length, for example, for objective comparison at follow-up examinations to document the progression of changes in the size and texture of a lesion, for instance. The more objective the measurement, the more consistent the application of treatment protocols will be, particularly when the patient is assigned to a new health care provider.

The Electronic Medical Record. In a clinic or doctor's office, multimedia computing not only can support live multiway conversations between physicians, patients, and specialists but can also facilitate off-line communication between the same health care team members. Telemedicine is uniquely positioned to demonstrate the value of a shared multimedia electronic patient record because the essential components are usually present in a telemedicine system:

- Primary diagnostic patient data are electronically captured as image, video, sound, and text
- Networked telemedicine sites can share database functions and access administrative and patient data located at the sites where they were acquired and saved or located at a centralized repository
- The telemedicine site is the point of care where findings are recorded and health care team members can access patient records, decision-support systems such as DXplain (Barnett et al., 1993), practice guidelines, and reference information such as the electronic version of the *Physician's Desk Reference.*

A multimedia electronic medical record is the basis for a type of telemedicine interaction known as Store and Forward (SAF). Teleradiology is usually implemented as an SAF applications since live video interaction is not essential and diagnoses and comments may be communicated as text and annotations on the images. For a second opinion, a group of pathology images might be communicated using an SAF application which displays a large number of small "thumb nail" images. Selecting one or a sequence of images would invoke a process to download high-resolution copies over a period of minutes to hours, while the consulting pathologist was attending to other business.

Internet Resources. The World-Wide Web (WWW) is an internet information resource through which information-producing sites offer hyperlinked multimedia information to public or restricted access. Graphical browser programs such as Netscape and Mosaic are specifically designed to access WWW resources and view their contents in text, graphics, images, and video. Getting started as a user or as a content provider at an information resource site can be overwhelming. A short useful article listing a number of procedures and resources for using the Web can be found in Vetter, Spell, and Ward (1994). Various Web sites offering medical information and educational content are listed in a report from the Office of Technology Assessment (U.S. Congress, 1995, pp. 207–11).

Standards

The standardization of telemedicine system functions is a first step toward interoperability and open access. Interoperability means that a telemedicine function performed on one brand of equipment will produce similar and predictable results if performed on another brand of equipment which supports the function. Open access refers to the existence of published technical specifications of a particular system or component which would allow a developer to design a system or component which will interoperate with it. All the technologies addressed in this chapter have applicable standards; where standards are in the process of development or where multiple standards exist, the flexibility to accommodate multiple standards is desirable. For example, component Y/C video gives superior performance compared to composite video, but some video sources such as ultrasound may have only composite video available. It is therefore necessary to allow a composite video source as an input to a telemedicine system even though the cameras may use another format.

When technical standards for health-information transfer and database design are implemented, they are often modified by local preferences to accommodate special needs. One has to look no further than the history of the development of the electronic medical record to find wide diversity in intent and solution. Because telemedicine may be viewed as a microcosm of the health care system it supports, it offers a unique opportunity to develop new approaches to standardizing the information content of medical records and the means for sharing that information for patient-care and health-systems operation and management.

Video and Still Imaging

Video-image quality depends on the performance and compatibility of the elements that comprise the system. Each element must be correctly specified and calibrated for optimum results in spatial resolution, color fidelity, and contrast range. The elements of an imaging system which must be considered, the variables which describe their performance, and specifications appropriate for medical imaging are listed in Table 4-3. These characteristics can be used to establish the criteria needed to verify that the available equipment at each site meets the minimum performance requirements and is correctly installed and configured, and to establish procedures for system operation and quality control.

Communication

The standards for terrestrial communications are defined by several odies within the International Telecommunications Union (ITU). While the names and descriptions of the applicable standards are beyond the scope of this chapter, relevant services provided by the Digital Data

Table 4-3. Imaging System Characteristics

Element	Performance Parameter	Specification
Light source	Color temperature	3000K
	Intensity	Sufficient for correct exposure.
Optics	Chromatic aberration	Apparent image shift (mm)
	Resolving power	Line pairs /mm
	Aperature; maximum	Manual iris control. f/n
Camera	Frequency response	6.0 MHz at -12 dB
	Gamma correction	Log response: OD 0.2 - 1.5 EIA-Reflectance chart
	Color balance for a given light source and peak luminance at 100 IRE	Manual setting within 2 per cent of SMPTE color bar values.
	Exposure	Manual iris control, AGC off.
	Output signal amplitude (mV)	Y: 714 peak, 54 setup, -286 sync C: 286 burst
	Video format	NTSC component (Y-C)
Digitizer	Sample rate	12 M samples/sec
	Image format	640 X 480
	Quantization	24 bits (8 bits each RGB)
Lossless compression	Image file compression ratio	1.1 to 3
Lossy compression (JPEG)		10 (depends on image content and "quality" parameter)
Display	Contrast ratio	20:1 (with minimal glare)
	Quantization	24 bits (8 bits each RGB)
	Intensity and wavelength of SMPTE color bars test pattern	Chromatic reading photometer
	Image resolution	640 X 480 minimum
	SMPTE medical image test pattern	0 - 5 per cent and 95 - 100 per cent contrast

System (DDS) and Integrated Services Digital Network (ISDN) are summarized in Table 4-4.

Videoconferencing

The Telecommunication Standardization Sector (TSS, formerly Consultative Committee on International Telephone and Telegraph, CCITT) has established a number of internationally recognized recommendations relating to teleconferencing (see Table 4-5). The goals of these standards are to assure a high degree of functional compatibility between like equipment supplied by different manufacturers and to standardize the interface protocols which access and control the communications network.

Table 4-4. Digital Communication Services

Type Service	Media/ Carrier	Cost	Data Rate	Availability and control	Switching	Terminal Equipment
Dedicated	4 wire/ T-1	Proportio-nal to rate and distance	56 Kbps - 1.54 Mbps	Continuous, manual access to DCS	None	CSU/DSU
Switched-56	2 or 4 wire/ T-0	Flat monthly + time	56 Kbps	As needed; dial-up	Circuit Switched	None or inverse multiplexer
2B+D	2 wire/ ISDN-BRI	Flat monthly + time	128 Kbps	As needed; dial-up		None or terminal adaptor
Frame Relay*	2 wire/ ISDN-BRI	Flat monthly + time	64 Kbps - 128 Kbps	Continuous; defined paths and rates	Packet Switched	CSU/DSU
	4/wire ISDN-PRI		384 Kbps - 1.54 Mbps			
Broadband	Fiber/ ATM	Not available yet	1.5 Mbps and higher	As needed; dynamic resource allocation	Cell Switched	ATM adaptor

* Data only; not suitable for video.

Patient Data

Health Level 7 (HL7) is a nationally recognized standard which defines a set of communication protocols for sharing clinical data among, for example, the pharmacy, order entry, charting, and billing components of a hospital information system (U.S. Congress, 1995, pp. 68–72). As a source and retrieval point for diagnostic images, heart sounds, and other patient information such as reason for consultation, findings, and dispo-sition of the case, a telemedicine system cannot operate efficiently in isolation from other patient information systems. Kuzmak and Dayhoff (1994) discuss some of the difficulties in successfully implementing stan-dards to build information systems which link standards compliant com-mercial products.

Table 4-5. ITU–TSS* Videoconferencing System Standards

Scope	ID	Description
Coordination of H-series recommendations	H.320	Narrow-band visual telephone systems and terminal equipment.
Communication	H.221	Protocol to multiplex audio, video, and control.
	H.233	Encription of H.221 data; no algorithm specified.
	H.242	Establish capabilities and negotiate procedures.
Communication interface	X.21	Synchronous interface on public data netowrks.
	X.25	Packet mode interface on public data networks.
Dialing	V.25(bis)	Dialing commands through a parallel (serial) interface.
Multipoint	H.231	Services for audio and video at channel rates up to 2Mbps.
	H.243	Establish communication among 3 or more terminals at rates up to 2 Mbps.
Video	H.261 (Px64)	Video coding algorithm for Px64 or Px56 Kbps data rate where P is 1, 2, 4, 6, 12, 24, 30.
Still frame	T.120	JPEG (Joint Picture Experts Group) compression.
Audio	G.711	64 Kbps, 3.4 KHz bandwidth, m law or A law
	G.722	64 Kbps, 7.0 KHz bandwidth
	G.728	16 Kbps, 3.4 KHz bandwidth

* International Telecommunication Union - Telecommunication Standardization Sector, formerly CCITT

Teleradiology

The American College of Radiologists and the National Electrical Manufacturers Association (ACR–NEMA) sponsored the development of a 13-part Digital Imaging and Communications in Medicine (DICOM) standard (ACR–NEMA, 1993) for interconnection of medical imaging devices, particularly for radiologic imaging equipment such as computed radiography (CR), computed tomography (CT), medical resonance imaging (MRI), and picture archiving and communication systems (PACS). Further, the ACR has published standards for " . . . the transmission of radiological images from one location to another for the purposes of interpretation and/or consultation" (ACR, 1994). In addition to personnel qualifications, licensing, and quality control, the standards include equipment guidelines for digitization of small matrix images digitized in arrays of 0.5k × 0.5k × 8 bits, large matrix images digitized in arrays of 2k × 2k × 12 bits, display capabilities, and patient database requirements.

Conclusion

The convergence of communications, video, and computer technologies has led to a wide range of telemedicine applications. The gradual migration of these applications into the daily routine of health care providers is controlled by three major events:

- *Application Development.* Designers must select appropriate technologies for each specific clinical application, develop easy-to-use interfaces, and control implementation and operational costs.
- *Standards Development.* Whether standards are developed *de facto* or through formal processes, they are needed to ensure interoperability of like applications, compatibility with existing and developing information systems, and the capability for extensibility through third-party value-added support.
- *Communication Infrastructure Development.* The National Information Infrastructure (NII) will drive down equipment and communications costs by opening up mass markets for digital video and digital communication services.

In a summary report on NII activities, Fenimore, Field, and Frank et al. (1995) describe the powerful driving force of high-definition television (HDTV) on infrastructure development. "The goal is to have a fully competitive marketplace in which any company may provide any service to any customer." Much of the infrastructure being developed to support HDTV services and applications will support telemedicine. In areas where market forces lag, such as affordable access to digital communication services in rural communities, government intervention may be required in the form of regulation, waivers, and public-private partnerships.

REFERENCES

American College of Radiology. 1994. *Standard for Teleradiology (Res.21-1994).* Reston, VA.

American College of Radiology, National Electrical Manufacturers Association. 1993. *Digital Imaging and Communications in Medicine (DICOM): Version 3.0.* Washington, D.C.: NEMA, Office of Publications.

AT&T. 1977. "The Digital Data System." In: *Networks and Services,* Vol. 3, *Telecommunications Transmission Engineering.* Indianapolis, IN: AT&T Technologies, pp. 474–506.

Barnett, G. O., E. P. Hoffer, M. S. Packer, K. T. Famiglietti, R. J. Kim, C. Cimino, M. J. Feldman, D. E. Oliver, J. A. Kahn, R. A. Jenders, and J. A. Gnassi. 1993. "DXplain—Demonstration and Discussion of a Diagnostic Decision Support System," Frisse, M. E. (ed.), in *Proceedings of the Sixteenth Annual Symposium on Computer Applications in Medical Care.* New York: McGraw-Hill, p. 882.

Chesters, M. S. 1982. "Perception and Evaluation of Images." In Wells, P. N. T. (ed.), *The Scientific Basis of Medical Imaging.* New York: Churchill Livingstone, pp. 237–80.

Collins, C. A., D. Lane, M. Frank, M. E. Hardy, D. R. Haynor, D. V. Smith, B. K. Stewart, J. E. S. Parker, G. D. Bender, and Y. Kim. 1994. "Design of a Receiver Operating Characteristic (ROC) Study of 10:1 Lossy Image Compression," in Kim, Yongmin (ed.), *Image Perception.* Bellingham, WA: SPIE–The International Society for Optical Engineering, Proc. SPIE, vol. 2166, pp. 149–58.

Conrac Division. 1980. *Raster Graphics Handbook.* Covina, CA: Conrac Corporation.

Fenimore, C., B. Field, H. Frank, E. Georg, M. Papillo, G. Reitmeier, W. Stackhouse, and C. Van Degrift. 1995. *SMPTE Journal* 104 (3): 148–52.

Goldberg, L. 1994. "ATM Switching: A Brief Introduction," *Electronic Design,* (December): 87–103.

Ji, T. -L., M. K. Sundareshan, and H. Roehrig. 1994. "Adaptive Image Contrast Enhancement Based on Human Visual Properties," *IEEE Transactions on Medical Imaging,* 13 (4): 573–86.

Keiser, B. E., and E. Strange. 1985. *Digital Telephony and Network Integration.* New York: Norstrand and Reinhold.

Kleinholz, L., and M. Ohly. 1994. "Supporting Cooperative Medicine: The Bermed Project," *IEEE Multimedia,* 1 (4): 44–53.

Kuzmak, P. M., and R. E. Dayhoff. 1994. "A Bidirectional ACR–NEMA Interface Between the VA's DHCP Integrated Imaging System and the Siemens-Loral PACS," In Jost, R. Gilbert (ed.), *PACS: Design and Evaluation.* Bellingham, WA: SPIE–The International Society for Optical Engineering, Proc. SPIE, vol. 2165, pp. 387–96.

Le Gall, Didier. 1991. "MPEG: A Video Compression Standard for Multimedia Applications," *Communications of the ACM,* 34 (4): 47–58.

Lee, B. G., M. Kang, and J. Lee. 1993. *Broadband Telecommunications Technology.* Norwood, MA: Artech House.

Minoli, D. 1991. *Telecommunications Technology Handbook.* Norwood, MA: Artech House.

Motorola Codex. 1992a. *The Basics Book of ISDN.* Reading, MA: Addison-Wesley.

Motorola Codex. 1992b. *The Basics Book of Frame Relay.* Reading, MA: Addison-Wesley.

Ohta, N. 1994. *Packet Video: Modeling and Signal Processing.* Norwood, MA: Artech House.

Parsons, D. M., and Y. Kim. 1994. "Quality Control Assessment for the Medical Diagnostic Imaging Support (MDIS) System's Display Monitors," In: Kim, Y. (ed.), *Image Capture, formatting, and Display.* Bellingham, WA: SPIE–The International Society for Optical Engineering, Proc. SPIE, vol. 2164, pp. 186–97.

Shure. 1994. *Data Sheet ST4300.* Evanston, IL: Shure Brothers Inc.

Trueblood, J. H., S. E. Burch, K. Kearfott, and K. W. Brooks. 1993. "Radiographic Film Digitization," In: Hendee, W. R., and Trueblood, J. H. (eds.), *Digital Imaging*. Madison, WI: Medical Physics Publishing, pp. 97–122.

U.S. Congress, Office of Technology Assessment. 1995. *Bringing Health Care Online: The Role of Information Technologies*. OTA–ITC-624. Washington, D.C.: U.S. Government Printing Office.

van Overveld, I. M. C. J. 1994. "The Effect of Gamma on Subjective Quality and Contrast of X-ray Images," In: Kim, Y. (ed.), *Image Perception*. Bellingham, WA: SPIE–The International Society for Optical Engineering, Proc. SPIE, vol. 2166, pp. 96–104.

Vetter, Ronald J., Chris Spell, and Charles Ward. 1994. "Mosaic and the World-Wide Web," *Computer*, 27 (10): 49–57.

SECTION III
CLINICAL APPLICATIONS
OF TELEMEDICINE

Chapter 5

TELERADIOLOGY

Joseph N. Gitlin

General Overview

Interest in teleradiology as a practical cost-effective method of providing professional radiology services to "underserved" areas began some 30 years ago when telemedicine was considered an alternative to recruiting medical specialists for isolated communities. At the same time, radiologists who provided services to multiple hospitals in sparsely populated areas saw an opportunity to increase their productivity by decreasing their time spent in "circuit-riding." Initial efforts to provide image interpretation services via teleradiology depended on digitizing analog images produced at a hospital or local clinic. This was accomplished by focusing a television camera on x-ray film and transmitting the data over conventional telephone lines to a medical center. There, a specialist viewed the images on a television screen and discussed the interpretation with the general practitioner, paramedic, or nurse practitioner at the patient's location. The early telemedicine systems were often useful when they facilitated local patient care and avoided the unnecessary transport of patients to distant hospitals. However, evaluation of the use of these systems for the primary diagnosis of radiographic images identified many technical deficiencies in the hardware, software, and telecommunication links (Gayler et al., 1979). Image resolution was inadequate; transmission was slow; storage and retrieval of data were expensive; and formal reporting was delayed.

Since there were potential medical and economic benefits to military personnel at small installations and ships at sea that had no radiologists, the Department of Defense supported several pilot programs to evaluate the feasibility of utilizing new technology to solve the hardware and software problems identified in the earlier studies. Equally important in these studies were defining the level of functionality needed to support teleradiology for the primary diagnosis of radiographic images, and

111

specifying the systems' requirements to fulfill the concept of "filmless radiology."

By 1985, the efforts of government, academia, and industry had advanced the state of the art to the point where teleradiology was technically feasible, but it was not widely accepted by radiologists and other physicians, nor had its cost-effectiveness been demonstrated (Gitlin, 1986). Limited success had been achieved in marketing inexpensive systems for "on-call" screening of images by radiologists at home, and the remote interpretation of CT and MR cases by radiologists in group practice had begun. In most hospital applications, teleradiology was generally considered a subset of the broader image management concept referred to as Picture Archiving and Communication Systems (PACS). The major technical advances applicable to teleradiology were improvements in components, such as film digitizers, storage and retrieval systems, and workstation monitors. Equally important were advances in telecommunications which made it possible to transmit diagnostic images at data rates in excess of one megabit per second.

Thanks to the telecommunications and computer industries, by 1990 it appeared that both teleradiology and image management systems had all the hardware and software needed to proceed with addressing the issues of clinical acceptance and cost-effectiveness.

The clinical acceptance and the cost-effectiveness of teleradiology depend on many factors. They include radiologists' confidence in the accuracy of image interpretations using electronic displays, efficiency and productivity when using workstations, the use of lossy data compression algorithms, reimbursement for consultation and second opinions, differing interstate licensure requirements, competitive ethics, and economic considerations related to system purchase and maintenance. Many hospital administrators and radiologists have recently deferred decisions about new equipment until the shape and impact of health care reform become clearer.

Most important in clinical acceptance of teleradiology is its utility in facilitating the primary diagnosis of conventional radiographic examinations performed at distant sites. Many well designed comparative studies of screen and film interpretations have shown that the accuracy of screen readings is significantly lower than film viewing and that the reader's confidence in the interpretation is adversely affected by the electronic workstation. Improvements of contrast and spatial resolution of the images and increased luminance of the display monitors are needed.

Users also need more intensive training and available manipulation functions when viewing images on the workstation.

The most important economic issues are the relatively high cost of image-display equipment with adequate functionality, and the recurring costs of transmission media capable of the speeds necessary for prompt consultation on data intensive medical images. The latter issue may be resolved soon by the Federal High Performance Computing and Communications (HPCC) Program, which is supporting the development of high-speed networks to transmit medical images to improve health care (Office of Science and Technology Policy, 1994).

While many issues remain to be resolved before teleradiology is routinely available in health care delivery, progress has been made and interest in the technology is increasing among pathologists, emergency medicine physicians, HMOs, and third-party payers. Where implementation of teleradiology depends on the installation of image-management systems in hospitals and medical centers, acceptance of the technology will be facilitated by the recent demonstrations of the network version of the American College of Radiology—National Electrical Manufacturers Association (ACR–NEMA) Standard, Version 3.0 (DICOM), and its proposed adoption as an international standard. It is anticipated that all manufacturers of imaging equipment and related devices will be able to provide ACR–NEMA compatible products. Featured as part of the centennial celebration of Roentgen's discovery of x-rays, teleradiology is an important example of science and technology's contribution to health care.

Radiology is the only medical specialty created by technology. While all modern medical practice depends to some degree on one or more technologies, only radiology is founded on a technological discovery. The year of 1995 was the centennial of that discovery by Wilhelm Roentgen and, coincidentally, the 150th anniversary of Roentgen's birth (Moliter, 1995). In 1895, Roentgen was professor of physics and director of the Physics Institute at the University of Würzburg in Germany. A contemporary edition of the *Encyclopedia Britannica* published in the early 1900s described the circumstances of his discovery, in the late evening of November 8, 1895.

(W)hile experimenting with a highly exhausted vacuum tube on the conduction of electricity through gases, he notices that a paper screen covered with barium platinocyanide, which happened to be lying near, became fluorescent under the action of some radiation emitted from the tube, which at the time was enclosed in a box of black cardboard. Further investigation showed that this radiation had the power of passing through various substances which are opaque to ordinary light, and also of affecting a photographic plate.

In such serendipitous circumstances as a messy laboratory bench— with the right ingredients "which just happened to be lying near"—are momentous scientific breakthroughs achieved. For his discovery of what the scientific community called "Roentgen rays" (but which Roentgen insisted on referring to as x-rays), he was awarded the Rumford medal of the Royal Society in 1896. Five years later, Roentgen received the Nobel prize in physics, the very first ever awarded.

It might well have been the Nobel prize for medicine. In the century since 1895, we have come to take for granted the revolutionary diagnostic power of the Roentgen ray, focusing our fascination on succeeding developments in fluoroscopy, arteriography, and computed tomography (CT). In the case of CT, British engineer/inventory Godfrey Hounsfield along with American physicist Allan Cormack also received a 1979 Nobel prize—by applying modern computer technology to Roentgen's undemonstrated mathematical theory of tomography.

Background

Some applications of telecommunications in the health field have been in use longer than others. We have had, for example, about 30 years' experience with teleradiology, whose literature dates from the early 1970s. "Telehealth" projects sponsored by the National Center for Health Services Research (NCHSR) were reported in *The Telehealth Handbook* (May 1978). In 1972, NCHSR employed the MITRE Corporation to study the telecommunications technology that would be appropriate to support the delivery of health care services in rural areas. MITRE concluded that telephone-based technologies should be exploited much more vigorously than had been the case in past government-funded research. The handbook discussed a range of applications including not only expensive, exotic technologies, but also rather inexpensive, everyday technologies readily available in many parts of the country. The handbook's purpose was to alert the health community to the possibilities that telehealth offered to increase access to affordable, high-quality health care services, while encouraging the adoption of only those technological approaches that were appropriate and sound.

While many of the projects demonstrated the benefits of telemedicine to patients, providers, and society, the results also indicated technical difficulties and economic problems, particularly with teleradiology.

Among those projects with radiology content described in the *Telehealth Handbook* were:

- The ATS-6 Satellite Advanced Health Care and Education—Alaska Health Experiment was conducted during 1974–1975 in Anchorage, Alaska, using satellite transmissions and 2-way black-and-white television for primary care applications. This project was designed to assess the potential of a satellite system for the presentation of health, education, and information programs. All four sites included in the project had local exchanges in addition to ATS-1 satellite radios which linked them to Tanana Hospital, Anchorage Medical Center, and other ATS-1 sites, Services included teleconsultation, the transmission of x-rays by the video capability, and EKG tracings over the audio channel. The demonstration also included the computerized patient record system, Health Information System (HIS), developed by the Indian Health Services (Boor et al., 1975; Wilson and Brandy, 1975).

- The Navy Remote Medical Diagnoses System (RMDS) was demonstrated in San Diego, California, in 1972, using slow-scan television and telemetry for primary care applications. RMDS was an interactive slow-scan television system for linking ship-to-ship, ship-to-shore, and land-to-sand points to provide a variety of diagnostic services. Patient x-rays from naval vessels at sea were examined via UHF, HF, and satellite, using narrowband slow-scan transmissions. EKG and electronic stethoscope devices were used to send physiological data using UHF and HF links (Rasmussen and Silva, 1976; Rasmussen et al., 1977).

- The Massachusetts General Hospital/Logan Airport project, conducted in Boston, Massachusetts, in 1968, demonstrated the use of microwave transmissions of 2-way audio/video/data for primary care applications. This system had the capability and was used to link a nurse clinician at Logan Airport with the physicians at Massachusetts General in Boston, providing telediagnosis and teleconsultation services. The system was designed to provide fuller utilization of the skills of the health professionals via telecommunication's capacity to overcome the problems of accessibility and travel time in an urban setting (Murphy, 1972).

- From 1973–1975, the Remote Radiographic Transmission for Diagnostic Interpretation project demonstrated the use of microwave, cable 2-way audio, 1-way data (x-rays) and telephone for primary care applications in Omaha, Nebraska. This project was designed to provide on-demand interpretation of x-rays transmitted between the small rural community of Broken Bow, which had no radiologist, and the Department of Radiology of the University of Nebraska Medical Center. The purpose was to implement and evaluate a slow-scan system for transmitting x-rays (Chan and Massick, 1975).

- The Playas Telehealth System demonstrated the use of microwave, 2-way audio/video, telephone and facsimile for primary care and administrative applications in 1975 at Playas Lake, New Mexico. This system was designed to provide comprehensive health care to over 500 workers and their dependents who live at the Phelps Dodge Corporation copper smelting town of Playas, New Mexico. Playas Clinic, operated by physician extenders, was linked to the Med Square Clinic in Silver City where physicians were available to supply a wide range of services via telecommunications to the Playas Clinic. The services included consultation, diagnosis, and administrative support. Playas and surrounding community residents had a broad base of health services available to them, including primary care, emergency care, laboratory/x-ray services, and pharmaceuticals (University of New Mexico, 1975).

- The Rhode Island Rural Health Demonstration Project, conducted between 1976–1979 at Block Island, Rhode Island, demonstrated the use of slow-scan television for primary care applications. This project was designed to provide comprehensive care to the populations of the medically underserved towns of Gloucester and Scituate, Richmond and Charleston, and Block Island. The first two areas were served by mobile vans offering a variety of services. Block Island, medically isolated from the mainland, was linked to a mainland hospital via slow-scan television to permit voice communication and transmission of x-rays, EKGs, and body views for specialty consults (Rappaport and Skinner, 1978).

- The Canadian Telemedicine Experiment U-6 was conducted in London, Ontario, Canada, from October 1976–February 1977, using satellite transmission of one-way video, two-way audio, facsimile, and 1-way data for professional supervision and support, as well as primary care. The Hermes Communication Technology Satellite linked a remote base hospital via one-way TV plus interactive audio and facsimile to the University Hospital. Further, a nursing station was linked via interactive audio and facsimile to the base hospital to support primary care at the nursing station. The camera at the base hospital was controlled remotely from the University Hospital, a TV link was used for support of practicing physicians and specialists at the base hospital in the areas of radiology, anesthesia, psychiatry, cardiology, pathology, hematology, physiotherapy, dentistry, pharmacy, respiratory technology, nursing support, infection control and administration (Carey and Russell, 1977; Roberts et al., 1978).

The U.S. Public Health Service interest in teleradiology was to provide for the prompt interpretation of radiographic images related to the health care of an estimated 28 million Americans who resided in rural areas underserved in such care (Gitlin, 1986). These people, who represented about half of the country's total rural population, were poorer and sicker and had a significantly shorter life expectancy than their urban counterparts. Distance, poor roads, and climatic extremes impeded their access to services considered routine in urban areas. Rural areas have long been plagued by a shortage of health care providers and an attendant disparity, compared with urban areas, in access to quality health care. The shortage of physicians in rural areas is slow to change, and access to health care services in geographically isolated communities continues to challenge the medical community and the government.

The Public Health Service Act was amended in November, 1978, and the Health to Underserved Rural Areas (HURA) Program was replaced and modified by the Primary Care Research and Demonstration (PCR&D) Program. The approaches of the former HURA Program and that of the PCR&D Program were essentially the same, but the target population was expanded to include those in urban as well as rural areas.

Earlier investigations supported by the National Center for Health Services Research (NCHSR) had led to the identification of promising communications and computer technologies. Slow-scan television was one of the technologies selected for demonstration because it appeared to have a high cost-benefit potential. Slow-scan television enables a rural health provider to send pictures over a telephone line to a receiver located at a site where specialized medical knowledge is available. The initial efforts used a wide variety of applications, including pictures of patients with dermatologic conditions, dental problems, and injuries. The transmission of radiographic images emerged as one of the most important applications after practical experience was gained at several clinics linked to medical centers. In 1978, the Bureau of Radiological Health (now the Center for Devices and Radiological Health) was invited to evaluate the potential of the system to provide routine diagnostic radiology services, and a series of special studies was undertaken that continued through 1985.

Field Trials

A teleradiology system was evaluated at The MITRE Corporation's Telehealth Laboratory in McLean, Virginia, during the summer of 1980. Focusing on significant differences between teleradiology and conventional films, four evaluative questions were addressed: (1) the average accuracy of findings and impressions; (2) the average confidence that radiologists have in the accuracy of their findings and impressions; (3) the ability of radiologists to avoid false negative and false positive findings and impressions; and (4) whether the technical controls for image enhancement available to users in the teleradiology mode improve the accuracy of findings and impressions when these controls are used extensively.

The results showed that the accuracy of interpretations from radiographs consistently exceeded those from teleradiology images, but not by a wide enough margin to eliminate teleradiology's promise as a solution to health care delivery problems in remote areas. The results also suggested that the next step in the development of teleradiology would be to test its performance and acceptability in a field setting.

For the longer term (i.e., following the field test), the possibility of incorporating new technical developments in image processing, such as digital radiography into the teleradiology system directly, should be investigated. If feasible, digital radiography, for example, could eliminate the need to make a video image of a radiograph (and the inevitable loss of information this step implies), thereby improving the accuracy of teleradiology images relative to conventional alternatives. Both radiography and image processing are developing rapidly, and their possible contributions to teleradiology systems development should be reviewed periodically.

The Department of Defense has also supported the development of teleradiology to help solve problems related to the management of military radiologist resources. Allman et al. (1983) expressed the following opinions:

> Teleradiology cannot solve all resource management problems, but it certainly can help in at least four areas that have plagued the military medical establishment. First, by improving the distribution of workload from remote treatment facilities to locations staffed by radiologists; second, in helping to overcome the shortages of radiologists by improved career incentives; third, by improving the consultative services of radiologists; and fourth, by its potential for improving image handling, distribution, storage, retrieval and analysis in the future.

In 1980, the Army, the Navy, and the Air Force joined the studies through the Uniformed Services University of the Health Sciences and the Tri-Service Medical Information System in an effort to improve radiologic services to personnel, their dependents, and retired personnel. Diagnostic radiology service at medical care facilities without radiologists was usually delayed and involved the transfer of films to distant medical centers or weekly visits by a qualified radiologist from military hospitals or by contract specialists. The Army Medical Research and Development Command envisioned a teleradiology system that would not only serve military personnel during peacetime but also lead to "filmless radiology" systems that would improve medical care under combat conditions. With the participation of the MITRE Corporation, university-based radiologists, and other consultants, the efforts of the four uniformed services (that is, the Public Health Service, the Army, the Navy, and the Air Force) resulted in two field trials of different teleradiology systems, in 1982 and 1984.

The results of the Teleradiology Field Trial should be considered in the context of the population served by the four participating clinics and the heavy weight given to discrepant cases and special studies included in the detailed analyses. The military personnel, their dependents, and the rural population in central Virginia may differ significantly from other groups of interest, and the accuracy of interpretations reported might be much different from that resulting from another mix of radiographic procedures. Nevertheless, the results appeared encouraging for the future of teleradiology. An important indicator was the relatively small number of discrepancies between video and film reports in both the 1982 and 1984 trials. The magnitude of the number may be related to the large proportion of normal cases in the study populations as well as to the original radiographs and the transmitted images.

Despite the superior accuracy of film interpretations demonstrated in these trials, the teleradiology concept showed promise as a method of providing radiographic services to certain populations. Shortcomings of image quality were understood more fully as a result of the evaluations reported here. With this understanding, it was possible to correct these shortcomings and increase the diagnostic accuracy of video images. It was also hoped that the results, together with corresponding technologic improvements, would contribute to the advancement of digital radiology.

Digital Imaging Network System

Based upon the findings and results of the 1982 and 1984 Teleradiology Field Trials, the Picture Archiving and Communication System (PACS) appeared feasible and a Digital Imaging Network project was designed.

In 1986, the U.S. Army Medical Research and Development Command contracted with The MITRE Corporation to investigate the use of filmless radiology in both its fixed facility and battlefield medical systems. To support this investigation, commercially available Digital Imaging Network System (DINS) equipment was competitively procured and installed at two university medical centers. Equipment provided by AT&T was installed at Georgetown University and equipment provided by Philips Medial Systems, at the University of Washington. In parallel, a prototype system was developed, using commercially available workstation technology, to support concept exploration of DINS in the battle-field environment.

The objective of DINS was to replace film-based radiological image management with filmless image management. In the filmless environment, images are acquired digitally, transferred over computer networks, stored on magnetic or optical computer storage media, and displayed on video monitors. However, the technologies and issues involved in implementing filmless radiology in the peacetime and combat medical arenas constantly evolve. The project had to maintain flexibility regarding current technologies to assure that those evaluated were relevant to the Army's plans for the future.

The DINS project can be credited with significant accomplishments in medical applications of the system:

- First full-scale use of DINS in a radiology department with links to other parts of the hospital and to remote sites
- First evaluation of filmless radiology in the combat medical environment
- First evaluation at Army medical centers of Computed Radiology (CR) employing phosphor plate technology
- First evaluation of teleradiology as a peripheral to a hospital image-management system as distinguished from stand-alone teleradiology systems
- First demonstration of teleradiology linking overseas U.S. Army bases and continental United States facilities.

Relationship to Telemedicine

Telemedicine may be broadly defined as the use of information technology to deliver medical imaging services and related information from one location to another. However, authors differ about what the definition should include (Bashshur, 1995a). Most agree that it includes applications in areas such as pathology and radiology, as well as consultations in specialties such as neurology, dermatology, cardiology, and general medicine. While some consider certain forms of medical education within the definition, others would exclude the use of video to transmit purely didactic classroom lectures in which no direct interaction between student and teacher occurs. Whatever the definition, telemedicine implies a closer link between the telecommunications infrastructure and the health care system that includes the entire range of teleservices.

The goals, organization, funding, and technology of current telemedicine projects vary. This diversity is shown below:

- Initial urgent evaluation of patients, triage decisions, and pretransfer arrangements
- Medical and surgical follow-up and medication checks
- Supervision and consultation for primary care encounters in sites where a physician is not available
- Routine consultations and second opinions based on history, physical exam findings, and available test data
- Transmission of diagnostic images
- Extended diagnostic work-ups or short-term management of self-limited conditions
- Management of chronic diseases and conditions requiring a specialist not available locally
- Transmission of medical data
- Public health, preventive medicine, and patient education.

The practice of diagnostic radiology and by extension, teleradiology, is an important component of clinical decision support (CDS) in the delivery of health care. CDS can be broadly and simply defined as the use of information to help a clinician diagnose and/or treat a patient's health problem. Two kinds of information are involved: (1) information about the patient; and (2) information about the kind of health problem afflicting the patient and alternative tests and treatments for it. CDS is by no means a new phenomenon; such information traditionally has been available from several sources. However, those sources, including medical imaging, have limitations that often diminish their reliability or their accessibility at the point of care.

The time pressures of clinical practice do not allow clinicians to study the patient's entire health history or to review the latest clinical knowledge about every nonroutine health problem they encounter. Consequently, one major goal of CDS is to locate needed information and make it available to the clinician in readily usable form at the point of care as quickly as possible, and in a manner that minimally interferes with the care process. Moreover, the potentially severe consequences—for both the patient and the clinician—of incorrect clinical decisions require that the information retrieved be as accurate as possible.

As impressive as their applications are, the usefulness of clinical decision support systems can still be hampered by incomplete, inaccurate, or inaccessible information-problems that advanced information technologies could help overcome. However, the capabilities of many of the information technologies employed in CDSs remain limited and their costs remain high, posing substantial barriers to their widespread use. Several technological advances are needed, especially in teleradiology, for faster, easier and more accurate collection, entry, and retrieval of patient information, and more readily accessible information about the health problem. The needed advances include:

- advanced human-computer interface technologies, particularly speech recognition, for easier and possibly hands-free input and retrieval of information
- more extensive use of structured data entry, such as onscreen forms and menus and prepared blocks of text, to ensure complete data collection and reduce keying errors
- higher capacity and more flexible electronic storage devices, such as updatable CD–ROMs
- higher-resolution computer displays
- more powerful and flexible graphics software
- improved technologies for capturing and storing digitized radiographic images, full-motion videos, and sound recordings, and faster methods of retrieving such information
- faster and more flexible methods of online query using relational databases
- higher-capacity telecommunications equipment and transmission lines
- more complete coverage of the research literature by online bibliographic databases.

As Wyatt (1994) put it: (clinicians) need a system that is easy to use: computer terminals must be ubiquitous, system response must be immediate (not seconds), necessary data should always be online, accessible, and confidential, and very little training should be required. In addition, system downtime must be at an absolute minimum, and data should be retained for as long as possible without diminishing system response times. Systems that meet all the needs of clinicians may have to be developed in-house rather than adapted from commercial products.

The technologies used for collecting, distilling, storing, protecting, and communicating data are widely used throughout American industry. In the health care industry, however, their application has been limited to scattered islands of automation, usually within discrete departments of hospitals. Among these departments, radiology is one of the most frequent users, reflecting on teleradiology as the most widespread application within telemedicine.

The use of telecommunications to transmit medical images is quite well developed and widespread. A teleradiology system acquires radiographic images at one location and transmits them to one or more remote sites, where they are displayed on an interactive display system and/or converted to hard copy (Batnitzky et al., 1990). Transmissions might include CAT scans, MRIs, or x-ray images. CAT scans and MRIs originate in digital form, but a film digitizer must be used to convert conventional radiographs from film to digital form. Teleradiology systems often employ a wide area network.

Teleradiology systems transmit images from one hospital to another, from an imaging center to a hospital, or from an imaging center or hospital to a radiologist's office or home. Each requires different technologies and communication links, and each site has different requirements (Dwyer, III, et al., 1993). For example, a radiologist who is on call may review an image in his or her office or at home, but later will also review the original image before making a final diagnosis. In this instance, a lower image resolution, requiring less expensive equipment, may be acceptable. A higher quality image is required if the radiologist is making a final interpretation without seeing the original image, as with a request for a second opinion or a hospital that contracts for its radiological interpretations.

In its study for HCFA, the Center for Health Policy Research found that, with some exceptions, radiology using telecommunications is feasible (Grigsby et al., 1993). For providers, training in the use of the equipment appears to be a critical component, particularly learning to manipulate and interpret images using a video image on a monitor. Preliminary research suggests, however, that most radiologists prefer conventional films and view boxes to teleradiology because reading them is less time-consuming and perhaps because they are more familiar with them (Grigsby et al., 1994).

Digital Imaging

Medical imaging offers one of the most fertile grounds for the application of advanced computer and communications technologies. All-digital systems, known as picture archiving and communications systems (PACS), now offer imaging of sufficient quality for primary diagnosis in radiology, although their high costs are a barrier to diffusion. The University of Virginia operates a PACS system that it plans to link to two remote sites, at distances of four and ten miles, using a T1 telecommunications line. Further expansion is planned based on the experience with the two sites (Healthcare Telecom Report, 1994).

Imaging is an essential component of the health care delivery system (Allman et al., 1992). All medical specialties include radiologic imaging in the diagnostic work-up to some degree. Most diagnostic images are produced from film/screen combinations and are interpreted using film viewed on light boxes. The images are stored, transported, and communicated from one physician to another on film. X-ray film has served physicians faced with diagnostic problems well for almost 100 years. When this film-based system works, it works extremely well. However, over the years, problems in film-based systems have emerged which adversely affect the delivery of health care. Quality issues such as timeliness of reports; adequacy of images; accessibility of images to the radiologist and the referring physician; availability of comparison films for treatment planning or teaching conferences; transfer of films to other facilities; and availability for quality assurance, risk management, and medical audit purposes need to be addressed.

Radiology information systems have helped to define the management issues related to handling and distributing film-based images but have not solved the problems. Although efforts have been made to improve the consistency and quality of images, technically poor studies, particularly from emergency room and portable devices, still abound. Delays may affect the clinical value of diagnostic examinations. Efforts to establish tight control over films fail for many reasons, and films are still lost, or more accurately are not available to individuals who require them at convenient times. The perceived and real deficiencies of the film-based system affect the quality of care and the professional stature of the radiology department. Film management affects: (1) patient satisfaction, (2) length of patient hospital stay, and (3) complications resulting from delay in accessibility of films or reports.

The expansion of technology over the past 20 years has greatly increased the quantity of medical images, a growing number of which are generated in digital format. Nevertheless, physicians still use paper to request examinations and radiologists still interpret nearly all studies from film. Films are stored in bins in a file room, and reports are scattered in a variety of locations that defy counting. Efforts have been made to exploit the ability of computers to manipulate images and data and to interface imaging systems with radiology and hospital information systems. Recent developments make it possible to convert all images to electronic digital records regardless of their original form. It is, therefore, reasonable to consider managing imaging services electronically.

In the transition from film to digital imaging, it is essential to involve radiologists in all stages of a design and in the selection of components. Within the radiology department, various imaging modalities need to be integrated for an efficient operation, for example, linking CT, MR, and computed radiography units to shared laser printers, storage devices, and networks. Investments in equipment which improves productivity and decreases personnel overhead will obviate large numbers of staff to handle, distribute, and store images in the digital department.

One should also consider the effect of new technology on other departments. Improving patient care is always a first priority that brings specialty interests together. For example, the emergency room patient may need a radiologist to interpret images at any time of the day or night. A PACS teleradiology subsystem can make a radiologist available 24 hours a day, seven days a week. The radiologists may be available any place accessible to a network. This brings most hospital locations and the radiologist's home into working relationships. The best practice demands accompaniment of all images transmitted outside the radiology department by a radiologist's interpretation.

Digital imaging can ensure that portable and other images will always be of a high quality rather than hit-or-miss, as is often the case. Digital networking insures that off-site locations will always have a radiologist available for interpretation. Other departments which will benefit by being connected to a PACS include plastic surgery, neurosurgery, and oncology where staging, surgical planning, and treatment evaluation will be enhanced by ready access to images from multiple modalities. Many other specialty areas will also be able to receive needed services.

The "digital file room" will be at the heart of any image management system. When selecting the hardware and designing the network, it is essential to provide access to current and previous images by primary interpreters and by practitioners in other medical services. Primary radiologists need a high level of image resolution for interpretation. For review in other departments, a lower level may be acceptable and data compression may be used to reduce the time for image transmission and space required for image storage. A responsive mass storage device and an excellent database management system are critical.

Networking defines the distribution of images in and outside the radiology department to facilitate communication and consultation between locations, other radiologists, and other physicians. Although a single network might accomplish this, it may be managerially and economically advantageous to consider multiple networks with different primary purposes. For instance, the networks within the radiology department must be tightly controlled by the radiology manager. Images obviously must be available for the radiologist at the appropriate workstation. The internal network is one of the medical devices essential to professional performance. On the other hand, images for review or consultation in other departments may be more appropriately managed and distributed by a hospital network manager. The hospital imaging network is part of the information service to the professional community at large. Although the two networks have much functionality in common, their role in professional performance is qualitatively different. These needs will vary from place to place.

Digital imaging, particularly computed radiography, may offer consistent image quality at lower patient dose. The need for repeat examinations may be reduced because of increased image linearity and dynamic range. Several utilities are available for image processing and enhancements, something which radiologists have been striving to accomplish for years. Magnification and reduction, black-white reversal, edge enhancement, and other types of image manipulation may improve visualization of tissue characteristics. Digital storage and transmission offer ready image availability. Images will be almost instantly available at display stations through a variety of strategies for retrieval of data. Image replication without degradation means images transferred to radiology for interpretation may be released simultaneously for review by other physicians caring for the patient. Images may be networked to common storage devices and to shared printers, thus reducing the need for duplicating equipment at multiple locations. Images will no longer be lost. ("Of course, it's not lost," Belgarath retorted. "I just haven't pinpointed exactly where it is at the moment." "Belgarath, that is what the word lost means" (Eddings, 1988).

For a variety of reasons, image manufacturers in the past have avoided standardization not only with other vendors but within their own product line. It has been difficult or impossible to connect different imaging devices and display modules to each other through standard interfaces because of unusual connection requirements or non-standard signals. Obviously, interfaces are required for image networking and for integration of PACS with radiology and hospital information systems. Many professional organizations and equipment vendors are currently working on standard hardware and communications protocol characteristics for PACS. The ACR–NEMA standard presently represents the most promising approach to interconnectivity. Integrally associated with teleradiology are a number of attributes of electronic digital radiology in general. These contribute strongly to the appeal of rendering service by teleradiology. Some of the particularly appealing qualities are:

- The ability to keep detailed records of technical attributes, examination events, and personnel with respect to time. This provides a new level of technical oversight and quality control which can be focused on the teleradiological procedures in particular. These measures can be folded into the continuing quality improvement interests of the department, clinic, or hospital in general.
- The capability of using the great latitude of digital radiography to adjust image formation. Digital radiography enables manipulation of images to achieve specialized diagnostic objectives, and it furnishes a means to exercise tight control over image characteristics.
- The ability to handle images as intangibles. Electronic image storage and retrieval provide a different orientation toward image library functions from that of traditional film management with its tangible, physical images. Performing radiology at or with images from remote sites has always faced problems of identifying, recording, and keeping track of the foreign films. Correlating these images with records from other sources has been a major challenge. With electronic imaging systems, particularly if integrated with an electronic hospital information system, the possibility of accessing all needed information for sophisticated analysis and interpretation of radiologic studies offers particular appeal to the responsible radiologist.
- Finally, the ability to access electronically various segments of the patient record, and various decision support services. These may be either part of the hospital complex or distant services accessible through networks such as Internet.

Relationship to PACS

The future of teleradiology is closely related to the successful implementation of picture archiving and communication systems (PACS) in hospitals, medical centers, and other health care facilities where radiologists are available to provide decision-support information to clinicians. Teleradiology differs from PACS in transmitting medical images usually over longer distances than within health care facilities, but all the technical considerations pertinent to PACS apply to teleradiology to some extent. What is clear from the experience to date is that the multiple uses of required equipment for both applications can substantially reduce the capital and operational costs of "stand-alone" teleradiology systems. As PACS become more widespread, teleradiology will become easier to justify economically. Therefore, when considering the technical requirements for teleradiology, it is appropriate to discuss them in the context of PACS technology.

When the future of diagnostic radiology or medical imaging is discussed by practitioners, scientists, administrators, and entrepreneurs, many of them present a picture of radiologists and other physicians viewing

Figure 5-1. Multiple monitors facilitate comparison of images from recent and previous examinations.

images on multiple electronic display stations. These images will have been produced by a wide variety of modalities such as computed tomography (CT), magnetic resonance imaging (MRI), and conventional x-ray equipment, each of which conforms to an international standard and is linked to the display stations via a high-speed network. In between the image-generating equipment and the viewing stations are high-density storage devices, computers, and interfaces to other information systems. The referring or attending physician is portrayed as having immediate access to patient's images and interpretive reports through an inexpensive personal computer connected to the imaging network.

In this scenario, the proponents assume that the necessary technology to support such practice exists, or will be available in the near term; that the product will be cost-effective regarding improved patient care, increased efficiency, and replaceable costs; and that physicians, nurses, technologists, and other health care providers will accept the changes required to implement the new systems.

But the claims and expectations of proponents of image management systems should be considered realistically. Each of the major components of picture archiving and communication systems (PACS) should be examined for feasibility, economics, and acceptance. On the one hand, dramatic changes of great benefit to patient care have been introduced to diagnostic radiology by digital imaging. However, modalities such as CT and MRI require large expenditures for acquisition and maintenance, and when used routinely, are associated with substantial data-volume problems. The concern with the management of the large volume of digital information has been central to the concept of PACS and to efforts to justify the systems in cost-effectiveness and professional acceptance.

The impact of the development of computed tomography has been compared with that of the discovery of x-rays by Roentgen in 1895. In her excellent article "The New Light," Nancy Knight presents a comprehensive summary of the reaction to Roentgen's discovery in the context of prevailing use and scientific advances of that era. She points out that, because of the simplicity of x-ray apparatus, access to the new technology was soon available to all interested parties, amateur or professional. This is not true of modern imaging modalities, and we are somewhat less sanguine today about the future. We should therefore try to evaluate the costs and the benefits of image management systems as objectively as possible. The large amount of digital data and information related to diagnostic imaging cannot be ignored. They must be managed in ways that will maximize patient-care benefits and minimize related costs.

Most workers have found it useful to classify the components of PACS or digital imaging networks within four functional categories. H. K. Huang's textbook, *Elements of Digital Radiology,* uses this approach very effectively to describe the components of electronic PACS. These include: (1) image acquisition, (2) storage and retrieval, (3) local and wide area transmission networks, and (4) display and interpretation. Each of these categories is comprised of hardware and software that may be considered individually as well as for their interaction within and among the categories.

One issue associated with image acquisition is data compression (i.e., whether to use an algorithm that produces relatively low compression ratios, but permits access to the original data, or a program that produces relatively high compression ratios that result in the loss of some of the data). The outcome affects both storage and retrieval functions as well as data-transmission requirements. Another important issue within the acquisition category is the value of implementing the ACR/NEMA digital image communication standards that will assure recognition of output from the various image generators, each of which produces unique format and content that would be indecipherable to other components of a PACS system without such a standard.

Within the category of storage and retrieval, several major issues arise when considering the inclusion of radiographic films that were produced by conventional x-ray examinations. Film digitization techniques, spatial and contrast resolution, storage media and image access time: each poses important questions when the conversion from analog to digital form is attempted. With regard to digital images generated by such modalities as CT and MRI, there are issues concerning the status of the images to be stored (e.g., pre- and postprocessed (or both), and whether to store all the images or just those that in the judgment of the radiologist illustrate the interpretation).

Decisions regarding networks also require careful consideration if the system is to meet the stringent requirements of medical imaging. While there is no question about the need for "high-speed" media to transmit medical images, the specific "throughput" capacity of a network often *is* an issue. Furthermore, the potential benefits of a separate, dedicated high-speed network for images installed in parallel with a lower-speed line for other data must be carefully considered in terms of economic and technical capacity.

Perhaps the most important issue related to digital imaging from the practicing radiologist's point of view is the commitment to utilize electronic displays for primary diagnosis rather than film or other hard copy. While the adaptation to "softcopy" display for interpretation may be cost-effective, the primary user (i.e., the radiologist) must find it acceptable. Before this issue is resolved, much needs to be learned about the human/machine interface requirements and a generation of radiologists will have to be retrained. The utility of image-processing techniques also is important when considering electronic display and interpretation. If "softcopy" is to become routine in primary diagnosis, its value will depend on the positive contribution of image processing to the accuracy of the interpretation as well as the prompt availability of comparison and correlation images. A great deal remains to be done to demonstrate the value of these techniques to the diagnostic process.

Immediate access to electronic images by referring and attending physicians poses another issue for the medical imaging community. Should images be made available without the interpretive report or should they be withheld until the report is available? One solution is to produce the report immediately following the diagnostic procedure and include it with the data representing the images. Several technical developments offer this possibility, including speech recognition modules linked to personal computers. The integration of automated reporting at a diagnostic workstation with display and image processing deserves considerable attention.

Goals and Objectives. The central questions facing many radiology department and hospital administrators are (Forsberg and Sawehak, 1994): Do I have excellent radiology coverage now? Can my radiologist interpret all modalities? And finally, is the radiologist available whenever needed? In light of these questions, there are four basic areas where teleradiology services can improve the quality of medical care:

(1) Interpretation of CT, MRI, or other digital studies: Teleradiology may allow the addition of recently developed digital modalities to a medical care facility's services even if there is no available radiologist trained in that modality or if the additional workload cannot be handled with existing staff.

(2) Call or after-hours coverage: Teleradiology allows coverage of hospital radiology departments and emergency rooms 24 hours a day, every day.

(3) Second opinion: Teleradiology allows for immediate "difficult case" consultation, and facilitates continued medical education, thereby eliminating the solo practitioner's isolation. Subspecialty consults by academic radiologists in major medical centers are easily obtained using this technology.

(4) Extra coverage: The staffing needs of many practices evolve in increments difficult to cover. Teleradiology can provide ancillary daytime coverage without the expense of adding another full-time radiologist when only additional part-time support is needed.

Twenty-four-hour comprehensive radiology services with access to subspecialists should increase the hospital's revenues. Access to up-to-date radiologic technology increases technical revenues, decreases patient out-migration, and enhances recruitment of physicians, all of which improve the economic status of the hospital. These issues become even more critical as the health care system is reformed. Hospitals may choose to add modalities when professional expertise is available. High-quality radiology services aid in the recruitment of other physicians who depend on these services. More timely radiographic interpretations result in more judicious patient transfers, shorter hospitalizations, and, when appropriate, aid in patient retention. As confidence in the radiology department grows, more utilization occurs. In the future capitated-environment, high-quality radiology services will attract managed care contracts.

The hospital perspective of the benefits of teleradiology emphasizes the technology's contributions to the support of expanding services. But while teleradiology has been touted as a panacea for underserved rural and urban America, it cannot solve all the problems in the delivery of radiology services.

Teleradiology, in its most rudimentary forms, has existed for over 30 years. One of the earliest demonstrations was by Jutra in 1959. But it was not until the last decade, with improvements in computer, video board, and monitor technology that the use of teleradiology could become widespread. Carey and others were among the first to recognize the power of telemedicine to improve health care delivery (Carey et al., 1989). They showed the positive impact of teleradiology on family practice patient care (Franken et al., 1989). Image transmission has proven useful in hospital settings as intensive care units link directly with radiology departments. One recent study demonstrated that, in an intensive care unit with digital viewing capability, the ordering physician was 9.5 times more likely to have access to examination results within one hour (Humphrey, 1993).

This dramatic reduction in turnaround time is a direct result of advances in teleradiology. Before the arrival of the digital revolution and the refinement of teleradiology technology, radiology coverage could be provided only when the physician was on site, leaving many hospitals with undesirable gaps of coverage throughout the day as well as "after-hours." Traditional radiology coverage consisted of full-time, on-site coverage or circuit-riders. If on-site coverage was unavailable, radiographs were sent via mail or courier to a radiologist. New modalities could be added at a hospital only if the incumbent radiologist possessed the professional expertise to interpret those modalities' images or, if not, at least the desire and time to develop those skills and abilities. Teleradiology can provide all medical facilities, regardless of size or location, access to twenty-four hour coverage and access to radiological subspecialists as well. The results are more comprehensive radiology department coverage and better patient care. Although outcome and economic analyses must be performed to quantify the impact of teleradiology fully, clearly, there is a place for this technology that improves quality of care, that will keep costs the same, and in some circumstances, even decrease them.

The most alluring attribute of electronic medical imaging may be the prospect of viewing radiographs instantaneously from remote sites (Allman et al., 1994): no delay, no lost films, not having to join a borrower's queue, just summon and behold. With electronic imaging, geography no longer poses a barrier to quick, easy, cost-effective service. For anyone who has ever covered a busy radiology practice, teleradiology's ability to surmount the obstacles of serving a rural hospital, an office across town, a satellite installation, or the emergency room at night, has an immense appeal.

In addition to benefits to practice, economic imperatives encourage the development of teleradiology. Radiologists and administrators recognize market value for both buyer and seller. The seller has the prospect of increasing the quality of service to patients by electronically extending the radiologist's consulting service to areas where constant physical presence is simply not possible, cost-effective, or convenient. This expands the market in which providers of health care can compete. From the buyer's perspective, this "opening up" of the market frees "hostage" institutions to seek competitive bids for coverage, thereby reducing contract costs. Success in reconfiguring these service patterns should be accompanied by better quality of care and better access to the health care system by the patient.

When the radiologist can be in all places at all times, what are some of the expected benefits?

First on the list must be the potential improvement of patient outcomes.

This will be brought about in many ways, but foremost is timely access to skilled service. Specialty diagnostic teams will reach many remote sites convenient to the patient instead of requiring the patient to travel long distance to receive consultative services. This outreach not only augurs well for better medical results but also for better accommodation and satisfaction of the patient.

By reducing the time to conduct services as well as to deliver them, teleradiology has the potential to increase the efficiency and productivity of many members of the radiology team. Radiologists, technologists, assistants, and clerks will all acquire better control of the processes in their charge and, thereby, will be able to turn out a larger volume of high-quality work.

The use of telecommunications to deliver health services has the potential to reduce costs, improve quality, and improve access to care in rural and other underserved areas of the country. Although the extent of this potential is largely speculative at this time, researchers are beginning to address telemedicine's impacts. According to one researcher:

> . . . telemedicine may be unique in having the potential for introducing low-cost, high-efficiency components that may, under certain conditions, increase access to care while possibly limiting increases in cost by enhancing health outcomes (Bashshur, 1995a).

If research results prove to be largely positive, telemedicine is likely to become fairly routine over the next 10 to 20 years (Grigsby et al., 1994). Whether and how telemedicine affects the quality of care delivered has not yet been proven. However, it is possible to speculate that some aspects of the electronic medical encounter might provide better care from the patient's perspective. Telemedicine could provide faster, more convenient treatment (Holand and Pedersen, 1993). Receiving the services of a specialist without having to leave one's community also provides better continuity of care. Similarly, allowing a patient to remain in the local hospital with family and friends for support could improve the quality of the experience for the patient and could, in fact, contribute to a faster recovery. These benefits would minimize the disruption of the patient's life and reduce the amount of working time lost. Follow-up care seems well suited to telemedicine, and might be carried out more effectively and efficiently by electronic means, thereby avoiding the costs of time and travel for an office visit.

Telemedicine can also offer high-quality services to the health care provider. Having timely, convenient access to the most up-to-date information, continuing medical education programs, decision support systems, and consultations with specialists in large medical centers should increase the provider's options and improve the accuracy of his or her diagnoses and effectiveness of treatment. Moreover, the development of clinical practice guidelines for telemedicine could lead to better care. However, whether telemedicine consultations improve the quality of care will be known only when patient outcomes have been determined.

Standards

The proliferation of teleradiology, especially in conjunction with hospital-based image management systems, requires the integration of existing and new architectures, products, and services. Health care data standards are essential to make these diverse components work together effectively. Two organizations have assumed responsibility for the coordination and promotion of health care data standards development in the U.S.: the Health Care Informatics Standards Planning Panel (HISPP) of the American National Standards Institute (ANSI), and the Computer-based Patient Record Institute (CPRI). HISPP coordinates the work of standards development groups for health care data interchange and health care informatics. CPRI's mission is to promote acceptance of the vision set forth by the Institute of Medicine Study report "The Computer-based Patient Records: An Essential Technology for Health Care" (Institute of Medicine, 1991).

Of the many standards available to the health care infrastructure, communication standards have the most direct applicability to the proposed project. These include the following (Blair, 1995):

Digital Imaging and Communication (DICOM): This standard was developed by the American College of Radiology and National Electronic Manufacturers Association (ACR–NEMA). It defines the message formats and communication standards for radiology images. DICOM is supported by most Picture Archiving and Communication Systems (PACS) vendors and has been incorporated into the Japanese Image Store and Carry (ISAC) optical disk system, as well as Kodak's PhotoCD. ACR–NEMA has applied to be recognized as an accredited organization by ANSI.

Health Level Seven (HL7): This standard is used for intra-institutional transmission of orders; clinical observations and clinical data, including test results; admissions, transfers and discharge records; and charge and billing information. HL7 is being used in more than 300 U.S. health care institutions, including most leading university hospitals. HL7 has been recognized as an accredited organization by ANSI.

The Institute of Electrical and Electronic Engineers Inc. (IEEE) P1157 Medical Data Interchange Standard (MEDIX): IEEE Engineering in Medicine and Biology society (EMB) is developing the MEDIX standards for the exchange of data among hospital computer systems. Based on the International Standards Organizations (ISO) standards for all seven layers of the Open Systems Interconnect (OSI) reference model, MEDIX is working on a framework to guide the development and evolution of a compatible set of standards. IEEE is recognized as an accredited organization by ANSI.

The American College of Radiology (ACR) has published a "Standard for Teleradiology" that defines goals, qualifications of personnel, equipment guidelines, licensing, credentialing, liability, communication, quality control, and quality improvement for teleradiology. While not all-inclusive, the standard should serve as a model for all physicians and health care workers who utilize teleradiology.

The ACR publication lists the goals of teleradiology as:

- providing consultative and interpretative radiological services in areas of demonstrated need;
- making services of radiologists available in medical facilities without on-site radiological support;
- providing timely availability of radiological images and radiologic image interpretation in emergent and non-emergent clinical care areas;
- facilitating radiological interpretations in on-call situations;
- providing subspecialty radiological support as needed;
- enhancing educational opportunities for practicing radiologists;
- promoting efficiency and quality improvement; and
- sending interpreted images to referring providers.

With regard to licensing, credentialing, and liability, the ACR stated:

- Physicians who provide the official, authenticated interpretation of images transmitted by teleradiology should maintain licensure appropriate to delivery of radiologic service at both the transmitting and receiving sites. When providing the official, authenticated interpretation of images from a hospital, the physician should be credentialed by that hospital medical staff in accordance with their bylaws. These physicians should consult with their professional liability carrier to ensure coverage in both the sending and receiving sites (state or jurisdiction).
- The physician performing the official authenticated written interpretations must be responsible for the quality of the images being reviewed. The ACR Rules of Ethics state: "It is proper for a diagnostic radiologist to provide a consultative opinion on radiographs and other images regardless of their origin. A diagnostic radiologist should regularly interpret radiographs and

other images only when the radiologist reasonably participates in the quality of medical imaging, utilization review, and matters of policy which affect the quality of patient care."

- At the teleradiology receiving site, the retention period of that jurisdiction must be met as well. The policy on record retention should be in writing.
- The physicians who are involved in practicing teleradiology will conduct their practice in a manner consistent with the bylaws, rules, and regulations for patient care at the transmitting site.
- The use of teleradiology does not reduce the responsibility for the management and supervision of radiologic medicine. Procedures should be systematically monitored and evaluated as part of the overall quality improvement program of the facility. When feasible, monitoring should include the evaluation of the accuracy of the interpretations as well as the appropriateness of the examination. Incidence of complications and adverse events should be reviewed to identify opportunities to improve patient care. The use of teleradiology shall be documented. Periodic reviews will be made for appropriateness, problems, and quality of transmitted data. The data should be collected in a manner which complies with statutory and regulatory peer-review procedures to protect the confidentiality of the peer-review data.

The DICOM Standard

The combined efforts of the American College of Radiology (ACR) and the National Electrical Manufacturers Association (NEMA) have led to the development of the Digital Image Communication in Medicine (DICOM) standard, which provides a format for communication among medical imaging devices. This standard has been developed over a period of time, and the numerous demonstrations of practical capability in recent years have confirmed the viability of this approach. This is important in teleradiology, because it provides another way to link the teleradiology equipment to the imaging modalities at the image source. The use of this direct digital connection not only ensures image fidelity, but also allows access to the full resolution of the original image, such as the wide contrast range in CT scanners' images.

The medical imaging community has worked hard on the development of image communication standards for PACS. ACR–NEMA Version 2.0 has been used for some communication between imaging equipment and other devices, such as display workstations. The latest version of the standard, ACR–NEMA Version 3.0 (DICOM), includes network communication capability.

Simply stated, DICOM is a standard for the communication of medical images and associated information. It differs from ACR–NEMA Versions 1 and 2 in several major respects. Most important is that the basic design of the Standard was changed. Versions 1 and 2 relied on an implicit model of the information that is used in radiology departments. The data elements were grouped according to the experience of the designers, and though the mapping was imperfect, the message structure did allow the necessary information to be transmitted. In contrast, DICOM relies on explicit and detailed models of how the things (such as patients, images, and reports) involved in radiology operations are described, and how they are related. These models, called entity-relationship (or E–R) models, are a way to be sure that manufacturers and users understand the basis for developing the data structures used in DICOM.

In a decision of major importance for the Standard, it was agreed that developing an interface for network support would require more than just adding patches to Version 2. The entire design process had to be reengineered, and the method adopted was that of object-oriented design. In addition, a thorough examination of the types of services needed to communicate over different networks showed that defining a basic service would allow the top layer of the communications process (the application layer) to talk to a number of different network protocols. These protocols are modeled as a series of layers, and so are often referred to as "stacks." The existing Version 2 stack that defined a point-to-point connection was one. Two others were chosen, based on popularity and future expansion. These were the Transmission Control Protocol/ Internet Protocol (TCP/IP) and the International Standards Organization Open Systems Interconnection (ISO–OSI). Figure 5-2 is a diagram of the communication model developed. The basic objective was that a given medical imaging application (which is outside of the scope of the Standard) could communicate over any of the stacks to another device that used the same stack. With adherence to the Standard, it would be possible to switch the communications stacks without having to rewrite the computer programs of the application.

After three years of work, WG VI, with suggestions from industry and academia, completed ACR–NEMA DICOM (also called DICOM 3.0 and DICOM without a version number appended). It is a much larger standard than Versions 1 or 2, but also supports many more features than either of the prior versions.

While this standard can be used to connect a CT scanner to a teleradiology workstation, at the image transmission end, it is not directly

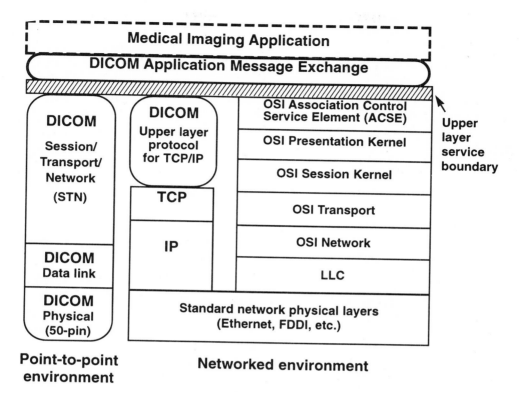

Figure 5-2.

applicable to the teleradiology link itself. That is, a teleradiology system may use a DICOM standard connection to the CT scanner, but include a proprietary method of transmitting the image over telephone lines using modems to the receiving workstation. This use of proprietary communication techniques prevents medical facilities from purchasing a transmitting station from one company and receiving workstations from other companies. This is particularly significant because it greatly limits the upgrade possibilities and may require that the user replace all the components in order to upgrade at a later date, even from the same equipment vendor.

It would be possible to use the image-communication capability provided under the DICOM standard over the telephone connection, although transmission might not be as fast as with some of the proprietary protocols. However, if a teleradiology workstation were capable of using either the proprietary or the DICOM method to transmit images, the user would have a wider range of upgrade options in the future.

Many decisions must be made in selecting the method of interfacing to a teleradiology system. If an interface to a digital imaging modality, such as CT or MRI, is required, then the user may have a choice of using a video frame grabber to capture the images, or using a direct digital interface. While the video connection may be faster and more flexible, the dynamic range of the acquired images is limited, and the viewer will not have control of the window width and level of the transmitted images.

The Role of ACR and NEMA in Supporting the Standard

The approval of DICOM 3.0 for publication and the demonstration of implementations as commercially available products at the 1994 Radiological Society of North America (RSNA) meeting were important milestones in the continuing development of the standard. These milestones represented the successful accomplishment of the major goals adopted by the ACR and NEMA when the joint effort was defined in 1983. After more than 10 years of intensive effort by both the ACR, which represents the principal users of imaging equipment, and NEMA, which represents the major suppliers, the network version of the standard became available. It is anticipated that this version will be formally adopted as a national and international standard in 1996, after submittal to the appropriate governing agencies in Japan, Europe, and North America.

The demonstrations of the standard at RSNA '94 included the public domain implementation developed at the Mallinckrodt Institution of Radiology with support from RSNA, and commercially available interpretations by more than 40 companies linked to RSNAnet at the annual meeting. The large number of products that comply with the standard should facilitate the utilization of digital image management systems and picture archiving and communication systems (PACS) by hospitals and other health care organizations. The demonstration was particularly appropriate in view of the emphasis on advances in medical imaging associated with the 1995 centennial celebration of Roentgen's discovery of x-rays.

The impetus for the development of the standard was the recognition of the value of digital imaging devices such as computed tomography (CT) (known as computerized axial tomography in the 1970s) in the diagnosis of a wide variety of diseases and injuries. The immediate clinical acceptance of CT and the availability of other digital imaging modalities such as nuclear medicine, ultrasound, and magnetic resonance imaging generated the perceived need for integrated electronic systems that could accommodate all types of imaging equipment produced by different manufacturers. This, in turn, led to the recognition that a standard was required to ensure effective communication among the imaging devices and related computer equipment.

While the need for an interface standard was long recognized, effective action to create the standard did not occur until 1983, when members of the ACR combined efforts with representatives of manufacturers through the newly formed ACR–NEMA Standards Committee. This committee, through joint action, was extraordinarily effective. In just two years, the committee and its working groups created an industry-standard interface originally known as the ACR–NEMA Digital Imaging and Communications standard (defined in NEMA publication no. 300-1985). By the RSNA annual meeting in 1987, most manufacturers had included the ACR–NEMA standard in their lists of equipment specifications (Lodwick, 1988).

The ACR–NEMA standard interface allows digital medical images and related information to be communicated between imaging devices, regardless of manufacturer or image format. Development of this interface was believed to be the first step necessary in the development of standards for PACS. Without data exchange, the backbone of digital image communication, other efforts to encourage the development of PACS would fail.

At the first PACS meeting (held in 1982), the problem of acquiring images and related data from different manufacturers' equipment was already apparent. The desire to acquire digital data directly was strong, but vendors were cautious about revealing how their software worked. The potential users of PACS, namely, the radiology community, asked the ACR to help find a solution. Engineering experts among the manufacturers also began to realize that little would be gained by keeping image formats proprietary. The result was the formation of the ACR–NEMA Digital Imaging and Communications Standards Committee early in 1983 (Horii, 1990).

After two years of work, the first version of the standard, ACR–NEMA 300-1985, was published and distributed at the RSNA annual meeting in 1985. As with many first versions, errors were found and improvements suggested. In response, the committee began working on changes to improve the standard. In 1988, ACR–NEMA 300-1988 (or ACR–NEMA

Version 2) was published. It used substantially the same hardware specification as Version 1, but added new data elements and fixed a number of errors and inconsistencies.

By 1988, many users recognized that an efficient interface between imaging devices and a network was required. While this could be done with Version 2, the standard lacked the parts necessary for robust network communication, and a solution to these problems meant that the standard would have to undergo major changes while retaining compatibility with earlier versions.

The ACR–NEMA DICOM standard has gained recognition by manufacturers and users as a significant factor in the future of medical imaging. It must be remembered that, because the standard is not mandatory, its implementation by equipment suppliers is optional. Support for the standard has grown with migration from the "black box" to an integrated interface (Spilker, 1989). The standard is seen as a dynamic development, changed and improved in response to the needs of the medical community with the support of the manufacturers. A goal of the committee is to make future versions of the standard compatible with previous versions and, whenever possible, develop new versions that can be "understood" by existing implementations.

It is anticipated that ACR and NEMA will continue to support further development of the standard and that collaboration with other medical disciplines and international agencies will enhance its utility around the world. Although much remains to be done, a great deal of credit is due the individuals and organizations that have contributed to the progress to date. Roentgen's centennial year should see many implementations of DICOM and the establishment of an international "consortium" that will have the responsibility and resources to continue the dynamic process of developing future versions of the standard. These efforts often seem far removed from the delivery of care to patients, but if medical imaging technology is to continue its contributions to cost-effective diagnosis and treatment, an evolving standard is essential.

The Concept of an Open System

The DICOM standard is patterned after the Open System Interconnection (OSI) of the International Standards Organization (ISO). The key feature of OSI is communication between heterogeneous systems. It was developed as a generalized model based on experience with ARPA-

NET and CYCLADES (Zimmermann, 1980). "Openness" is established when participating parties agree on a communication protocol. The "message" that is transmitted between the communicating partners ("nodes") is expressed in a specified form. The DICOM standard specifies this form through the transfer syntax that defines the coding of the information. The message itself is accompanied by instruction elements appropriate to the communication channel (for instance, which communication "stack" will be used).

A nontechnical parallel would be a communication between two mathematicians with no common conventional language, in which one asks a question, and the other replies, by means of symbols. They would use mathematical coding as a common language. Alternatively, they could each find a translator who speaks a common conventional language. This would require double translation in each direction; faulty "coding" (misunderstandings) would probably occur (Hindel, 1994).

Aside from the enormous amount of detail and complexity of the OSI standard, its essential features are the transportable message in well-defined form, the capability of the sending node to generate this message, and the corresponding capability of the receiving node to decipher or parse this message. The sending and receiving nodes need not use the same operating system or the same application program. They can be, and in many cases will be, heterogeneous.

By 1988, many users wanted an interface between imaging devices and a network. While this could be accomplished with ACR–NEMA Version 2, the version lacked the parts necessary for robust network communication. For example, one could send a device a message that contained header information and an image, but one would not necessarily know what the device would do with the data. Since Version 2 was not designed to connect equipment directly to a network, solving these problems meant very large changes to the Standard. Very early the Committee had adopted the idea that future versions of the Standard would retain compatibility with the earlier versions, and this placed some constraints on its work.

The scope of medical imaging extends well beyond radiological images. Endoscopists, pathologists, dentists, and dermatologists (to name just four specialty areas) all produce images as part of their practice. Recently, representatives of the professional societies of these groups met with a special Working Group of ACR–NEMA to begin planning how they could take advantage of the DICOM work. The approach of an object-oriented design is making this process relatively straightforward. The professional society members can provide the expertise about constructing appropriate information objects, and the ACR–NEMA groups can then help to translate them into the DICOM standard. The first task undertaken will be the definition of information objects for color imaging in endoscopy (including pulmonary, gastrointestinal, genitourinary, and orthopedic applications).

Interest in medical imaging both within and outside of radiology, and the general interest in electronic medical records have fostered the development of a group that has representation from all the medical informatics standards bodies and professional groups. This group is the Healthcare Informatics Standards Planning Panel (HISPP) sponsored by the American National Standards Institute (ANSI). ANSI is the U.S. representative to the ISO, and is the formal conduit for information exchange with The European Standardization Committee (CEN), the body that has responsibility for all medical informatics standards in Europe. The ISO also provides official communication with the Japanese standards organizations. As standards develop, coordination through ANSI HISPP and its international counterparts is vital if the end products are to be compatible. A look at the international nature of the imaging equipment business should serve to illustrate the importance of standards considerations that go beyond national boundaries.

Technology and Implementation

Medical imaging is an ideal domain for integration of information technologies. The various imaging machines are expensive and so highly specialized that no single vendor can impose proprietary standards on vendors of different types of machines. The imaging machines can share the same type of displays and data manipulation computers, however, and this has encouraged the development of broad data-exchange standards. Radiology and nuclear medicine are consultative disciplines, but since the images need not be interpreted at the site where they are collected, consultations can be carried out at a distance over telemedicine links. Radiologists usually examine images displayed with the highest possible resolution, but primary care physicians who rely on their interpretations may also wish to have access to lower resolution copies of images to explain the interpretations to patients. Finally, an economic incentive exists for fully developing digital image storage: medical images are among the most commonly misplaced or unavailable records. Some 40 percent of all x-ray films are not retrievable, making it necessary to repeat some imaging procedures and extend hospital stays (Horii et al., 1992).

In implementing a telemedicine project, it is important first to define its objectives, and then select the appropriate technology to meet those needs (Puskin, 1992). The focus should be on the application's contribution to better clinical decision making and patient care. Depending on what the project is designed to achieve, the technology might range from a touchtone telephone to a sophisticated multimedia system. This points to the need for careful planning and a clear understanding of the project goals. Experience suggests the need for a flexible, open system that can be adapted to advances in technology. The complexity and sophistication of the technology selected will obviously depend on its intended functions. Transmission speed, image resolution, storage capacity, mobility, and ease of use are important considerations.

The system's users must be involved in its design from the beginning. Once implemented, onsite technical assistance is necessary to ensure that problems are immediately addressed and rectified. The system should be located conveniently to providers. Users will need adequate training, and a facilitator will ensure that consultations run smoothly. As desktop multimedia telemedicine platforms become available to the health care provider's office, issues of cost, location, convenience, and the operational and maintenance difficulties of telemedicine will diminish significantly.

More than 7,000 teleradiology systems had been sold by 1994 by two of the largest manufacturers, Image Data, San Antonio, Texas and Icon Medical Systems, Campbell, California. In the first six months of 1994, approximately 15 interfacility teleradiology programs in North America provided teleradiology services to about 90 remote sites and interpreted approximately 22,000 studies (*Teleradiology 1994,* 1994).

A full description of all the technical factors involved in selecting a teleradiology system is beyond the scope of this chapter, but is well described in a booklet published by the American College of Radiology (ACR, 1995). The sections which follow give an overview of the types of capabilities provided by typical teleradiology systems, and serve as an introduction to the detailed specifications provided in the ACR booklet.

Image Acquisition

The process of sending a radiological image to a remote viewing site begins with capturing an adequate representation of the image and transferring it to a digital computer. This image-acquisition process can be accomplished by any of a variety of methods, depending on the hardware available, and the type of image to be transmitted. Some systems will use more than one method, such as video-image capture for CT images and a laser film scanner for radiographic film images.

The simplest method for capturing images is to digitize a film by scanning it with a light beam, and recording the amount of light which passes through the film. This is also the most flexible method, because it can be used to capture almost any type of image, if the image has been recorded on film. Most radiology departments routinely print the majority or all of their images on film. Many radiographic images, such as chest or abdomen x-rays, are already on film, and CT or MRI scans are usually photographed or "printed" on films for viewing on an alternator. Even images that are cine sequences, such as real-time ultrasound or cardiac cineangiography, can be printed as a series of static images.

Film digitizers are flexible because they can be used with any and all radiographic modalities in the hospital or clinic, and do not require specialized connections, as might be required to link a video-image digitizer to a CT scanner. The laser film digitizers are the easiest of the film digitizers to use because they scan the film with a narrow light beam which gives high spatial resolution, and provides an excellent ability to resolve details in both the lightest and the darkest areas of the film, often resolving 1,000 shades of gray or more.

Other methods for scanning medical images on film include the use of a Charge Coupled Device, called a CCD, which may be thought of as a row of tiny photocells that are scanned across the film, while a uniformly bright light illuminates the image. Alternatively, a video camera can be focused on a film mounted on a viewbox and used to digitize the image, but this technique limits the image quality significantly.

A video-image capture device is typically a computer circuit board placed in a computer. This board measures and digitizes the video signal voltage waveform going to an image display, such as a CT scanner console video monitor. These devices may be used to capture any video image, including a television camera pointed at an x-ray film mounted on a viewbox. However, they tend to be more troublesome to operate

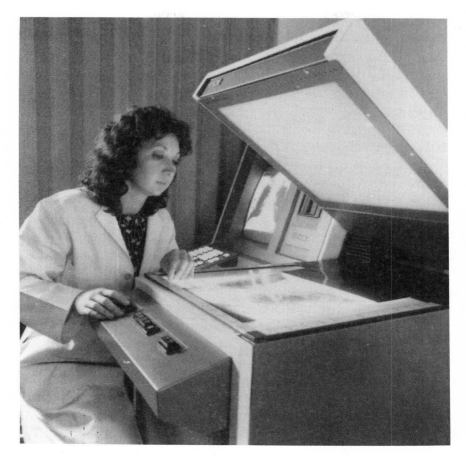

Figure 5-3. Teleradiology's potential to provide prompt radiological expertise may prove a boon to small or remote clinics and hospitals. At the originating clinic, an x-ray is placed in a scanner to be converted into a digitized image that can be stored in a computer.

since they require careful attention to lens setting, such as focus and f-stop, in addition to the usual difficulty in providing uniform illumination of the x-ray film. A more reliable method of interfacing to digital radiographic modalities is the use of an all-digital connection between the image modality computer and the teleradiology system computer, using the DICOM standard.

Compression

Compression technologies are used extensively to store and transmit digital medical images. They have become very fast and inexpensive as a result of the intensive development efforts by computer and communications companies seeking efficient ways to transmit digitized images and video for publishing and broadcasting. There are many different methods for compressing data, some of which have been identified by the DICOM standard. Methods of data compression that achieve relatively low ratios but regenerate the original image precisely are said to be "lossless," while methods that result in relatively high ratios and do not exactly reproduce the original images are termed "lossy." While some lossy methods may not provide a perfect rendition of the image, the resulting image may be adequate for the purpose of the teleradiology system.

The process of compressing images for more compact storage or faster transmission is associated with both uncertainty and controversy. In some circumstances data compression must be totally reversible, so that the original image can be reproduced exactly. Some compression algorithms result in an image that can be diagnostically equivalent to the original image. While compression may reduce the size of a CT image by only a factor or three or four, a chest x-ray may be compressed by a factor of 20 or more, without the loss of clinically significant information. Wavelet compression of 50-to-1 shows promise as being acceptable for CT, MRI, CR, and digitized analog radiographs.

Compression techniques that result in any loss of detail may be inappropriate for primary diagnosis of such images. Medical images are used in many other less demanding contexts, however. For instance, the referring physician may wish to discuss the image with the radiologist over the telephone while viewing it on a desktop computer monitor incapable of displaying the subtleties of the image, even if they were present in the file. JPEG or other compression technologies are often used in situations like this to reduce the size of an image file by a factor of 10 or more. Minimizing the size of image files is even more important when they are to be transmitted to a remote site over a telemedicine link or modem because compression can drastically reduce the cost and the elapsed time necessary to transmit the image.

While the amount of data compression to be used may be decided on a case-by-case basis, it is more likely that one method will be applied to all images, making it easier to use the system on a routine basis. Even though the amount of data compression cannot be adjusted for each clinical situation a gross error is less likely to occur in the amount of data compression if these factors are selected automatically.

Figure 5-4. At the medical center, the computerized x-rays are displayed on video monitors. The terminal's keyboard can be used to enhance the image, to highlight an area of interest, for example. For cases needing prompt diagnosis, consultation between the medical center and clinic can be done by phone.

Storage

Data storage is accomplished variously in different systems, and must often be tailored to the needs of a specific user of these facilities. Usually for teleradiology short-term storage of images is of concern, not the long-term storage required of a PACS archive.

Although the typical teleradiology system does not require a great deal of image storage, in some circumstances substantial image storage is extremely useful. At the transmitting site, sometimes one wishes to capture or acquire a series of images but not transmit them immediately.

Also in some circumstances there may be a problem accessing the available telecommunications line, e.g., when the receiving end may be giving a busy signal. The most frequent need for local storage arises when the transmitting site wishes to acquire images faster than they can be transmitted or to transmit them later for the convenience of the receiver.

At the receiving end, adequate storage is very useful to review recent images a day or two later, in addition to comparing several images from the current examination or comparing the current with previous related examinations. The specific type of image storage is not critical, but the number of images that may be stored does depend on the size of the storage facility, often a hard disk, and on the amount of data compression applied to the images before storage. In some systems, the images are "decompressed" as they are received and are stored on the hard disk in restored format. While this limits the number of images that may be stored, the images may be recalled and displayed more rapidly because the additional decompression step is not required before viewing.

Table 5-1. Storage Requirements for Imaging Techniques
(adapted from K.G. Baxter et al., 1991)

Modality	Image Size (pixels)	Dynamic Range (bits)	Average number of images per exam	Average storage requirement per exam (MBytes)
Computed tomography	512 × 512	12	30	15.0
Magnetic resonance imaging	256 × 256	12	50	6.5
Digital subtraction angiography	1,000 × 1,000	8	20	20.0
Digital fluorography	1,000 × 1,000	8	15	15.0
Ultrasound imaging	512 × 512	6	36	9.0
Nuclear medicine	128 × 128	8	26	0.4
Computed radiography	2,000 × 2,000	10	4	32.0
Digitized film	4,000 × 4,000	12	4	128.0

The storage requirements for textual and numerical data in patient records are dwarfed by the storage space requirements shown in Table 5-1 for medical images such as x-rays, magnetic resonance images, and computed tomography scans. These new imaging technologies challenge the ability of information systems to store and process data, but they enable the development of new generations of highly localized surgical and radiation therapies that otherwise would not be possible. The computational resources necessary for medical imaging will continue to grow with the increasing use of high-resolution spatial imaging, in which multiple images are assembled into a three-dimensional model, and with the development of functional imaging, where processes such as the rate of oxygen metabolism in a particular body structure are studied by assembling multiple copies of the same image over time.

Medical images are typically stored and manipulated on large hard drives, similar to those in desktop computers, and then transferred to digital tape, magneto-optical disks, recordable CD–ROMs, or COLDs (Computer Output to Laser Disks) for archival storage. The latter three technologies use light beams to store and record information on durable plastic or magnetic disks. Although the disks themselves are likely to last for many decades, it is not clear that the equipment necessary to read the disks will be manufactured throughout the life of the medical information.

Although telemedicine is often conceived in terms of dynamic, inter-active video consultations, "store and forward" technologies are at least as promising. In these systems, static images or audio-video clips are transmitted to a remote data-storage device, from which they can be retrieved by a medical practitioner for review and consultation. The advantage of store-and-forward technology is that it obviates the need for simultaneous availability of the consulting parties. The low bandwidth requirements of store-and-forward systems also tend to make them much less expensive. Store-and-forward technology is the basis of at least one teledermatology research project currently sponsored by the National Library of Medicine in Bethesda (Perednia and Allen, 1995).

Radiology and pathology are especially suited to a store-and-forward format. They have some unique requirements, however, most notably for higher resolution images than are used for most types of clinical consultation. Most teleradiology and telepathology programs are currently separate from the LATV-mediated telemedicine programs.

Image Transmission

In rural settings, image transmission is usually done over ordinary dial-up telephone lines, using computer modems to transmit the images. These modems are frequently part of the teleradiology system, and the manufacturer ensures that the modems on the receiving end are properly matched with the transmit modems in speed of transmission, data compression, error detection, and correction procedures. Image transmission can also be accomplished at higher speeds over a local area network, e.g., within a hospital or by an Integrated Services Digital Network (ISDN) circuit provided by the local telephone company.

High-bandwidth communications may have a profound effect on the structure and process of health care services analogous to the changes that automated teller machines brought to the banking industry. Already, demand management systems are in place within some integrated HMOs; nurses in centralized locations use telephones to give advice for managing particular health problems to patients located throughout the HMOs service area. Consultative practices such as radiology and dermatology are slowly beginning to use high bandwidth communications over dedicated phone lines to practice telemedicine.

Digital communication links are more robust than analog ones; they are more immune to noise and require less frequent reamplification of the signals as they travel between locations. If a hospital has a digital PBX or if it wants to connect computers at different sites, its administrators might choose to bring a digital line right into the facility. The capacity or bandwidth of such a connection is typically measured by the number of normal voice connections it can carry: a DS-0 connection (64,000 bits per second) can carry a single voice connection; a DS-1 or T1 connection (1.544 million bits per second) can carry 24 phone conversations; a DS-3 or T3 connection (45 million bits per second) can carry 672 voice conversations, and so on. Various fractional levels between these capacities can be ordered as well. A typical x-ray image could be transferred over a DS-0 connection in about eight minutes, over a DS-1 connection in about 20 seconds, over a DS-3 connection in about 0.7 seconds, and over the high-capacity cross-continent "backbones" in a few milliseconds. The bottleneck in telecommunications is the slowest connection.

Different computer networks utilize different packet formats and protocols; the packets on different networks might be different sizes or have different addressing schemes. It is inconvenient for a computer launching a packet to have to know what sort of packet is expected by the receiving computer, which might be thousands of miles away. Sometimes this problem is solved by having all parties on the network agree on a common standard such as the TCP/IP protocol (Transmission Control

Protocol/Internet Protocol) used for communication among computers in the worldwide Internet. Still, it is often necessary to transfer packets from one network to a different network with a different protocol. One way to do this is to employ a sort of diplomatic pouch called frame relay: packets traveling on the high-capacity telecommunications network are wrapped in standard envelopes or frames and then unwrapped at the receiving end into whatever packet format is required. The frames may have various lengths depending on the size of the packet inside. Frame relay communications work well for transferring data files such as x-ray images, but they are less effective for transferring video streams because unpredictable delays experienced by differently-sized frames can lead to pauses and jumps in the video playback.

A different and faster approach is to reformat all packets into minimalist cells that are a kind of least-common denominator. That is the approach behind asynchronous transfer mode communications (ATM). Because all ATM cells are exactly alike, the routing equipment that shuttles them around the world can be designed to be extremely fast, and the transmission delays for a series of cells will be relatively constant. Video streams can be reassembled from ATM cells with few noticeable delays. High-speed frame relay and ATM communications will be necessary for any large-scale networking that involves sharing large amounts of health information.

ATM has been chosen by the telecommunications standards committees around the world as the underlying switching and multiplexing technology that would be used within the very high speed (broadband) communications systems that are evolving. It offers the promise of eliminating the barriers between local and wide area networks, providing seamless interconnection of Local Area Networks (LANs) across the country and the world. For those knowledgeable about the issues related to acquiring communications circuits that could be used in the transfer of video images and information related to medical diagnoses, ATM is the start of a process predicted to remove one of the typical barriers to telemedicine acceptance by eliminating transmission speed as an issue in acquiring services. ATM offers the capability for flexible bandwidth on demand. Neither circuit switching (commonly used to support voice telephone applications) nor packet switching (a data-oriented service arrangement offered by commercial telephone carriers) is considered as capable and effective as ATM switching in satisfying the needs of users who want to process video, data, and voice simultaneously, with minimal delay (Ferrante, 1994).

Viewing the Images

The process of displaying images for interpretation involves the use of specific display technology that should be optimized for use in a given teleradiology application. The demands made on a system used for primary diagnosis are more stringent than those required for reviewing examinations. The display quality is one of the factors which determines the accuracy of the image interpretation, and the nature of the diagnostic information that may be extracted from the images. Most images transmitted using teleradiology are black and white, but on occasion there may be a need for color, such as when nuclear medicine images are transmitted. The use of a color display in a teleradiology terminal may produce some problems, because the color display screen is comprised of tiny luminescent dots alternating between red, green, and blue. The interaction of these colored dots with the pixels in the black-and-white medical image may blur details at the edge of the image and a slight reduction in the resolution capability of the overall system.

Luminance is another important factor in image display and interpretation. This refers to the amount of light emitted by the CRT display surface. It affects the human response through both the acuity of the eye and the detection of luminance differences. Commonly available gray scale monitors have a maximum luminance value of 50 foot-lamberts (ft-L) and some of the newer monitors have luminance levels over 160 ft-L, but are still not as bright as conventional film viewboxes, which usually have luminance of over 500 ft-L. As an alternative to electronic display, in some teleradiology systems the images are printed on film at the receiving end, so they can be viewed on conventional film alternators or viewboxes, and filed with other radiographic examinations.

Part of the functionality of a teleradiology workstation is provided by the software, and includes features such as brightness and contrast adjustments, magnification, and edge enhancement. Most teleradiology viewing applications require much less image processing than is available on the typical CT or MRI scanner. Of greater interest is the speed of image display, the number of images which can be displayed simultaneously, and the convenience of paging through a series of images.

Upgradability

Commercial teleradiology systems vary in their upgrade potential from relatively simple systems that are part of a narrow product line to entry level systems designed to add more capability, eventually becoming a significant portion of a total PACS system. Users may wish to upgrade only one aspect of the system operation, or purchase a combination of capabilities to increase functionality. As system use grows, faster methods of telecommunication may be required, and the ability to upgrade to a faster modem, or the use of specialized telecommunication capabilities, such as dedicated telephone lines, may be desired.

Frequently, users need to store images for several days, or to accommodate a large number of images at one time. This may require either the expansion of the existing disk storage capability, or the addition of a different method of image storage. Users may need to add display capability to view more images simultaneously, or to interpret images with higher resolution. In some circumstances, the initial teleradiology system provides for connection only to a limited number of image modalities, and the addition of a wider variety of imaging devices may be required in the future. Unfortunately, very few systems are compatible with teleradiology equipment manufactured by other companies. If an upgrade cannot be accomplished by the original suppliers, users may need to change manufacturers, and replace much or all of the old equipment to accomplish an upgrade. Teleradiology is not appropriate if the system does not provide images of sufficient quality to perform the indicated task. The image quality should be sufficient to satisfy the needs of the specific circumstance.

Interpretation of Images

Considerable interest exists in the development of "filmless" radiology departments. In 1990, the U.S. military sponsored pilot studies at several medical facilities, including small institutions served by teleradiology. The ability to interpret images displayed on electronic monitors with an accuracy equivalent to that achieved with conventional radiographs is essential to the success of such a change in practice.

Recent publications have reported that under certain conditions the accuracy of radiograph and screen interpretations are equivalent (Carey et al., 1989; Goldberg et al., 1993). Relative success has been reported by radiologists applying relatively low-resolution teleradiology systems to remote interpretation of nuclear radiologic examinations (Arnstein, 1990). Their findings may support the belief that remote primary diagnosis for emergency department coverage can be supported with teleradiology. Although the goal of achieving a filmless radiology department has merit, studies of the accuracy of interpretation indicate that radiograph and screen interpretations of emergency department radiologic cases should not be regarded as equivalent. The Johns Hopkins Medical Institutions have investigated the feasibility of providing teleradiology service for several years, and have reported on the accuracy of interpretations based on viewing at electronic workstations. These studies demonstrated that interpretation of digital images using the available teleradiology systems was significantly less accurate than interpretation of radiographs. With the increasing importance of emergency department services for patient care, teleradiology likely will be needed to supplement the medical imaging services provided to emergency medicine. An acceptable teleradiology system, however, must at least support the current level of diagnostic accuracy and confidence achieved by interpretation of conventional radiographs, if the quality of patient care is to be maintained.

In 1994, the Hopkins teleradiology system was upgraded with a smaller scanning spot size on a film digitizer and with an electronic workstation with higher resolution. Current studies also differ from previous efforts: the relative frequency of different types of radiographic examinations in the emergency department, such as cervical spine and chest, was represented among the cases chosen for the study. However, the selected cases differ from typical emergency department experience in the relatively high degree of diagnostic difficulty and the frequency of cases with positive findings. Difficult cases were selected to provide a more stringent test of the display capability of the electronic workstation.

The positive cases were selected to include clinically important diagnoses, such as pneumothorax, pneumonia, lung mass, and fracture, as opposed to diagnoses that would have little impact on patient care, such as degenerative changes in the spine. Emergency medicine physicians were included in the study to be certain that important diagnoses and decisions were being evaluated, and to determine if the same results would be achieved. Within the group of study cases, a subgroup was selected that consisted of examinations considered critical in requiring prompt, accurate diagnoses. For example, a pneumothorax requires prompt diagnosis, whereas detection of a lung cancer could be delayed for several hours or perhaps days, with little or no effect on the clinical outcome.

The results of the Hopkins studies revealed that radiologists and emergency medicine readers were significantly more accurate in case interpretation when viewing conventional radiographs than when viewing images displayed on the teleradiology workstation. The ability to demonstrate statistically significant differences between these viewing modes was due in part to the selection of sufficiently difficult test cases to challenge the performance of the teleradiology system and the interpreters. Results were nearly equivalent for radiograph and screen interpretation in low-difficulty cases, such as fairly obvious fractures, whereas the differences were relatively large in cases of moderate and high difficulty, such as pneumothorax.

Findings in several studies, including the Hopkins work, indicate that higher spatial resolution is important in the identification of some subtle abnormalities, such as a pneumothorax and nondisplaced fractures (Wegryn et al., 1990; Frank et al., 1993). This technical problem may be solved by the use of recently developed systems, which have display monitors with higher resolution and have greatly increased data transmission and storage capabilities. However, other factors also contribute to overall performance of teleradiology systems. For example, the availability of a user-friendly workstation is certain to be an important factor. A level of functionality is needed that matches the convenience, efficiency, and accuracy of the combination of radiograph, view box, and hot light.

One possible reason for the better performance of readers with conventional radiographs than with images on workstations may be interpreters' experience. Until we have a generation of physicians trained primarily in screen interpretation, the difference in interpreters' experience with radiographs and with the systems being evaluated may influence the results. This bias, to the extent that it exists, is difficult to eliminate. It may be necessary to conduct a study over an extended period of time to determine the nature and magnitude of a "training effect." A Quebec study demonstrated significant improvement in reader performance after three months of regular use of a teleradiology system (Page et al., 1981).

The effect of reader inexperience with the teleradiology workstation can be mitigated by using laser-printed radiographs, which can be handled in a conventional and familiar manner. The use of such radiographs for interpretation in Carey et al. (1989), may have been a factor in the apparent equivalence of interpretations from original radiographs and workstation images. Other comparisons of interpretation of laser-printed radiographs and screen images have yielded mixed results (Slasky, 1990; Cox, 1990; Razavi, 1992; Thaete, 1994). On the other hand, laser-printed radiographs cannot be used in practice if the advantages of filmless radiology are to be realized.

Some authors discounted the importance of inaccurate interpretations of radiographs by noting that, in many cases, no untoward results occurred or that the patient treatment was based on clinical observations or on some other concurrent condition (McLain and Kirkwood, 1985; Quick and Podgorny, 1977; Overton, 1987). This implies that the radiographic examinations had little clinical usefulness in such cases. In an emergency department setting, some radiographic findings are more important than others to patient treatment. For example, a lung nodule, although highly important for long-term patient care, usually does not require emergency treatment, so the consequences of a nodule missed on an initial radiographic review may be minor from a clinical standpoint. On the other hand, some radiographic findings, such as pneumothorax, have immediate impact on patient treatment. Such findings may occur infrequently in some settings; nevertheless, accurate diagnosis must be the foremost goal in radiographic interpretation.

A serendipitous result of the Hopkins studies is noteworthy. Radiologists performed less accurately with the workstation than with conventional radiographs, but their interpretations of screen images were more accurate than the interpretations of radiographs by emergency medicine physicians. The difference between findings by radiologists and emergency medicine readers may suggest that remote consultation with a radiologist using the teleradiology system can contribute to patient care in the emergency setting by improving the accuracy of film interpretation. Nonetheless, the difference in diagnostic accuracy in the results of radiologists and emergency medicine physicians must be interpreted cautiously. In this study, emergency medicine physicians were asked to perform the role of the radiologist, not that of the emergency medicine physician, since the complete clinical history and physical examination were not made available to the readers. Responses to the follow-up questionnaire indicated that emergency medicine physicians depend more on clinical history during interpretation of radiographs than do radiologists. This was not taken into account in the study. In a 1989 study, orthopedic surgeons were found to depend more on clinical history than did radiologists (Berbaum et al., 1989). Further studies are necessary to determine the influence of these factors on radiographic interpretations, particularly in cases critical to emergency care.

Given the current state of technology, medical images critical to the care of emergency department patients should be interpreted on plain radiographs. Results with the teleradiology workstation as configured are not acceptable for primary diagnosis of radiographic examinations seen in the emergency department. However, recent improvements in related technology may resolve the equipment deficiencies noted in this study. They may support the development of a system that performs satisfactorily and is clinically acceptable for the interpretation of digitized radiographs associated with the practice of emergency medicine.

Computer-Based Patient Records

Computer-based patient records serve as repositories for clinical information and as records of communications and transactions; their analogs in traditional health information systems are paper-based patient records or charts, usually kept in folders along with films at each site of care. Although computer-based patient records may be localized in a single data file, they might also be widely distributed in computers throughout an institution or among several institutions. In either case, the perceived location of the record is on a computer screen in front of the person using it at any particular moment.

It is possible to design stand-alone, computer-based, patient-record systems. Some are in use, but much of the advantage of computerizing health information is lost if other systems and processes within the provider institution are unable to interact with information in the record. Maintaining a stand-alone patient record, for example, could require caregivers or clerical personnel to retype test results produced by a computerized lab analyzer to get them into the record, or to retype administrative information from the record for use in creating financial statements. To avoid these inefficiencies, computer-based patient records are usually embedded in various other information systems. These systems include not only computer hardware and software and networks, but also the "people, data, rules, procedures, processing and storage devices—and communication and support facilities" involved in managing the record system and distributing data and information throughout the provider organization (Office of Technology Assessment, 1995a). For hospitalized patients, computer-based patient records are typically linked to clinical information systems that track clinician-patient encounters.

They also may be linked to administrative, laboratory, nursing, and pharmacy information systems. However, most health care encounters occur outside of hospitals. The large amount of health information generated in primary care and home care settings could be captured in computerized patient-record systems embedded in information systems appropriate for private and group practice doctors and public health workers.

Ideally, within a single institution, the distinctions between these various information systems should be transparent to users so that they become parts of a seamless enterprise information system. In practice, however, the components for each type of information system are usually procured separately, and their integration can be plagued by a lack of design coordination and technical standards.

Costs and Economics

The cost of telemedicine needs to be considered in relation to its contribution to improving the health of the population by preventing disease, treating illness, and ameliorating pain and suffering, in comparison with alternative systems (Bashshur, 1995b). A recent report prepared for the Health Care Financing Administration (HCFA) that included an extensive literature review of telemedicine research found no studies that provided an adequate overview of its cost-effectiveness (Grigsby et al., 1993).

The cost of a teleradiology system varies widely depending on the amount of capability desired, the number of teleradiology workstations to be purchased, and the amount and type of preexisting equipment available for use. If a PACS system is available with the capability of transmitting images remotely, the cost of the teleradiology equipment may be limited to the purchase of one or more remote teleradiology workstations. If there is a requirement to store a large number of examinations, to handle images with high resolution, or to transmit them very quickly, then the cost will increase.

The value of a teleradiology system is difficult to measure because it is not amenable to typical cost/benefit analysis. Very often, the benefits are improved patient care, improved service to referring physicians, and more rapid response to clinical needs. Because these benefits tend to be intangible, it may be difficult to quantify them and apply them to an evaluation of the life-cycle cost of the proposed teleradiology system.

Too frequently, the true cost of a teleradiology system is seen only after the purchase of a system that does not satisfy the requirements or is incapable of growth over time (Rowberg, 1994).

A 1990 OTA report noted that one of the greatest problems rural hospitals face is the outmigration of residents to urban areas for care (Office of Technology Assessment, 1990b). Many hospitals in small communities have been forced to close because their bed census dropped so low that they became uneconomical to operate. The economic impact on a small community when its hospital closes is enormous. In addition to reducing access to care, such closures have a major impact on employment opportunities. The viability of small hospitals might improve if telemedicine allowed more patients to receive consultative services locally, rather than being referred to large medical centers.

The costs of developing a telemedicine project can be high. These costs include telecommunication charges, equipment costs, technical support, training, and administrative support. Many small communities operating on their own will be unable to afford these costs. Telemedicine lends itself to cost-sharing as a way of financing projects as well as assembling the expertise necessary to make it successful. Within communities, systems too costly for health applications alone can be shared with educational, local government, social, and community services to make the investment feasible.

With the exception of the 20-year-old telemedicine program at Memorial University of Newfoundland, St. John's, none of the programs begun before 1986 has survived. Although data are limited, the early reviews and evaluations of those programs suggest that the equipment was reasonably effective at transmitting the information needed for most clinical uses and that applications were for the most part satisfied. However, when external sources of funding were withdrawn, the programs disappeared, indicating that the single most important cause of their failure was the inability to justify these programs on a cost-benefit basis. Other issues, such as limited physician acceptance, played a less definitive role in their demise (Perednia and Allen, 1995).

System Costs

A teleradiology system acquires radiographic images at one location and transmits them to one or more remote sites, where they are displayed and/or converted to hard copy (Batnitzky, 1990). These systems often employ wide area networks (WAN). Their goal is to provide improved radiologic services at all sites on the network. Experience in the use of teleradiology systems has demonstrated the need for a laser film digitizer, an optical disk, and a high-quality display and/or laser film printer at each site. At 1990 prices, single-site hardware purchase costs averaged $196,000, plus an additional 20 percent for yearly network services. Hardware purchased for a consultation or central referral facility approximated $344,000.

Based upon an example derived from a teleradiology network in Kansas, the following estimates were selected as typical of a system in which a community hospital is linked to an academic medical center to provide backup coverage (Dwyer, 1993). In this example, 34 examinations totaling 141 films are transmitted daily, after digitizing at the community hospital, to the academic center for interpretation. The hardware, software, and maintenance costs over a two-year period totaled $87,000 at the community hospital and $144,000 at the academic center, as shown in Tables 5-2 and 5-3.

Table 5-2. Estimated Teleradiology Costs for Community Hospital

Components	Costs
Laser Film Digitizer (2K × 2K × 12 bits)	$25,000
Grayscale Display Station (1K × 1K, 2 Monitors)	$15,000
Host Computer System PC	$12,000
System Software	$ 8,500
Network Access Controller	$15,000
Maintenance Costs (2 years)	$11,550
Total	$87,050

Even though hardware and software costs are declining, the initial capital required is substantial and operational costs, including maintenance, must be justified. In addition, the costs for the communication service between the two facilities amounted to $1,600/month. These costs are shown in Table 5-4.

The tariffs of this point-to-point service provide the 24-hour user with cost benefits when the system is fully utilized. An alternative for low-frequency users is a dial-up configuration charged only for transmission time.

The future of teleradiology depends on the constantly improving

Table 5-3. Estimated Teleradiology Costs for an Academic Radiology Department*

Components	Costs
Laser Film Digitizer (2K × 2K × 12 bits)	$25,000
Grayscale Display Station (1K × 1K, 2 Monitors)	$50,000
Host Computer System PC	$21,000
System Software	$15,000
Network Access Controller	$15,000
Maintenance Costs (2 years)	$17,600
Total	$143,600

*No PACS Infrastructure

Table 5-4. Communication Cost for Point-to-Point DS-1 Service* (1.544 M b/sec)

• Site access charges for both sites	$700/month
• T1-Carrier costs ($19/month/mile × 48 miles)	$912/month
Total	$1,612/month

*Plus a one-time installation charge of $1,600

technology of digital imaging modalities, film digitizers, digital data networks, interactive gray-scale display workstations, image-processing algorithms, and hard-copy recorders. Interactive gray-scale display workstations enable interactive manipulation and display of gray-scale images. A digital data network is a combination of computer technology and communication systems, the goal of which is to transmit digital data successfully, such as digitized radiographs, text, and compressed video. Departments are adding computed radiography (digital storage phosphor plates) to existing digital imaging modalities, such as computed tomography (CT), magnetic resonance (MR) imaging, digital subtraction angiography (DSA), digitized fluorography, nuclear medicine, and ultrasound.

Clinical acceptance and utilization of digital radiology is accelerating. Digital data systems are being implemented with standards such as the end-to-end integrated services digital network (ISDN). The ISDN is based on digital transmission and switching technologies. It is used to construct an integrated digital network for long-distance telecommunications of voice and data. Fiberoptics offer software-controlled virtual wide area networks with selectable data transmission rates. A fiberoptic channel is a communications medium made of fine glass or plastic fibers that can transmit light pulses, such as from a laser, and transmit information at several gigabits per second (billions of bits per second). Interactive gray-scale workstations are implemented with $1,024 \times 1,024 \times 12$-bit and $2,048 \times 2,048 \times 12$-bit displays. The incorporation of laser film digitizers and laser film printers enables teleradiology systems to provide film-to-film transmission.

Acceptance of teleradiology systems by radiologists depends on an understanding of digital image acquisition and display parameters. The cost of using teleradiology requires justification and evaluation of system and patient care.

The Cost of Communication

A wide variety of types of telecommunications systems may be used to support teleradiology applications. Each has costs which vary as widely as the benefits and features they provide. The simplest connection is a dial-up modem connection over ordinary telephone lines. These are easily established, and there is little problem with telephone lines in most circumstances. While the modems selected limit the speed of image transmission, the transmission times can be under one minute per CT image with data compression, and even long-distance transmission of images may be feasible.

The simplest method of transmission is unidirectional (e.g., images being transmitted from a hospital to an on-call radiologist). With more elaborate systems, it may be possible for a radiologist to discuss a case with another physician, with each person having the ability to move a cursor over the image, and have the cursor automatically move on the screen of the distant workstation, so that real-time voice consultation can be supplemented by interaction with the images, perhaps including brightness and contrast changes for image viewing, and text or graphical annotation over the images. More sophisticated systems may provide video conferencing, so that the radiologist may also interview the patient at a distance, and have a two-way video conference with the referring physician, as they discuss the approach to diagnosis and management of a difficult case.

When a large number of images needs to be transmitted rapidly, the transmission network must be configured with faster methods of digital image communication. Within a city, T-1 telephone lines may be used. These are dedicated connections leased from the telephone company or other carrier, and provide 1.5 million bits per second of digital communication capability, approximately 100 times that of a voice grade telephone connection using a modem. Such a connection over the distance of a few miles is likely to cost between one and two dollars per hour, and can be effective if it is used many hours per day. However, if the user pays for such dedicated lines 24 hours per day, for intermittent use, it may not be cost-effective.

In many rural areas, the communication infrastructure is unable to support the bandwidth necessary to carry the signals for telemedicine using two-way interactive video (Tangalos, 1993). In addition, the costs of connections between local and long-distance telecommunication carriers can pose a significant barrier to telemedicine projects. Under the existing tariff structures, telephone calls placed to locations inside the local access transport area boundaries are often more expensive than those placed outside the same service area.

In rural areas, hospitals, schools, government, and other community groups can aggregate demand and share a network to help spread the system costs (Puskin and Sanders, 1995). This can be accomplished, however, only when residents are involved in planning and developing a system and have a sense of ownership in it. In an earlier study, OTA suggested that Rural Area Networks would allow rural communities to customize networks to their own needs, while achieving economies of scale and scope (Office of Technology Assessment, 1991). By sharing in the creation of such a network, rural communities would be able to enjoy some of the benefits of their urban counterparts. A system of "bandwidth on demand," in which users pay only for the time they use on the system, would greatly reduce the costs of telemedicine and obviate the need for a dedicated communications line. Such service could be provided using an advanced switching technology such as asynchronous transfer mode, which can support many different kinds of services. To ensure that systems are interoperable, technology standards will be essential for communication providers (Office of Technology Assessment, 1992). Many of these issues are currently being addressed in the context of the Administration's NII initiative, particularly by members of the Information Infrastructure Task Force (IITF).

Faster methods of communication may be provided by using microwave links, satellite connections, fiber optics, or newer high-speed digital switching technology, such as the Asynchronous Transfer Mode (ATM) devices being installed in some telephone company facilities. Connections over longer distances are more affordable if some form of switched service is used, so that the user pays for the connection only while images are actually being transmitted. Such services are widely available using T-1 rates, in fractional steps, even over long distances.

Legal and Regulatory Issues

Telemedicine raises some difficult legal and regulatory issues as well. Remote diagnosis and treatment across state lines could bring different laws and regulations into play. The present legal scheme does not provide consistent, comprehensive protection of privacy in health care information, whether it exists in a paper or computerized environment. Clearly the privacy implications for telemedicine will continue to receive careful scrutiny. Physician licensing becomes an issue because telemedicine facilitates consultations without respect to state borders and could conceivably require consultants to be licensed in a number of states. This would be impractical and is likely to constrain the diffusion of telemedicine projects. Telemedicine may, in fact, decrease the threat of malpractice suits through improved record keeping and databases, and the fact that taping the consultations will automatically provide proof of the encounter. However, it may also raise other liability issues, such as the lack of a "hands-on" examination by the consultant.

There are several reasons why the technologies and standards underlying applications must be understood for purposes of setting public policy. First, technological changes are challenging the relevance and enforceability of the existing body of state and federal law. Several states virtually preclude the development of computer-based patient records by specifying in pen and quill legislation the required storage media for patient records (Office of Technology Assessment, 1995a). These laws were no doubt meant to ensure the singularity and permanence of patient records, but they were probably written without an appreciation of the compactness, duplicability, and durability of optical disks.

Standard Identifiers for Individuals, Providers, and Payers

Interstate electronic commerce for health information would be facilitated by a system of standard identifiers. Because each provider or provider group (as well as payers and other users of health information) maintains its own identification number scheme and assigns its own numbers, patient records are not uniquely identified once they leave the institutions where they have been created.

Some argue that the benefits of fully electronic records are more easily obtained if each individual could be uniquely identified. If each person had a universal patient identifier, it would be easier to link the health information maintained at different institutions, for example. In addition to identifying patients, health care providers and specific sites of care also need to be identified. While there are a number of recommendations for developing numbering schemes *de novo,* some industry organizations recommend modifying or expanding existing identification number schemes to get unique identifiers in place more quickly (Norman, 1993).

Privacy and Confidentiality. Privacy in health care information has been protected in two ways: (1) in the historical ethical obligations of the health care provider to maintain the confidentiality of medical information, and (2) in a legal right to privacy, both generally and specifically, in health information. Confidentiality involves control over who has access to information. Other terms frequently used in discussion of the protection of privacy are integrity and security. Integrity assures that information and programs are changed only in a specified and authorized manner, that computer resources operate correctly, and that the data in them are not subject to unauthorized changes. A system meeting standards for access allows authorized users access to information resources on an ongoing basis (Pfleeger, 1989). Security refers to the framework within which an organization establishes needed levels of security of information to achieve, among other things, the confidentiality goals.

The use of telecommunications to deliver medical care may pose additional risks to the privacy of patients and their records. For example, the creation of a videotape of a consultation might pose a new privacy threat for the patient unless appropriate safeguards to control access to it are built into the process. The issue of who has access to this information will need to be considered and resolved in advance. Depending on the nature of the examination, the patient may also have concerns about who is actually present in each location during the consultation. Nonmedical personnel, such as a technician or facilitator, may be needed to assist in the consultation.

If a videotaped consultation becomes part of the patient's medical record, it would be treated like other videotaped information on the patient (e.g., an angiographic procedure, for example). In these cases, the usual privacy laws would apply. State laws governing the transmission and retrieval of patient medical records vary, and officials are concerned about user verification and access, authentication, security, and data integrity.

Inconsistent Regulatory Environment

State governments generally have licensing authority over health care providers and require them to maintain medical records. Nearly every state regulates what media are permissible for storing medical records. In many states, the language is reasonably "technology neutral" and the use of catchall phrases such as "other useable forms" or "other appropriate processes" has been taken to mean that computerized record storage is permitted. In some states, however, legislation has served as a barrier to the development of automated patient records by specifying the permitted media (e.g., microfilm or paper) and excluding disks, tapes, and other computerized storage media. Other states require clinicians' signatures in ink on particular forms, implying a paper original to which the signature can be affixed. Some states specifically permit the use of computers for some functions but forbid it for others, thus hindering the development of a complete computer-based record. There are other paradoxes and inconsistencies in legislation as well, with some states permitting electronic signatures for some purposes but requiring retention of a paper or microfilmed record (Tomes, 1994).

Only a few states specifically authorize computerized medical records. Indiana statutes, for example, authorize the use of computerized records that maintain confidentiality. They specifically state that the recording of hospital medical records by the data-processing system is "an original written record" and authorize the courts to treat information retrieved from such systems as originals for purposes of admissibility into evidence. Record keeping rules for nonhospital providers—nursing homes and physicians' offices, for example—are often covered by different state statutes or regulations and can be very different from those that apply to hospitals in the same state. Implementing a complete electronic patient record in a multisite provider organization, that might include hospitals and nursing homes, can be complicated if these requirements differ widely. The regulatory inconsistencies among states can create difficulties for health care organizations that are attempting to develop common patient-record systems for sites in more than one state.

Electronic Signatures

Signatures are necessary to attest to the completeness and authenticity of a medical record. Generally, each entry in a record is signed or authenticated by the person responsible for that entry. An electronic record can be signed electronically, and this is permitted in many states; once again, however, electronic signatures are treated differently from state to state. Some states are silent about the specific means or technology to be used for the signature, or say that industry and professional standards should dictate the form of the signature. This seems to permit the use of electronic signatures in those states because the Joint Commission on Accreditation of Healthcare Organizations, American Hospital Association, and other industry groups have published guidelines related to electronic signatures. Some states (e.g., Pennsylvania, Alaska, and California) specifically authorize the use of an electronic signature activated by a computer key known only to the authorized user.

Physician Licensing

Physicians must be licensed by the states in which they practice. Telecommunication facilitates consultations without respect to state borders and could conceivably require consultants to be licensed in a number of states. This would be impractical and is likely to constrain the broader diffusion of telemedicine programs. For instance, in July, 1994, the State of Kansas passed legislation requiring that out-of-state physicians who provide consultations using telemedicine be licensed in Kansas.

The licensing problem for telemedicine could be addressed by the implementation of national licensing standards or the classification of physicians practicing telemedicine as consulting physicians. For a start, such a national license could be provided to physicians who provide consultations to underserved populations. A precedent exists for physicians serving in the military, the Department of Veterans Affairs, the Indian Health Service, and the Public Health Service (Sanders and Bashshur, 1995). Another way to address the problem of licensing is to place the overall responsibility for the patient's care in the hands of the referring physician and view a consultant in a different state as making recommendations only. A novel approach unique to telemedicine would be to consider that the patient is being "transported" electronically to the consultant, thus obviating the need for the specialist to be licensed in the patient's home state.

The issue of credentialing arises with telemedicine in terms of the use of consultants. In the Medical College of Georgia program, the Joint Commission on Accreditation of Hospitals (JCAHO) has determined that hospital credentials are not a problem for the consulting physician as long as all physician orders are written by the referring physician.

Liability. The liability implications of telemedicine are unclear. At least two aspects of telemedicine could pose liability problems. One is the fact that, in a remote consultation, the specialist does not perform a hands-on examination, which could be regarded as delivering less than adequate care. The second aspect is that the use of compressed video, in which repetitious information is eliminated as the data are converted from analog to digital and back, may raise the issue of diagnosing with less than complete information (Office of Rural Health Policy, Health Resources and Services Administration, 1994). On the other hand, telemedicine may, in fact, decrease the threat of malpractice suits by providing better record keeping and databases, and the fact that taping the consultations will automatically provide proof of the encounter. Tapes could also help to prove the innocence of providers who are falsely accused.

Consulting physicians could reduce their liability by adhering to practice guidelines for various telemedicine applications. Such guidelines would need to be established by national health professional associations.

Reimbursement. Implementation of telemedicine is likely to proceed with or without government support as providers recognize its benefits to their practices. However, federal government support will be required if it is to benefit those who need it the most—people living in rural and inner-city areas where market forces are unlikely to provide the services needed. If Congress wishes to encourage the diffusion of telemedicine to help solve the disparities in health care availability, it can have the most impact in the areas of research funding and reimbursement for telemedicine consultations. The two are closely connected; formulating a standard reimbursement policy depends on obtaining satisfactory answers to many of the questions raised about telemedicine's efficacy and cost-effectiveness.

Because the data that would support a uniform reimbursement policy for telemedicine consultations are not yet available, HCFA is moving slowly and deliberately in accumulating the necessary information on which to base a sound decision. This seems a prudent strategy. Experimenting with reimbursement in a small number of demonstration sites will provide valuable insights that will eventually enable the agency to craft a careful policy based on actual results (Office of Technology Assessment, 1995a).

Summary

Technology's trends suggest that, within the next few years, health care providers will be able to see patients at remote sites by using a desktop workstation or laptop computer in a mobile, wireless configuration. Clinicians will be able to select interactive video and store-and-forward modes as needed. Simple, intuitive software shells will allow seamless access to pertinent patient records, radiographs, pathology slides, pharmacy information, and billing records. Instant access to on-line libraries of medical information, diagnosis and treatment algorithms, and patient instructional materials will be available. Referral to specialists and allied health personnel will be done by computer-based scheduling. Patient information will be stored in archives that can be accessed by authorized medical personnel anywhere in the world. Privacy and security will be provided by encrypting data and restricting user access by means of passwords.

While the use of telecommunications in delivering health services has great potential, it also raises a number of issues that need to be resolved if telemedicine is to thrive. In general, patient consultations using telemedicine are not reimbursable (except for teleradiology and telepathology). This will have a negative effect on its diffusion until HCFA promulgates a national policy. One of the reasons for HCFA's reluctance is the paucity of research supporting the safety, efficacy, clinical utility, and cost-effectiveness of telemedicine (Office of Technology Assessment, 1995).

As large health care organizations develop regional alternatives to the classic competitive model, the ability to transmit patient data in every form—text, image, graphics, and sound—will make it possible to integrate patient information with care teams regardless of location. The development of teleradiology along with other electronic innovations will lead to the design of local and wide area networks that will provide the infrastructure of true "telemedicine." But application of this technology, like most innovations, will exact a price in the need to shift attitudes and to recast habitual work patterns. Electronic imaging, by its nature, reduces assembly time of the study, making it available for interpretation almost immediately. With this comes the possibility of smoothing patient flow, reducing patient workup time, and inserting a completed radiological interpretation into the dynamic decision stream. Following on the heels of possibility, of course, comes the demand that it be realized.

Digital imaging has made major contributions to patient care since its initial applications, and particularly since the wide acceptance of computed tomography in the 1970s. As the number and type of digital imaging equipment have proliferated, the benefits have been extended to more people and the relatively high costs have been justified by providers of health care and manufacturers of equipment. As of today, most of the benefits of digital imaging have been achieved through its technical characteristics to produce views of the human body that could not otherwise be attained except through more invasive procedures.

The major unanswered questions are related primarily to the cost-effectiveness of such systems and professional acceptance of the components, particularly the image display workstation. With regard to the former, the real costs of an image management system include not only the hardware, software, and related maintenance, but substantial changes in staffing, training, and relationships between the medical imaging specialists and primary care physicians. Interfaces between teleradiology components, the image management system, and other information networks are essential to the successful implementation of PACS and must be given high priority if the system is to be viewed as something other than an additional resource for the radiology department. For the full potential of image management systems to be realized, the availability of both images and interpretive reports to the referring physicians must be achieved in a timely manner.

With regard to professional acceptance, while the diagnostic radiologist plays a key role, particularly with regard to the human/machine interactions at the image display workstation, the needs of technologists, administrators, and clerical personnel within the organization must be considered. Of equal importance are the functions and responsibilities of clinical and administrative staff outside the department, who will view the system in terms of its contribution to patient care and its effects on the institution's financial status and marketing strategy. In teaching hospitals and university medical centers, the contributions of the image management system to training and research will also be critical to professional acceptance. Clearly, if successful implementation is to be achieved, all of the key personnel must be involved in the planning stage and in the decisions that determine the sequence of implementation.

REFERENCES

American College of Radiology. 1994. "ACR Standard for Teleradiology," Resolution 21.

Allman, R. M., R. F. Kilcoyne, and R. Shannon. 1994. *The Teleradiology Attraction: Understanding Teleradiology.* Harrisburg, PA: The Society for Computer Applications in Radiology.

Allman, R. M., D. J. Curtis, J. P. Smith, S. L. Brahman, and E. C. Maso. 1983. "Potential Contribution of Teleradiology to the Management of Military Radiologist Resources," *Military Medicine,* 148: 897–900.

Allman, R. M., R. F. Kilcoyne, and R. Shannon. 1992. *The PACS Attraction, Understanding PACS: Picture Archiving and Communication Systems.* Harrisburg, PA: The Society for Computer Applications in Radiology.

Arenson, R. L. 1986. "Foreword," In: Arenson, R. L. (ed.). "Use of Computers in Radiology," *The Radiological Clinics of North America* 24: 1: 1–3.

Arnstein, N. B., D. C. P. Chen, and M. E. Siegel. 1990. "Interpretation of Bone Scans Using a Video Display: A Necessary Step toward a Filmless Nuclear Medicine Department," *Clin Nucl Med,* 15: 418–23.

Bashshur, R. L. 1995a. "On the Definition and Evaluation of Telemedicine," *Telemedicine Journal,* 1 (1): 20–23.

Bashshur, R. L. 1995b. "Telemedicine Effects: Cost, Quality and Access," *Journal of Medical Systems,* 19 (2): 81–91.

Batnitzky, S., et al. 1990. "Teleradiology: An Assessment," *Radiology,* 177: 11–17.

Baxter, K. G., et al. 1991. "Wide Area Networks for Teleradiology," *Journal of Digital Imaging,* 4 (1): 51–59.

Berbaum, K. S., E. A. Franken Jr., and G. Y. El-Khoury. 1989. "Impact of Clinical History on Radiographic Detection of Fractures: A Comparison of Radiologists and Orthopedists," *American Journal of Radiology,* 153: 1221–24.

Blair, J. S. 1995. "Overview of Standards Related to the Emerging Health Care Information Infrastructure," *CRC Press, Inc.,* January.

BNA's Health Care Policy Report, vol. 2, Feb. 28, 1994.

Boor, J. L., G. Braunstein, and J. M. Janky. 1975. "ATS-6: Technical Aspects of the Health/Education Telecommunications Experiment," *IEEE Transactions on Aerospace and Electronic Systems,* AES-11 (6): 101532.

Carey, L. S., B. D. O'Connor, D. B. Bach, et al. 1989. "Digital Teleradiology: Seaforth-Londong Network," *Journal of the Canadian Association of Radiology,* 40: 71–74.

Carey, L. S., and E. Russell. 1977. *Canadian Telemedicine Experiment U-6.* London, Ontario, Canada: University of Western Ontario.

Cerva, J. 1982. *Teleradiology System Design. MTR-82W171.* McLean, VA: The MITRE Corporation.

Chan, S., and R. J. Massick. 1975. *33 Telecommunications Projects in Medical Education and Health Care.* East Lansing, MI: Office of Medical Education, Research and Development, Michigan State University.

Cox, G., L. T. Cook, J. H. McMillan, et al. 1990. "Chest Radiography: Comparison of High-Resolution Digital display with Conventional and Digital Film," *Radiology,* 176: 771–76.

Cox, G., A. Templeton, and S. J. Dwyer III. 1986. "Digital Image Management: Networking, Display, and Archiving." In: Arenson, R. L., (ed.) "Use of Computers in Radiology." *The Radiological Clinics of North America,* 24 (1): 37–54.

Curtis, D. J., B. W. Gayler, J. N. Gitlin, et al. 1983. "Teleradiology: Results of a Field Trial," *Radiology,* 149: 415–418.

Dwyer, S. J. III, et al. 1993. "Wide Area Network Strategies for Teleradiology Systems," *RadioGraphics,* 12 (3): 569.

Dwyer, S. J. III, A. W. Templeton, and S. Batnitzky. 1991. "Teleradiology: Costs of Hardware and Communications," *American Journal of Radiology,* 156: 1279–82.

Eddings, D. 1988. *King of the Murgos.* New York: Ballantine Books.

Ferrante, F. 1994. "Asynchronous Transfer Mode (ATM) and its Applicability to the Telemedicine Environment." Tutorial, Proceedings of the Symposium on Computer Applications in Radiology. Winston-Salem, NC, June.

"Filmless Radiology: The Design, Integration, Implementation, and Evaluation of a Digital Imaging Network," Annual and Final Report—June 1990.

Fitzmaurice, J. M. 1994. "Health Care and the NII: Putting the Information Infrastructure to Work: Health Care and the National Information Infrastructure," *AHCPR,* 41–47.

Forsberg, D., and M. Sawehak. 1994. "Options in Teleradiology," *Administrative Radiology,* 50–54.

Frank, M. S., G. R. Jost, P. L. Molina, et al. 1993. "High-Resolution Computer Display of Portable, Digital, Chest Radiographs of Adults: Suitability for Primary Interpretation," *American Journal of Radiology,* 160: 473–77.

Franken, E. A., et al. 1989. "Teleradiology for a Family Practice Center," *Journal of the American Medical Association,* 261: 3014–15.

Gayler, B. W., J. N. Gitlin, W. H. Rappaport, and F. L. Skinner. 1979. "A Laboratory Evaluation of Teleradiology," Proceedings of the Sixth Conference on Computer Applications in Radiology. June, 26–30.

Gitlin, J. N. 1986. "Teleradiology," In: Arenson, R. L., (ed.) "Use of Computers in Radiology," *The Radiological Clinics of North America,* 24 (1): 55–68.

Gitlin, J. N. 1994. "The Role of ACR and NEMA in Supporting the Standard," In: Hindel, R., (ed.) *Implementation of the DICOM 3.0 Standard,* 12–15.

Gitlin, J. N., and G. E. R. Weller. 1995. Teleradiology Tutorial CAR 95-Berlin. Department of Radiology, Johns Hopkins Medical Institutions, June.

Goldberg, M. A., D. I. Rosenthal, F. S. Chew, J. G. Blickman, S. W. Miller, and P. R. Mueller. 1993. "New High-Resolution Teleradiology System: Prospective Study of diagnostic Accuracy in 685 Transmitted Clinical Cases," *Radiology,* 186: 429–34.

Grigsby, J., et al. 1993. *Analysis of Expansion of Access to Care Through Use of Telemedicine and Mobile Health Services, Report 1: Literature Review and Analytic Framework,* Denver, CO: Center for Health Policy Research.

Grigsby, J., et al. 1994. *Analysis of Expansion of Access to Care Through Use of Telemedicine, Report 4: Study Summary and Recommendations for Further Research,* Denver, CO: Center for Health Policy Research.

Harrington, M. B., J. Cerva, B. Kerlin, et al. 1981. "A Laboratory Evaluation of the Technical Performance of the Teleradiology System: Summer 1980. MTR-81W201," McLean, VA: The MITRE Corporation, iii, xi–xiii.

Healthcare Telecom Report, Sept. 12, 1994.

Hindel, R. 1994. "Historic Development and Essential Features of the DICOM 3.0 Standard," In: Hindel, R., (ed.) "Implementation of the DICOM 3.0 Standard," *RSNA,* 16–19.

Holand and S. Pedersen. 1993. "Quality Requirements for Telemedical Services," *Telemedicine, Telektronikk,* 89 (1): 52.

Horii, S. C. 1990. "The ACR–NEMA Standards," *Administrative Radiology,* 69–75.

Horii, S. C. 1992. "A Comparison of Case Retrieval Times: Film versus Picture Archiving and Communications Systems," *Journal of Digital Imaging,* 5 (3): 138–43.

Horii, S. C., et al. 1995. "DICOM: An Introduction to the Standard," Revised May. Unpublished: 2–6, 22, 24.

Huang, H. K. "Elements of Digital Radiology."

Humphrey, L., et al. 1993. "Time Comparison of Intensive Care Units With and Without Digital Viewing Systems," *Journal of Digital Imaging,* 1: 37–41.

Institute of Medicine. 1991. *The Computer-based Patient Record.* Washington, D.C.: National Academy Press.

Lodwick, G. S. 1988. "PACS and the ACR–NEMA Digital Image Communications Standards," *Administrative Radiology,* May: 7–10.

Masel, J. P., and J. F. Grant. 1984. "Accuracy of Radiological Diagnosis in the Casualty Department of a Children's Hospital," *Australian Journal Paediatrics* 20: 221–24.

Mayhue, F. E., D. D. Rust, J. C. Aldag, A. M. Jenkins, and J. C. Ruthman. 1989. "Accuracy of Interpretations of Emergency Department Radiographs: Effect of Confidence Levels," *Annals of Emergency Medicine,* 18: 826–30.

McLain, P. L., and C. R. Kirkwood. 1985. "The Quality of Emergency Room Radiograph Interpretations," Journal of Family Practice, 20: 443–48.

Moliter, R. M. 1995. "A Century of Technology in Radiology," In: Devey, G. (ed.) "SCAR News," 6(1): 5.

Mowen, D. L., and B. D. Kerlin. 1980. *Rural Health Technology Demonstrations and Evaluation: An Overview and Annotated Bibliography,* McLean, VA: The MITRE Corporation.

Murphy, R. I. 1972. "Accuracy of Dermatologic Diagnosis by Television," *Archives of Dermatology,* 105: 833–35.

Nolan, T. M., F. Oberklaid, and D. Boldt. 1984. "Radiological Services in a Hospital Emergency Department and Evaluation of Service Delivery and Radiograph Interpretation," *Aust Paediatr J,* 20: 109–12.

Norman, D. A. 1993. *Things That Make Us Smart: Defending Human Attributes in the Age of the Machine.* Reading, MA: Addison-Wesley.

Office of Rural Health Policy, Health Resources and Services Administration, Public Health Service, Department of Health and Human Services, Reaching Rural, Washington, DC: 1994.

Office of Science and Technology Policy, Executive Office of the President. High Performance Computing and Communications: Toward a National Information Infrastructure, 1994, (Washington, D.C. 20506).

Overton, D. T. 1987. "A Quality Assurance Assessment of Radiograph Reading Accuracy by Emergency Medicine Faculty," abstract, *Annals of Emergency Medicine,* 16: 503.

Page, G., A. Gregoire, C. Galand, et al. 1981. "Teleradiology in Northern Quebec," *Radiology,* 140: 361–66.

Perednia, D. A., and A. Allen. 1995. "Telemedicine Technology and Clinical Applications," *Journal of the American Medical Association,* 273 (6): 483–88.

Pfleeger, C. P. 1989. *Security in Computing.* Englewood Cliffs, NJ: Prentice-Hall, Inc.

University of New Mexico Regional Health Program. 1975. *Playas Lake Telehealth System.* Albuquerque, NM: The University.

Puskin, D. S. 1992. "Telecommunications in Rural America: Opportunities and Challenges for the Health Care System," *Annals of the New York Academy of Sciences,* 670: 71–72.

Puskin, D. S., and J. H. Sanders. 1995. "Telemedicine Infrastructure Development," *Journal of Medical Systems,* 19 (2).

Quick, G., and G. Podgorny. 1977. "An Emergency department Radiology Audit Procedure," *Journal of the American College of Emergency Physicians* 6: 247–50.

Rappaport, W., F. L. Skinner, and B. Gayler. 1979. *A Laboratory Evaluation of Teleradiology,* McLean, VA: The MITRE Corporation.

Rappaport, W., and F. Skinner. 1978. *The Block Island–Rhode Island Hospital Telehealth System: A Progress Report,* McLean, VA: The MITRE Corporation.

Rasmussen, W. T., R. L. Crepezu, and F. H. Gerber. 1977. *Resolution Requirements for Slow-Scan Television Transmission of X-rays.* San Diego, CA: TR-150 Naval Ocean Systems.

Rasmussen, W. T., and J. Silva. 1976. *Navy Remote Medical Diagnosis System.* San Diego, CA: Naval Electronics Laboratory Center.

Razavi, M., J. W. Sayre, R. K. Taira, et al. 1992. "Receiver-Operating-Characteristic Study of Chest Radiographs in Children: Digital Hard-Copy Film vs 2K × 2K Soft-Copy Images," *American Journal of Radiography* 158: 443–48.

Roberts, P. (R–KS). 1994. Hearing before the Committee on Ways and Means, Subcommittee on Health, House of Representatives, U.S. Congress, Feb. 7.

Roberts, R., S. Skens, and G. Lyons. 1978. *Evaluation of the Moose Factory Telemedicine Project U-6.* Hamilton, Ontario, Canada: McMaster University.

Rosenthal, G. 1978. Foreword. *Telehealth Handbook – A Guide to Telecommunications Technology for Rural Health Care,* iii–iv.

Rowberg, A. H., S. J. Dwyer III, and B. K. Stewart. 1994. *Technical Factors in*

Teleradiology, Understanding Teleradiology. Harrisburg, PA: The Society for Computer Applications in Radiology.

Sanders, J. H., and R. L. Bashshur. 1995. "Challenges to the Implementation of Telemedicine," *Telemedicine Journal,* 1 (2): 115–23.

Scott, W. W., D. A. Bluemke, W. K. Mysko, et al. 1995. "Interpretation of Emergency Department Radiographs by Radiologists and Emergency Medicine Physicians: Teleradiology Workstation versus Radiographic Readings," *Radiology* 195: 223–29.

Slasky, S. B., D. Gur, W. F. Good, et al. 1990. "Receiver Operating Characteristic Analysis of Chest Image Interpretation with Conventional, Laser-Printed, and High-Resolution Workstation Images," *Radiology.* 174: 775–80.

Spilker, C. 1989. "The ACR–NEMA Digital Imaging and Communications Standard: A Non-Technical Description," *Journal of Digital Imaging* 2: 127–31.

Tangalos, E. G. 1993. "Telemedicine Outcomes: What We Know and What We Don't," paper presented at the Rural Telemedicine Workshop, Office of Rural Health Policy, Washington, DC, Nov. 3–5.

Telehealth Handbook: A Guide to Telecommunications Technology for Rural Health Care. 1978.

Templeton, A. W., S. J. Dwyer III, S. J. Rosenthal, et al. 1991. "A Dial-up Teleradiology System: Technical Considerations and Clinical Experience," *American Journal of Radiology,* 157: 1331–36.

Thaete, L. F., C. R. Fuhrman, J. H. Oliver, et al. 1994. "Digital Radiography and Conventional Imaging of the Chest: A Comparison of Observer Performance," *American Journal of Radiology,* 162: 575–81.

Tomes, J. P. 1994. *Compliance Guide to Electronic Health Records.* New York: Faulkner and Gray, pp. 14–19.

U.S. Congress, Office of Technology Assessment, Critical Connections: Communication for the Future, OTA–CIT-407 Washington, DC: U.S. Government Printing Office, January 1990a.

U.S. Congress, Office of Technology Assessment, Health Care in Rural America, OTA–H-434 Washington, DC: U.S. Government Printing Office, September 1990b.

U.S. Congress, Office of Technology Assessment, Rural America at the Crossroads: Networking for the Future, OTA–TCT-471 Washington, DC: U.S. Government Printing Office, April 1991.

U.S. Congress, Office of Technology Assessment, Global Standards: Building Blocks for the Future, TCT-512, Washington, DC: U.S. Government Printing Office, March 1992.

U.S. Congress, Office of Technology Assessment, Advanced Network Technology, OTA–BP–TCT-101 Washington, DC: U.S. Government Printing Office, June 1993.

U.S. Congress, Office of Technology Assessment, Issue Update on Information Security and Privacy in Network Environments, OTA–BP–ITC-147 (Washington, DC: U.S. Government Printing Office, June 1995).

U.S. Congress, Office of Technology Assessment, Wireless Technologies and the

National Information Infrastructure, OTA–ITC-622 Washington, DC: U.S. Government Printing Office, July 1995.

U.S. Congress, Office of Technology Assessment, Bringing Health Care Online: The Role of Information Technologies, OTA–ITC-624 (Washington, DC: U.S. Government Printing Office, September 1995).

Vetter, C. 1985. "Bringing X-ray Expertise to Medical Outposts," *FDA Consumer,* 19: 7.

Wang, Y., D. E. Best, J. G. Hoffman, et al. 1988. "ACR–NEMA Digital Imaging and Communication Standards: Minimum Requirements," *Radiology,* 166: 529–32.

Wegryn, S. A., D. W. Piraino, B. J. Richmond, et al. 1990. "Comparison of Digital and Conventional Musculoskeletal Radiography: An Observer Performance Study," *Radiology,* 175: 225–28.

Wilson, M. R., and C. Brady. 1975. "Health Care in Alaska via Satellite," AIAA Conference on Communication Satellites for Health/Education Applications. Denver, Colorado, July 21–23.

Wiltel Communications Corporation. 1995. *The Telecommunications Glossary.* Wiltel.

Wyatt, J. C. 1994. "Clinical Data Systems, Part 3: Development and Evaluation," *Lancet,* 344 (8938).

Zimmermann, H. 1980. "OSI Reference Model: The ISO Model of Architecture for Open Systems Interconnection," *IEEE Tran Comm* 28: 425–32.

Chapter 6

TELEPATHOLOGY

Ronald S. Weinstein, Achyut Bhattacharyya,
John R. Davis, and Anna R. Graham

Introduction

Telepathology: Definition and General Orientation

Telepathology is the practice of pathology from a distance. It involves
the viewing of pathology specimen images on a video monitor
(Kayser, 1992; Kayser, 1993; Krupinski et al., 1993; Murphy and Bird,
1974; Park, 1975; Riggs et al., 1974; Weinstein, 1986; Weinstein et al.,
1987a; Weinstein et al., 1989; Weinstein et al., 1990; Weinstein, 1991;
Weinstein, 1992). The images are acquired, either as single frames or in
real time, by a video camera mounted on a light microscope. Images are
transmitted over a telecommunications link to a computer workstation
for analysis by a telepathologist. A second video camera can be used to
transmit images of gross specimens and x-rays.

Currently, two significantly different technical approaches, dynamic
imaging and static imaging, represent competing technologies in telepathology (Table 6-1). **Dynamic imaging** telepathology systems provide a
real-time video imaging capability via a broadband telecommunications
linkage (Krupinski et al., 1993; Weinstein et al., 1987a; Weinstein et al.,
1987b; Eide and Nordrum, 1994). In their full implementation, dynamic
systems incorporate a motorized, robotically controlled light microscope
which enables the telepathologist to operate the microscope remotely via
a keypad, mouse, or other input device. The telepathologist can manipulate the microscope's specimen stage in the X- and Y-axes to move the
microscope slide laterally and in the Z-axis (vertical) to adjust focus.
Typically, the referring pathologist relinquishes control of the microscope to the telepathologist (Figure 6-1). The referring pathologist and
telepathologist communicate by telephone or video conferencing during
the consultation. Dynamic imaging is an attractive method because it

179

approximates the usual technique of pathological examination of glass slides of histology and cytology specimens.

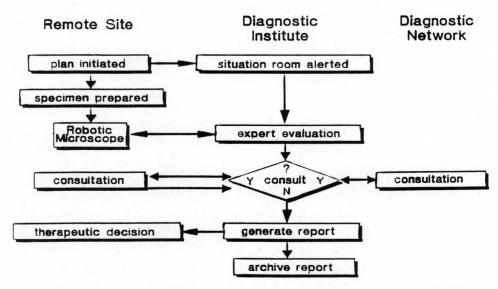

Figure 6-1. Telepathology strategy. According to this scheme, a subspecialist (expert) and a network consultant at different computer workstations can individually manipulate a robotic microscope, simultaneously view real-time video images of tissue sections and discuss and render diagnoses. The turnaround time for generating a final surgical pathology report, including expert opinions, is minimized. (From Weinstein, R.S., Bloom, K.S., and Rozek, L.S. 1987. *Archives Pathology and Laboratory Medicine,* 111: 646–652. Used with permission.)

Static imaging based on less expensive "store-and-forward" computer imaging technology is the other method (Krupinski et al., 1993; Linder and Masada, 1989; Weinstein et al., 1990; Black-Schaffer and Flotte, 1995; Weinstein, 1996). With static imaging, a pathologist at the referring site selects a few still video images (i.e., typically fewer than ten), stores them locally on a computer, and then uploads the images to the telepathologists' static imaging workstation. The consultant reviews the images, renders an opinion, and then communicates it to the referring pathologist by computer, telephone, or FAX. This approach requires less expensive equipment, and it has the advantage of image transmission over conventional telephone lines (Table 6-1).

A major limitation of static imaging is the use of a small number of images for each case. Complete sampling of tissue sections on histopa-

Table 6-1. Comparison of Static and Dynamic Telepathology Systems.[a]

	Static	*Robotic-Dynamic*
image system	still	live
motorized microscope	no (optional)	yes
robotic remote control	no	yes
images per average case	5	unlimited
specimen sampling	limited	comprehensive
image selection	referring pathologist	telepathologist
transmission time/image	45 seconds	1/15 second
average time/diagnosis	15 minutes	3 minutes
image compression	yes	yes
video conferencing	no	yes
network capacity	28.8 kbps[b]	1.54 Mbps
equipment cost	$30,000/$20,000[c]	$130,000/$40,000

[a]Apollo/Corabi Telepathology Systems (Alexandria, VA)
[b]Use of an ISDN (64 kbps) or T1 (1.54 Mbps) telecommunications link reduces transmission time per image. (Yu et al., 1994)
[c]referral site/consultation site

thology slides typically requires the capture of hundreds or even thousands of images (Weinstein et al., 1990). Another drawback is the virtual exclusion of the consulting telepathologist from the process of selecting the microscopic fields for imaging.

Although telepathology is in its infancy compared with teleradiology, several telepathology projects have been established since 1989. These are providing services on an ongoing basis (Allert and Dusserre, 1990; Bhattacharyya et al., 1995; Eide and Nordrum, 1992; Goncalves, 1993; Martin et al., 1991; Martin et al., 1992; Miaoulis et al., 1990; Miaoulis et al., 1992; Weinstein et al., 1995). The projects serve as important test beds for the technology.

Clinical Application of Telepathology

In the United States, pathology services are traditionally classified into two broad categories: anatomic pathology and clinical pathology. Anatomic pathology encompasses many of the diagnostic services for which light microscopy is the central technology. It includes surgical pathology, cytopathology, and autopsy pathology. Clinical pathology includes hematology, microbiology, cytogenetics, and blood banking, services that have light microscopy components to varying degrees. To date, the major uses of telepathology have involved video light microscopy, but other services such as the remote analysis of electrophoresis data and DNA ploidy are feasible (Phillips et al., 1995). Table 6-2 lists telepathology light microscopy services.

Table 6-2. Pathology Services Using Light Microscopy.

Histopathology	Microbiology
Cytology	Urinanalysis
Hematology	Cytogenetics

Conceptual/Theoretical Framework of Telepathology

Video microscopy, the core technology for telepathology, is used for other purposes in the modern pathology laboratory. This paved the way for the introduction of telepathology into practice. A decade ago, video light microscopy was the domain of researchers; today, video light microscopy applications are commonly used in clinical practice for automated cytogenetics and quantitative immunohistochemistry. Video microscopy has also been integrated into many pathology education programs. The fact that video microscopy is a standard tool should affect users' acceptance of telepathology. Medical students and residents are accustomed to viewing pathology images on video monitors and accept the suitability of the images for analyzing tissues and cells.

A number of conceptual and technical issues in telepathology need to be resolved. The most important one involves the use of static imaging for histopathology specimen viewing. For this application, static imaging represents a major shift from the traditional practice of surgical pathology. Pathologists do not perform static imaging in their everyday practice of surgical pathology, nor do they have surrogates control the acquisition of images of surgical pathology slides for viewing. Surgical pathologists perform their customary microscopy activities in real time, exerting total "hands-on" control over the microscope's functions, including stage movements, focus, magnification, and illumination. The pathologist who is rendering a diagnosis takes full responsibility for conducting the examination of glass slides that includes scanning all tissue sections, shifting between low, intermediate, and high magnifications as needed, and adjusting focus.

Division of responsibility for control of the microscope by the referring pathologist, and for interpretation of images by the telepathologists is novel and raises certain questions. First, the extent to which some aspects of the consulting pathologist's traditional role can be delegated to surrogates, such as the referring pathologist, needs to be resolved. In everyday practice, pathologists may have surrogates (e.g., residents or physician's assistants) dissect and sample gross specimens, but they are responsible for viewing the glass slides. Pathologists assume full responsibility for the totality of the histopathology examination. "Sins of omission" while examining histology slides by light microscopy are regarded as unacceptable. Although tissue sampling is acknowledged to be inherently incomplete, since it is usually not feasible to have serial sections of all tissue blocks or tissue blocks of every piece of surgically-removed tissue, the pathologist is responsible for overseeing specimen sampling, assuring a degree of sampling that meets local and/or national practice standards, and examining all tissue sections processed for light microscopy. This level of participation is compromised, especially with static imaging telepathology.

Not only are numbers of images and selection of static images at issue, the sequence in which they are presented to and viewed by the telepathologist can affect the outcome of a consultation. Normally, pathologists become adept at maximizing the information to be gleaned from the examination of histopathology specimens with a light microscope. They tend to solve certain diagnostic problems by identifying pertinent microscopic fields in their peripheral vision and bringing these areas to the center of their visual fields for further scrutiny through their light microscopes. A less experienced referring pathologist could miss the significance of crucial microscopic fields, fail to include images of them in the data set, and thereby deprive the teleconsultant of important information. Thus, with static imaging the referring pathologist seeking the consultation forfeits the benefits of the consultant's expertise in field selection, which can result in diagnostic errors (Bhattacharyya et al., 1995; Ito et al., 1994; Shimosato et al., 1992).

There are other issues and concerns relevant to static imaging telepathology. Static imaging is inefficient in comparison to either conventional light microscopy or to robotic-dynamic telepathology. (Shimosato et al., 1992; Krupinski et al., 1993) Despite the use of fewer images per case, consultations take longer to complete. Another issue involves the partitioning of medical-legal responsibility between the referring pathologist and the telepathologist rendering the diagnosis. This needs to be addressed for both static image and dynamic-robotic telepathology.

Currently, dynamic-robotic telepathology is quite expensive, requiring a motorized microscope and access to a broadband telecommunications channel (Table 6-1). This has discouraged its wide use. Nonetheless, efforts are ongoing to improve the capability of dynamic-robotic systems, reduce their cost, and create hybrid static-dynamic systems to incorporate some features of dynamic-robotic systems while utilizing less expensive telecommunications links (Nordrum et al., 1991; Oberholzer et al., 1993; Shimosato et al., 1992).

Evolution of Telepathology

Historical perspective is useful for understanding the design configuration of dynamic-robotic telepathology systems as they exist today, and might explain why efforts continue to improve robotic real-time telepathology systems, even though less expensive static image technology is available. Three independent groups—in the United States, Norway, and Japan—developed the early dynamic-robotic telepathology systems.

Development of the first dynamic-robotic telepathology system addressed some perceived needs of the National Bladder Cancer Group (NBCG) in the United States. The aim was to improve the quality of pathology diagnoses used to assign clinical trial patients to the cancer therapy protocols of the NBCG (Weinstein et al., 1987a; Weinstein, 1996). Traditionally, local institutional surgical pathology reports provided the NBCG with the basis for patient stratification and assignment to specific arms of cancer therapy studies. The Central Pathology Laboratory of the NBCG was responsible for retrospective validation of the surgical pathology reports, often a year or more after the completion of therapy.

Telepathology became an option for consideration when it was realized that the use of institutional surgical pathology reports potentially compromised clinical trials of bladder cancer therapies. Unexpectedly high interobserver variability rates in pathology diagnoses had been experienced within the Group which was of concern to the senior author of this chapter, who served as Director of the Central Pathology Laboratory at that time. A number of training measures for the referring hospital pathologists were instituted, but results, in terms of reducing interobserver variability, were uneven and often temporary. Technical solutions such as telepathology were considered. For this application, it was considered essential that the diagnostic accuracy of telepathology equal that achievable by an expert subspecialist using conventional light microscopy. Federal funding was unavailable for the telepathology project,

but the idea was of interest to investors in the United States and Japan. Subsequently, a dynamic-robotic telepathology imaging system was developed in order to link some of the collaborating institutions with the NBCG's Central Pathology Laboratory (Figure 6-1). The aim was to circumvent the problem of interobserver variability by having the Central Pathology Laboratory pathologists render the primary surgical pathology diagnoses before the local urologists, oncologists, or radiotherapists initiated therapy (Weinstein, 1996).

Ironically, the technology was never used according to the original intent because the NBCG terminated its activities before the implementation of telepathology could take place. However, by that time, a dynamic-robotic telepathology system had been engineered by Corabi International Telemetrics, Inc., to satisfy some of the NBCG's perceived needs (Weinstein et al., 1987a, 1987b). In 1986, a prototype of the Corabi system was used in a test-of-concept demonstration between El Paso, Texas and Washington, D.C., using a satellite for video image transmission and standard telephone lines for telemetric control of the robotic microscope and voice communications (Erickson, 1990; Weinstein et al., 1987a).

In Norway, establishment of comprehensive telemedicine services became a national priority in 1988. Remote intraoperative frozen section services were initiated in Northern Norway in 1990 (Eide and Nordrum, 1992; Nordrum et al., 1991; Weinstein, 1991). This application also required a high level of diagnostic accuracy. The University Hospital in Tromsø, Norway, was connected by a dedicated 2 Mbps point-to-point telecommunications link to two small, distant, island-based hospitals without pathologists located above the Arctic Circle. The telecommunications link required one-day advance notice to activate. The telepathology system used for the project incorporated many of the features of the dynamic-robotic system developed earlier by the Corabi engineering team in the United States. A "super mouse" was added for controlling the robotic microscope (Nordrum et al., 1991; Eide and Nordrum, 1994). In order to overcome the bandwidth limitations imposed by the telecommunication system, the Norwegian group used a hybrid video system combining low-resolution, dynamic-real time imaging for specimen scanning and higher resolution static imaging. The Norwegian telepathology system has been operational for six years and has been used to render intraoperative frozen section diagnoses on over 100 frozen section cases.

In Tokyo in 1990, the National Cancer Center Hospital opened a satellite hospital approximately 25 miles from the main hospital. A telepathology project was initiated to enable the Pathology Department

at the main hospital to provide intra-operative frozen section consultations to their smaller satellite hospital, as was done in Norway (Shimosato
et al., 1992). Unlike Norway, where the referring sites lack a pathologist,
the National Cancer Center Hospital planned to staff the referring
hospital with a pathologist. The staff at the National Cancer Center
Hospital teamed with Nikon Corporation to evaluate and compare the
diagnostic accuracy of dynamic-robotic telepathology and static image
telepathology, as well as a hybrid system combining dynamic and static
imaging. It was concluded that a hybrid dynamic-static imaging system
provides a reasonable level of diagnostic accuracy, but that it is not
equivalent to conventional light microscopy (Shimosato et al., 1992).

These three demonstration projects have shown that there are features
of robotic-dynamic telepathology systems that meet some of the needs of
pathologists. Because the telepathologist controls all microscope functions,
including stage movements in the X- and Y-axes, the crucial issue of
histopathology field selection for remote examination is circumvented.
The telepathologist can view entire tissue sections with the full range of
magnifications used in regular practice and can scan specimens in real
time with reasonable rapidity.

Although early designers of dynamic imaging telepathology systems
attempted to emulate conventional light microscopy, there are significant differences between dynamic-robotic telepathology and conventional light microscopy that are bothersome to pathologists. First are the
differences in image quality. Image quality is not identical when comparing video microscopy images and conventional light microscopy images.
Light microscopy images are of noticeably better quality as viewed
directly through a glass microscope lens (Figure 6-2).

Can video images be used by pathologists responsible for rendering
primary diagnostic opinions or second opinions? Ultimately, individual
pathologists are responsible for determining the adequacy of images for
rendering a diagnosis on a case-by-case basis. Data currently being
analyzed and published regarding the diagnostic accuracy of telepathology
may influence the acceptability of video-imaging technology among
pathologists in the future.

In retrospect, would the less costly static image telepathology have
satisfied the needs of the early telepathology projects? Findings of several studies indicate that although static image telepathology can be
useful, overall diagnostic accuracy fails to meet the standards of accuracy
expected of either a primary diagnostic service or a consultation service

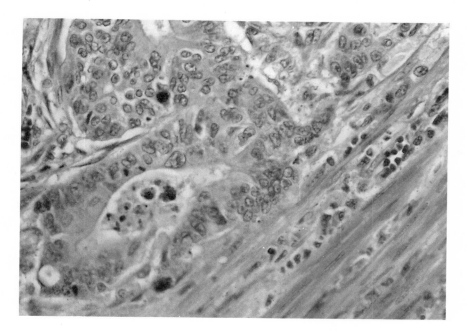

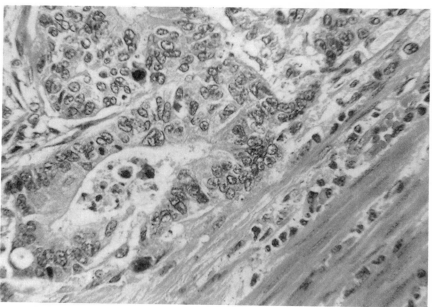

Figure 6-2. Comparison of images obtained with conventional light photomicroscopy and digital telepathology. (A) Photo-micrograph of invasive colon adenocarcinoma taken by conventional light microscopy. (Paraffin section, H&E; optical magnification ×400; print magnification ×460). (B) Telepathology digital image of same field (1024 × 768 pixels; optical magnification ×400; print magnification ×460). Photograph was taken from the screen of a Roche Image Analysis System with 35-mm camera. Although the video image (B) appears somewhat "sharper", this represents an artifact. System software and the image digitization process have produced overly crisp representations of cellular structures. (From Bhattacharyya, AK, JR Davis, BE Halliday, AR Graham, SA Leavitt, R Martinez, R Rivas, RS Weinstein. 1995. *Telemedicine Journal* 1:9–17. Used with permission.)

(Ito et al., 1994; Oberholzer et al., 1993; Shimosato et al., 1992; Weinstein, 1996). Limiting factors, especially constraints on tissue sampling, render the static imaging technology less accurate than the "gold standard," conventional light microscopy. The level of diagnostic accuracy of dynamic-robotic telepathology imaging comes closer to that achievable with conventional light microscopy (Eide and Nordrum, 1992; Eide et al., 1994; Krupinski et al., 1993).

Early studies showed that dynamic-robotic telepathology has its own set of limitations. It is somewhat less efficient than hands-on light microscopy, although much more efficient than static imaging. By substituting a video screen for an eye piece lens on a microscope for observing pathology slides, viewing times increase due, in part, to reductions in the size of microscopic fields as viewed on a video monitor. This results from the placement of the video camera above the microscope. In addition, the incorporation of a motorized robotic microscope into the system adds hardware-, software-, and telecommunication-time overhead. The inefficiencies of the robotic microscope are apparent to the remote microscope operator. Control of stage movements is cumbersome compared with the hands-on control of a conventional light microscope. Manipulating the robotic microscope involves activating an input device, transmitting a stream of commands to the robotic microscope, and waiting for robotic microscope telemetry data. Motorized microscope stages are often driven by stepmotors, one in the X-axis and the other in the Y-axis. Target sites on the glass slides are approached by small incremental X- and Y-axis (lateral) movements. This is unnatural to the system operator who is accustomed to unrestricted stage movements.

Focus controls also require transmitting a stream of information to the motorized microscope. The telepathologist verifies achievement of suitable specimen focusing on the video screen. Focus is reached by rotating a robotically controlled motorized knob that adjusts the height of the specimen stage in the Z-axis. Achieving focus with a motorized robotic microscope takes more time than with a manually controlled microscope.

Motorized microscopes respond slowly to electronic commands. This is an impediment to pathologists accustomed to using conventional light microscopes in a highly efficient manner. Pathologists become very facile at manipulating their light microscopes. To maximize efficiency, many pathologists strip their microscopes of the mechanical stage control device since they prefer to control the lateral movements of a mounted glass slide by lightly resting their fingertips on the glass slide.

This maneuver allows the pathologist to move glass slides in increments of tenths of millimeters in any direction to control lateral movements. The degree of hand-eye coordination and fine motor control of an experienced pathologist is often remarkable. In conventional practice, the experienced pathologist and a microscope perform as an integrated unit, akin to a musician and a violin. This produces routine efficiencies, and the pathologist comes to regard the examination of glass microscope slides as if it were second nature. These efficiencies are appreciated when input devices, computers, a motorized microscope, and a video imaging system are employed to emulate the processes of conventional light microscopy. Initially, the use of early robotic-dynamic telepathology systems seemed unnatural. Later models have incorporated improvements in design and provide a higher level of user friendliness.

History of Telepathology Field

The first reported transmission of a pathology video image was in the late 1960s as part of the Massachusetts General Hospital-Logan Airport project (Park, 1975). Early phases in the evolution of telepathology involved the design, construction, and evaluation of telepathology systems (Carr et al., 1992; Eide and Nordrum, 1992; Ito et al., 1994; Kayser et al., 1991a; Kayser and Drlicek, 1992; Keil-Slawik et al., 1991; Krupinski et al., 1993; Martin et al., 1991; Nordrum et al., 1991; Oberholzer et al., 1993; Schwartzmann, 1992; Shimosato et al., 1992; Weinstein et al., 1987a; Weinstein et al., 1987b; Weinstein et al., 1989; Weinstein et al., 1990; Yu et al., 1994). There were test-of-concept demonstrations (Becker et al., 1993; Kayser et al., 1991b, Nordrum et al., 1991; Riggs, 1974; Weinstein, 1987a) and evaluations of human performance using the technology (Bloom et al., 1987; Krupinski et al., 1993; Linder and Masada, 1989; Nordrum et al., 1991; Shimosato et al., 1992; Weinstein et al., 1987a). Since the first telepathology practices became operational, a number of different telepathology systems have been used for primary diagnostic services (Nordrum et al., 1991; Oberholzer et al., 1993; Shimosato et al., 1992), second opinions (Bhattacharyya et al., 1995; Ferrer-Roch, 1993; Goncalves, 1993; Martin et al., 1992; Miaoulis et al., 1990; Miaoulis et al., 1992; Weinstein et al., 1995; Wold and Weiland, 1992; Busch and Olsson, 1995; Kayser, 1995), and quality assurance programs (Bhattacharyya et al., 1995; Cronenberger et al., 1992).

Technical Considerations and Human Performance Studies

Telepathology images generated by available video systems are of
poorer quality than images viewed directly through light microscopes
(Weinstein et al., 1987a; Krupinski et al., forthcoming). Commenting on
the suitability of telepathology images in clinical practice, several authors
have noted video-image quality to be problem, especially with respect to
low magnification scanning of histopathology tissue sections and resolution of certain critical features of cellular morphology (Ito et al., 1994;
Kayser and Drlicek, 1992; Krupinski et al., 1993; Shimosato et al., 1992).

Theoretically, image-related technology issues in telepathology would
be eliminated if video equipment capable of producing images equivalent in quality to light microscope images became available at affordable
prices. The alternative is to use images of inferior quality than light
microscopic images, which appears to be a reasonable decision for many
telepathology applications. Important judgments have to be made on
the acceptability of compromises in image quality and the extent to
which this may affect patient care. Tradeoffs between poorer image
quality and increased accessibility to expertise may be necessary. Some
pertinent human performance studies have been conducted both in
experimental laboratories and clinical practices. These laboratory studies and clinical trials provide baseline data for the field of telepathology.

Weinstein and coworkers carried out early studies of video microscopy
in a vision laboratory and validated the use of the technology for diagnostic pathology (Bloom et al., 1987; Krupinski et al., 1993; Weinstein et al.,
1987a; Weinstein et al., 1992). They assessed the diagnostic accuracy of
pathologists using video microscopy. The video-microscopy system incorporated an analog Sony Model SHR-10 video camera (1050 scan lines)
mounted on a microscope (without eye pieces). Pathology images were
displayed on a 15-inch Sony Trinitron 950-line monitor. A study set
consisting of glass slides from 115 consecutive breast frozen section cases
was examined independently by six pathologists blinded to clinical
information for the cases.

To compare pathologist performance with conventional light vs. video
microscopy, each pathologist viewed the 115 cases twice, once by light
microscopy and once by video microscopy, at different diagnostic sessions.
A counter-balanced experimental design was used in which each pathologist saw half the cases with light microscopy and half with video microscopy in phase I of the study, and saw each case in the opposite viewing

condition in phase II. Phases I and II occurred at least a month apart. The size of the panel of pathologists permitted analysis of interobserver and intraobserver variability with respect to both diagnostic accuracy and viewing times. In all, 1,380 diagnoses were rendered by the pathologists at 140 separate diagnostic sessions. The "baseline truth" diagnosis was established by a breast pathology expert examining the same frozen sections with a light microscope.

To assess performance using Receiver Operator Characteristic curve (ROC) analysis, the pathologists rated each case with a five-level confidence scale where: 1 = definitely benign; 2 = probably benign; 3 = possibly malignant; 4 = probably malignant; and 5 = definitely malignant. This scale is somewhat different from the scale used in clinical practice but has the advantage of increasing the sensitivity of measurement of human performance. It also produces more reliable data for generating ROC curves (Krupinski et al., 1993).

The ROC studies showed that pathologist performance in the light and video microscopy conditions were virtually identical and nearly perfect. Interobserver and intraobserver variability was insignificant. Several pathologists actually performed better with video microscopy than with light microscopy (Krupinski et al., 1993).

Another aspect of the study was the measurement of viewing times. All six pathologists had significantly shorter viewing times for both benign and malignant cases in the light versus video microscopy conditions. Although the viewing-time difference was statistically significant for light and video microscopy, the difference would not represent a problem in clinical practice since over 95 percent of cases were diagnosed in less than four minutes in both viewing modes (Weinstein et al., 1992).

In another laboratory study, the diagnostic accuracy achievable with video microscopy was assessed using a much lower resolution video system (Eide et al., 1992). Video images were viewed on an 11-inch Sony monitor with 300-line horizontal resolution. Two pathologists examined 80 frozen section cases representing over 13 organs. "Truth" diagnoses were established by the same pathologists examining paraffin-embedded tissues, rather than the original frozen sections. The results showed no significant difference in accuracy for diagnosing benign and malignant lesions by light microscopy or video microscopy. Overall concordance with the "truth" diagnoses was somewhat lower for video microscopy due to a somewhat higher proportion of inconclusive diagnoses as well as a few false-positive and false-negative diagnoses. The authors speculated that the errors might be eliminated using a higher resolution video system.

Results of a third laboratory study, in which CD–ROM still images were viewed on a video monitor, indicated a lower level of diagnostic accuracy than that reported for real-time video microscopy (Weinberg et al., 1996). This study is relevant to the issue of the diagnostic accuracy of static image telepathology.

Clinical trials of telepathology have demonstrated levels of diagnostic accuracy similar to those achieved in the laboratory-based video micros-copy studies. Generally, higher levels of diagnostic accuracy have been reported for clinical trials using telepathology systems incorporating a dynamic-robotic component. Eide and Nordrum (1992), using their hybrid static-dynamic imaging system, had a diagnostic efficiency (true-positive plus true-negative diagnoses divided by the total number of cases) of 92 percent, which is essentially the same as what they achieved by light microscopy. Correct benign versus malignant diagnoses were rendered in 46 of 50 frozen section cases (Eide et al., 1992; Nordrum et al., 1991).

Shimosato et al. (1992) evaluated dynamic-robotic and static-robotic imaging telepathology systems and then compared the results. First, using a dynamic-robotic system with an HDTV camera and monitor (1125 lines of horizontal resolution) and fiber optical telecommunications, 158 lesions were examined (144 paraffin section lesions and 14 frozen section lesions) and an overall diagnostic accuracy of 90.5 percent was achieved.

The study was continued but switched to a static imaging system operating at HDTV standards, with images displayed on a monitor which provided the same resolution as the monitor used for the previous dynamic-robotic imaging study. Static images were transmitted over an ISDN channel.

Using this static-robotic imaging system, Shimosato achieved a diag-nostic accuracy of 88.1 percent for paraffin section cases, 68.8 percent for frozen section cases, and 96.3 percent for cytological smears. Some of the errors were judged to be "permissible" because they did not result in the choice of inappropriate therapy. Nevertheless, these errors should not be discounted in calculating diagnostic accuracy rates. The occurence of diagnostic errors was attributed to the inability to adequately examine glass slides at higher magnification and to significant sampling errors despite the use of robotic microscopy (Shimosato et al., 1992).

In addition to measuring diagnostic accuracy, the time needed to render a diagnosis was assessed. The average time with dynamic-robotic imaging was 3 minutes in comparison to 15 minutes for static-robotic imaging (Shimosato et al., 1992).

Oberholzer et al. (1993), in Switzerland, also tested a static-robotic telepathology system. A motorized microscope and telepathologist's workstation were linked by ISDN (64 kbps). The telecommunications system was not adequate to support true dynamic imaging but allowed designers of the system to incorporate some features of a dynamic-robotic system. A motorized robotically controlled video microscope transmitted images that could be electronically demagnified to serve as orientation points from which slide movements were remotely controlled on the motorized microscope stage. Optical magnification could be changed remotely, and a built-in autofocus feature was used to circumvent the lack of timely visual feedback to the telepathologist. Using this system, diagnostic opinions were rendered for 15 cases. Six cases were premalignant or malignant and nine cases were benign. Three of six malignancies were correctly diagnosed as were all nine benign tumors. With Oberholzer's low-resolution static-robotic system, it took from 25 to 35 minutes per case to render diagnoses (Oberholzer et al., 1993).

Ito et al. used a static image telepathology system without robotic microscopy to provide consultations on kidney transplantation cases. Correct diagnoses were rendered for 10 of 12 renal biopsies. One error was attributed to inadequate sampling, and inadequate static images accounted for the second error. The average time was 13 minutes per case. The authors commented that although their results showed that static imaging can be useful, a robotic-dynamic system might improve their diagnostic accuracy (Ito et al., 1994).

Current Range of Applications

Uses of telepathology are: (1) to provide urgent services at sites either without a pathologist or with a pathologist requiring back-up (i.e., frozen section services); (2) to provide immediate access to subspecialty pathology consultants; (3) to generate second opinions from peers; (4) to provide a pathology "telepresence" to assist pathologists in completing or refining a differential diagnosis; and (5) for continuing medical education, proficiency testing, and recertification of pathologists as well as other laboratory personnel. The first two applications have been studied in greatest detail.

Dynamic-robotic telepathology is clearly of value for rendering intraoperative frozen section diagnoses (Eide and Nordrum, 1994). The usefulness of telepathology for the second application, providing subspecialty pathology services, is less certain. It is noteworthy that the overall use of subspecialty pathology diagnostic services in the United States is not well documented, although such services are of obvious value in many cases. Little is known about the relationship of use of subspecialty pathology services to patient outcomes. Despite such lack of documentation, the trend toward subspecialization in diagnostic pathology has accelerated over the past decade. This is reflected in a proliferation of organ-oriented subspecialty pathology societies and journals (Weinstein et al., 1995). It is estimated that 10–25 percent of surgical pathology cases benefit from the input of a pathologist with subspecialty expertise. Certain types of cases (e.g., renal, skin, and lymphoma-related lymph node biopsies) are often routinely sent directly to subspecialists for evaluation. Currently, the consultation of a pathology subspecialist can be obtained locally or by sending glass slides and tissue blocks to a reference laboratory by express mail.

Based on the premise that access to subspecialty pathologists can be a justification for telepathology, independent networks with hubs in Dijon, France, Washington, D.C., and Tucson, Arizona have been established (Martin et al., 1994; Weinstein, 1995).

The Arizona-International Telemedicine Network (AITN) has quantitated the actual use of subspecialty pathologists in its telepathology practice (Figure 6-3). The AITN uses a case triage model in which a telepathologist, functioning as a generalist, has the option of signing out a telepathology case or referring the case to a subspecialty telepathologist, based on criteria used in their regular consultation practice (Figure 6-4).

AITN's experience was that referring pathologists were typically satisfied with the level of service provided by the general telepathologists and infrequently requested that cases be shown to subspecialists. Workflow analysis within the Network showed that a subspecialty telepathologist participated in rendering an opinion in less than 30 percent of cases, although they were available for consultation. This suggests that when telepathology services are organized so that all cases go directly to subspecialty pathologists without the intervention of a triage pathologist, "overuse" of subspecialty pathologists might be fostered (Weinstein et al., 1995). The AITN personnel concurred that users of their telepathology services did not require a subspecialist for the majority of cases sent to

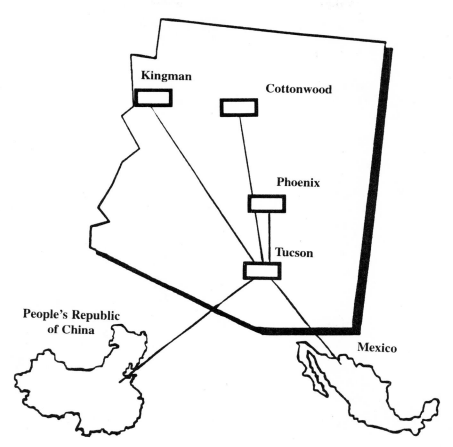

Figure 6-3. Locations of the five Arizona-International Telemedicine Network (AITN) referring sites and a diagnostic center in Tucson, Arizona. (From Bhattacharyya, AK, JR Davis, BE Halliday, AR Graham, SA Leavitt, R Martinez, R Rivas, RS Weinstein. 1995. *Telemedicine Journal* 1:9–17. Used with permission.)

the Central Telepathology Laboratory. Often the referring pathologist was more interested in obtaining a second opinion from a peer than receiving a consultation by a *bona fide* subspecialist. This was especially common for referring pathologists in solo practice. (Bhattacharyya et al., 1995)

A "telepresence" service is also offered by AITN telepathologists on an ongoing basis. The referring pathologist requests help to reach a final diagnosis, to develop a differential diagnosis, or to decide what additional studies would lead to a diagnosis. The AITN found that use of this service varied among different referral sites.

TELEPATHOLOGY WORKFLOW

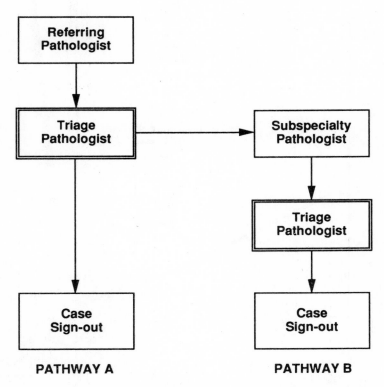

Figure 6-4. Case triage practice model: work flow in Central Telepathology Laboratory (CTL) of the Arizona-International Telemedicine Network. For majority of cases, the triage tele-pathologist renders the diagnosis (Pathway A); in a minority of cases, a subspecialist telepathologist communicates the diagnosis to referring pathologist via a triage pathologist (Pathway B). (From Bhattacharyya, AK, JR Davis, BE Halliday, AR Graham, SA Leavitt, R Martinez, R Rivas, RS Weinstein. 1995. *Telemedicine Journal* 1:9–17. Used with permission.)

Clinical Effectiveness/Efficiency

A few studies have addressed issues related to the effectiveness and efficiency of telepathology. As described above, in Norway, Nordrum and Eide have shown that intraoperative frozen section consultations can be performed for surgeons whose patients are at remote island hospitals. Consultations were considered accurate and performed within accept-able turnaround times for their clinical setting. The alternative was transporting patients by air to the mainland (Nordrum et al., 1991).

Kayser and Drlicek used a static-image telepathology system to provide expert intraoperative frozen section consultations. They reported that substantial assistance in the intraoperative classification of disease was achieved in 37 percent of cases (Kayser and Drlicek, 1992).

In Sweden, beginning in 1992, twenty-nine histopathology and cytology laboratories experimented with static image telepathology. After a two-year period, the technology was rejected because the low-resolution static imaging (512 × 512 pixel) used for the project was judged inadequate by pathologists. Many of the Swedish pathologists wanted higher resolution and dynamic imaging (Busch, 1992; Busch and Olsson, 1995; Olsson and Busch, 1995).

The AITN has addressed the issues of the effectiveness and efficiency of static-image telepathology, relative to conventional pathology practices. It confirmed that its telepathology case mix included many diagnostic dilemmas and complex cases. This emphasizes the difficult nature of many cases sent to telepathology consultants. Some of these problems cannot be readily resolved even by conventional light microscopy. Examination of the AITN cases by an independent, in-house review panel showed that telepathologists examining a few preselected static images on a video monitor tended to underestimate the complexity of some cases. This draws into question the appropriateness of static image telepathology for a consultation practice. Clearly, some types of cases, especially those involving unusual lesions with which the referring pathologist may be unfamiliar, benefit from static-image teleconsultation. However, based on the AITN's experience to date, the benefits of static-image telepathology for consultations on complex cases (as opposed to "unusual" cases representing uncommon lesions) is problematic. It may be more efficient and cost-effective to send tissue blocks and glass slides from complex cases to the consultant by express mail.

Matching the Concept with Reality— Constraints on Developing Telepathology

Provider Acceptance. Image quality is of obvious importance to telepathology service providers. There are also concerns regarding video-image sampling, especially with the static-image system. Several different sampling problems have emerged in static imaging studies. We found that telepathologists using static-imaging systems can be inadvertently misled by referring pathologists into underestimating the complexity of cases. The referring pathologist may select what seem to be representative microscopic fields for transmission and viewing, but the fields may not reflect the full range of morphologies present in the histopathology sections. The telepathologist, relying on the referring pathologist's judgment about microscopic field selection, may assume adequacy of sampling even when it is not justified. Cases have been documented in which the referring pathologist arrived at an incorrect diagnosis, selected images that supported the incorrect diagnosis, and then received "confirmation" of this incorrect diagnosis by the consulting telepathologist based upon review of the nonrepresentative static images (Bhattacharyya et al., 1995). This risk may be encountered whenever the referring pathologist has responsibility for image selection. It might be avoided by instituting explicit, objective sampling rules, such as sampling images at predesignated X- and Y-coordinates on a tissue section. However, such an approach is generally deemed impractical because of the structural heterogeneity of tissue sections.

We also found that static-image telepathologists seldom requested additional images even in complex cases. This was unexpected since the same pathologists frequently ask for deeper cuts of tissue blocks when they work up complex cases by conventional light microscopy. Their altered *modus operandi,* when functioning as static-image telepathologists, may be related to a paucity of visual clues in static images. This can be thought of in terms of the differences between the sampling procedures used by pathologists examining a slide through a light microscope while directly manipulating a glass slide on the specimen stage *versus* the telepathologist downloading images at a static-image computer workstation.

Viewing a histopathology glass slide through a light microscope and controlling slide movements, a well-trained pathologist systematically surveys the entire slide at low magnification. The pathologist makes mental notes of locations of morphologies of potential diagnostic importance. Having completed this survey for all tissue sections on the glass slide, the pathologist then returns to areas of special interest and begins to examine them individually, using a higher-power objective lens. Although this routine may be accomplished in a few minutes for each glass slide, the pathologist can view the equivalent of hundreds of static images in the process (Weinstein et al., 1990). As the tissue section on the glass slide moves beneath the objective lens, driven by finger movements, the pathologist constantly makes fine adjustments in the direction and speed of the slide movements, often based upon visual clues that appear in the pathologist's peripheral vision. The pathologist is continually and actively involved in controlling the sequence of images presented to his/her visual cortex. The images themselves often generate questions addressed simultaneously or sequentially by the pathologist studying the histology section with the light microscope. This often requires moving the glass slide to other microscopic fields. Thus, real-time, image-based feedback directs the microscope operator's neuromuscular activities. This activity can be emulated with a robotic-dynamic telepathology system.

The telepathologist seated in front of the monitor of a static image system foregoes the comprehensive examination of tissue sections. Opinions are rendered on a few preselected images that represent only a small fraction of the information content in the tissue sections for a case. Under these constraining circumstances, the telepathologist not only abdicates responsibility for image selection but also relinquishes much of the intellectual process that normally takes place while scrutinizing tissue sections in real time and rendering diagnoses. This seems to have a profound effect on the behavior of the pathologist using a static image telepathology system.

We have observed what may be referred to as an "icon mentality" when telepathologists view static images. The telepathologists appear to become mesmerized by single images and imagine mentally seeing considerably more detail than represented by the images themselves. In some cases, single images of inferior quality, or those lacking certain diagnostic criteria, have led to inappropriate diagnoses because the telepathologist imagined seeing details that were not present (i.e., "telehallucination"). In other cases, when confronted with a set of static images, telepathologists sometimes tend to overextend themselves in rendering a diagnosis based upon a nonrepresentative set of images assembled by the referring pathologist. This occurs without the telepathologist questioning the appropriateness of the image set. The stream of questions normally generated and answered when the pathologist moves a slide around on the stage of a light microscope is inadequately replicated when viewing static images. Frequently, in cases for which the telepathologist is unwilling to render an opinion based on the available static images, he or she requests that the glass slides be forwarded to the telepathology laboratory rather than ask for additional static images. This is far less likely to be encountered when pathologists use dynamic-robotic telepathology systems.

There are fewer concerns related to service provider acceptance of dynamic imaging telepathology systems. Reservations regarding the user-computer interface have been expressed, especially in relation to manipulation of the robotically-controlled motorized microscope (Carr et al., 1992; Nordrum et al., 1991). Recently, computer scientists have worked on this problem and have made significant advances in improving these interfaces. Image quality has proven to be of less concern to users of dynamic-robotic systems. Static imaging allows viewing tissue sections in two dimensions, whereas, dynamic imaging allows the pathologist to through-focus on tissue sections and gain a sense of the morphology as it exists in three dimensions. Furthermore, increasing the extent of tissue sampling adds to pathologists' confidence in the adequacy of video images, even at lower levels of resolution than are generally used for static imaging. The added information gained by using telepathology in a dynamic mode appears to significantly reduce telepathologists' anxiety regarding video image quality.

As with other telemedicine services, telepathology providers are also concerned about the lack of equipment standards and opportunities for reimbursement as well as various medical-legal issues (Allaert and Dusserre, 1995; Kayser and Schwartzmann, 1992; Schwartzmann et al., 1995).

Client Acceptance. Generally, numbers of cases accrued by telepathology consultation services, as well as other telemedicine services, have been somewhat below expectations. Only one client-satisfaction survey has been published for telepathology services (Olsson and Busch, 1995). Based upon survey results for the Swedish national clinical trial and our own observations within the AITN, we suggest that the somewhat low level of usage of telepathology services experienced by various networks to date may reflect many factors including: absence of standardized practice protocols; users' reservations about the technology; uncertainties about the appropriate criteria for case selection; perceived problems with the quality of teleconsultations; skepticism concerning accuracy; preferences for consultants other than those offered by the network; a reluctance to enter into novel pathologist-consultant arrangements; and concerns over human resource issues raised by telemedicine. At this time, the majority of telepathology-slide-consultation services rendering second opinions are using static-imaging technology, although several networks are installing robotic-dynamic telepathology systems. Justified skepticism about the adequacy of static images partially accounts for the low utilization of static-imaging services.

On the other hand, apparently working in favor of user acceptance is the socialization that can accompany telepathology consultation. For example, in the AITN case triage practice model, one of the three triage telepathologists discusses each case with the referring pathologist and transmits diagnostic opinions back to the site of origin. These triage telepathologists are well acquainted with the clients and have good rapport with them. When subspecialists are used for teleconsultation, their opinions are communicated by an AITN triage pathologist. Hence, the professional interactions tend to be consistent and positive. The AITN found that lack of adequate interpersonal interactions, which occasionally does occur, discourages the use of the technology and can become a source of client dissatisfaction. Nurturing interpersonal interactions between clients and service providers must be a high priority for any telemedicine network.

Telepathology system managers should be aware that there may be a downside to establishing excessively cordial interpersonal relationships between client pathologists and service providers. We have identified what can be referred to as the "error of acquiescence." This is defined as a diagnostic error resulting from a partial loss of objectivity when there is a motivation for the consultant to render a diagnosis concordant with that of the referring pathologist. This error may be obviated by "blinding" the telepathologist to the referring pathologist's diagnostic opinion.

Technological Limitations. Several technical limitations may result in diagnostic errors. The first step in processing surgical pathology specimens involves dissecting tissue specimens and selecting areas for tissue sectioning. Normally, this involves both visualizing and palpating the specimen to identify areas of abnormal appearance and consistency. The intraoperative frozen section service in Norway uses a video camera to allow the telepathologist to examine images of gross specimens (Nordrum et al., 1991). Tissue palpating is done by local technologists and surgeons who relay the information to the telepathologists. In the French national network, images of gross specimens and X-rays are sent with light-microscope images (Martin et al., 1995).

Image quality is a technological limitation. The problem may be minimized by carefully defining the limitations of the imaging technology and establishing practice protocols for its use.

The most significant technical limitation regarding imaging, per se, is the loss of the high optical resolution in low magnification video images (Kayser and Drlicek, 1992). In conventional pathology practice, low-magnification (i.e., low power) images are used for glass slide orientation and identifying potential areas harboring diagnostic fields. The amount of detailed information lost through video imaging is especially apparent at low magnification. The quality of the video images obtained at low magnifications virtually precludes the use of static images for surveying tissue sections. This problem can be circumvented by using a robotically-controlled microscope and real-time imaging. In this way, tissue sections can be sampled expediently at intermediate levels of magnification.

Robotic-dynamic telepathology systems have their own technical limitations which prevent the precise duplication of the full range of activities of the conventional light microscopes. Robotically-controlled microscopes and their motorized stages have discernible time lags in response to external commands. The equipment-time overhead coupled with the time consumed by the transmission of control signals and receipt of response information from the robotic microscope causes the operation of robotic microscopes to seem somewhat sluggish.

Cost. The cost of telepathology systems is related to the costs of video cameras and motorized microscopes. Although low-resolution cameras have become reasonably inexpensive, it seems unlikely that the image quality will completely satisfy all users. High-resolution imaging systems are costly and may be difficult to justify on the basis of small case loads.

Robotic-dynamic telepathology systems are becoming less expensive in part because of competitive pricing of motorized microscopes. A significant operating expense of dynamic systems has been the tariffs of telecommunications companies. These costs will decrease as the result of deregulation and increased competition within the telecommunications industry. The high level of interest of federal and state agencies in telemedicine may have an impact on telecommunications expenses and benefit the field.

Users of consultative services can obtain comparable services by sending glass slides or tissue blocks to a consultant by express mail. This is a well-established, expedient means for getting consultations, offering the referring pathologist a wider choice of consultants than will be available with telepathology in the immediate future. Potential advantages of telepathology consultations are access to panels of experts and shorter turnaround times. For the next few years, full-time staffing of small telepathology consultation services may be difficult to maintain. Once a number of larger telepathology consultation services are in place that use standardized telecommunications protocols and compatible equipment, the situation could be reversed. Cross-coverage by subspecialists employed by independent networks might become commonplace.

Future of Telepathology

Human Resource Issues

Subspecialty pathology has grown rapidly in the past decade. Subspecialists (e.g., renal pathologists, hematopathologists, dermatopathologists, neuropathologists, etc.) are generally concentrated in university medical centers, leaving many communities without subspecialists. Levels of usage of subspecialty services vary enormously. In general, subspecialists and their employers are eager to increase their patient base; this explains some of the current interest in telepathology.

Given the current economic environment, obtaining second opinions from subspecialty pathologists may be discouraged by health maintenance organizations and other carriers. On the other hand, the proliferation of large regional and national pathology reference laboratories and the consolidation of healthcare-provider institutions under large corporate umbrellas are affecting many aspects of the practice of pathology. Such organizations may find it beneficial to incorporate telepathology into their group practices to take full advantage of system-wide expertise and to enhance the level of professional satisfaction among their pathologists.

The use of telepathology is not likely to have a major impact on human resource requirements. Robotic real-time telepathology could negate the need for an on-site pathologist in some small hospitals.

Organizational Issues

The organization of telepathology practices was addressed by the AITN when it developed the case triage model for handling pathology consultation cases. This practice model uses non-specialty pathologists to triage cases. They sign them out personally or, in a minority of cases, refer them to a subspecialist. The organization of the French national telepathology practice is similar to that of the AITN. An alternative practice model has been deployed by the Armed Forces Institute for Pathology (AFIP) in Washington, D.C. With the AFIP practice model, telepathology cases are immediately transferred from a case acquisition center by a clerk to an organ-based subspecialty pathology division. The organization of these subspecialty divisions are themselves hierarchical with trainees, junior and senior staff members in the larger divisions. It is assumed that an individual pathologist with a focused interest and immediate access to senior subspecialty experts will see all cases that fall within his/her organ-based jurisdiction. It remains to be seen if a higher level of diagnostic accuracy can be achieved using the AFIP specialty-based approach. Recent studies by the French network suggest that a combined approach may be suitable for larger practices (Viellefond et al., 1995).

Pathologists in solo practice lack easy access to second opinions and subspecialists. One approach would be for pathologists in solo practice to network with a reference laboratory or university department. Another approach would be to organize "telegroups" consisting of clusters of pathologists in similar solo or small group practices. Pathologists in these groups would be cognizant of the demands and constraints of a small hospital practice. When organized into such telegroups, the pathologists might be in a better position to provide cross-coverage for a broader spectrum of pathology services. Some members of "telegroups" might wish to develop individual subspecialty interests and become the telegroup's "expert," warranted by the increased case load. It is envisioned that subspecialty expertise may no longer be found exclusively in universities and other large practices. Telepathology would provide the means for solo practitioners and small groups to create larger "virtual group practices" including pathology subspecialists.

Other Issues

There is an urgent need to develop standards for telepathology. The College of American Pathologists has established a committee to develop telepathology equipment standards and telecommunications protocols. There is an ongoing European telepathology project within the framework of the European Community Program on Telematics and Healthcare. An aim of their PILOT project (Pathology International and Local Operational Telematics) is to establish intercommunications between European and other networks on a worldwide basis. A set of standard communications protocols is currently being developed for robotic microscopes and telepathology workstations.

Summary and Conclusions

Summary

Telepathology is a relative late comer to the field of telemedicine. Applications of the technology remain somewhat ill-defined, and the best technical solutions are a matter for continued debate. It would be premature to predict which of the available technological approaches (i.e., dynamic versus static imaging) will gain greater acceptance in the future. Test projects to date have produced useful information on the diagnostic accuracy of the technology, specimen viewing times, and system use. However, several networks have ceased operation because of low utilization or user interest, or have remained in operation while accruing relatively small numbers of cases. Case accrual may increase once the accuracy of the technology is well established in the literature, limitations of the technology are better defined, and licensure and reimbursement issues are resolved.

Telepathology and Medical Practice

Telepathology has the potential to alleviate the maldistribution of subspecialty pathology expertise in the United States and elsewhere. Potentially, this would be beneficial to patient care and would help to promote a higher level of service by subspecialty pathologists. Currently, pathology standards are set, in part, at the local or regional level. National standards of care could be applied by ensuring that all patients have access to subspecialty pathology expertise.

Better access to routine pathology services by certain segments of the population could be achieved with telepathology. The Norwegian experience with intraoperative frozen sections services demonstrated that surgical and pathology procedures could be performed without transporting patients. Many ambulatory care facilities and small hospitals lack an on-site pathologist. Immediate diagnoses could be rendered by off-site telepathologists for procedures such as fine-needle aspirate biopsies. This could help optimize specimen sampling, improve diagnostic accuracy for such procedures, and provide an important justification for implementing the technology.

REFERENCES

Allaert, F., and P. Dusserre. 1990. "Images Transmission. The Office of the Pathologist," *Gestion Hosp,* 298: 597–600.

Allaert, F., and P. Dusserre. 1995. "Medical Liability and Telemedicine," *Archives D'Anatomie et de Cytologie Pathologiques,* 43: 200–205.

Becker, R. L., C. S. Specht, R. Jones, M. E. Rueda-Pedraza, and T. J. O'Leary. 1993. "Use of Remote Video Microscopy (Telepathology) as an Adjunct to Neurosurgical Frozen Section Consultation," *Human Pathology,* 24: 909–11.

Bhattacharyya, A. K., J. R. Davis, B. E. Halliday, A. R. Graham, S. A. Leavitt, R. Martinez, R. Rivas, and R. S. Weinstein. 1995. "Case Triage Mode for the Practice of Telepathology," *Telemedicine Journal,* 1: 9–17.

Black-Schaffer, S., and T. J. Flotte. 1995. "Current Issues in Telepathology," *Telemedicine Journal,* 1: 95–106.

Bloom, K. J., L. S. Rozek, and R. S. Weinstein. 1987. "ROC Curve Analysis of Super High Resolution Video for Histopathology," *SPIE Proc Visual Image Process,* 845: 408–12.

Busch, C. 1992. "Telepathology in Sweden: A National Study Including All Histopathology and Cytology Laboratories," *Zentralblatt Pathologie,* 138: 429–30.

Busch, C., and S. Olsson. 1995. "Future Strategy for Telepathology in Sweden: Higher Resolution, Real Time Transmission in a Multipurpose Work Station for Diagnostic Pathology," *Archives D'Anatomie et de Cytologie Pathologiques,* 43: 242–45.

Carr, D., H. Hasegawa, D. Lemmon, Cl Plaisant. 1992. "The Effects of Time Delays on a Telepathology User Interface," *Proc Annual Symposium Computer Applications in Medical Care,* 256–60.

Cronenberger, J. H., H. Hsiao, R. J. Falk, and J. C. Jennette. 1992. "Nephrology Consultation via Digitized Images," *Annals of NY Academy of Science,* 670: 281–92.

Eide, T. J., and I. Nordrum. 1992. "Frozen Section Service via the Telenetwork in Northern Norway," *Zentralblatt Pathologie,* 138: 409–12.

Eide, T. J., and I. Nordrum. 1994. "Current Status of Telepathology," APMIS, 102: 881–90.

Eide, T. J., I. Nordrum, and H. Stalsberg. 1992. "The Validity of Frozen Section Diagnosis Based on Video-Microscopy," *Zentralblatt Pathologie,* 138: 405–07.

Erickson, D. 1990. "Do You See What I See? Pathologists Lead the Way for Long Distance Diagnosis," *Scientific American,* (July): 88–89.

Ferrer-Roca, O. 1993. "New Technologies: The Introduction of Videophones in Pathology." *Arq Patol,* 25: 51–58.

Goncalves, L. 1993. "Telepathology in Portugal," *Arq Patol,* 25: 7–9.

Ito, H., H. Adachi, K. Taniyama, Y. Fukuda, and K. Dohi. 1994. "Telepathology Is Available for Transplantation-Pathology: Experience in Japan Using an Integrated, Low-Cost, and High-Quality System," *Modern Pathology,* 17: 801–05.

Kayser, K. 1992. "Telepathology: Visual Telecommunication in Pathology: An Introduction," *Zentralblatt Pathologie,* 138: 381–82.

Kayser, K. 1993. "Progress in Telepathology," *In Vivo,* 7: 331–33.

Kayser, K., M. Oberholzer, and G. Weisse. 1991. "Telekommunikation in der Pathologie-erste Ergebnisse der Expertenkonsultation in der Biopsie-Diagnostik," *Verh Dtsch Ges Pathol,* 75: 441.

Kayser, K., M. Oberholzer, G. Weisse, I. Weisse, and H. V. Eberstein. 1991. "Long Distance Image Transfer: First Results of its Use in Histopathological Diagnosis," *Acta Pathol Microbiol Immunol Scand,* 99: 808–14.

Kayser, K., and P. Schwarzmann. 1992. "Aspects of Standardization in Telepathology," *Zentralblatt Pathologie,* 138: 389–92.

Kayser, K., and M. Drlicek. 1992. "Visual Telecommunication for Expert Consultation of Intraoperative Sections," *Zentralblatt Pathologie,* 138: 395–98.

Kayser, K., M. Drlicek, and W. Rahn. 1993. "Aids of Telepathology in Intraoperative Histomorphological Tumor Diagnosis and Classification," *In Vivo,* 7: 395–98.

Keil-Slawik, R., C. Plaisant, and B. Shneiderman. 1991. *Remote Direct Manipulation: A Case Study of a Telemedicine Workstation.* College Park, MD: University of Maryland Center for Automation Research, CAR–TR-551, pp. 1–6.

Krupinski, E. A., R. S. Weinstein, K. J. Bloom, and L. S. Rozek. 1993. "Progress in Telepathology: System Implementation and Testing," *Adv Pathol Lab Med,* 6: 63–87.

Krupinski, E. A., R. S. Weinstein, and L. S. Rozek. Forthcoming. "Experience-Related Differences in Diagnosing Video-Displayed Medical Images," *Telemedicine Journal.*

Linder, J., and C. T. Masada. 1989. "Frozen Section Diagnosis by Videotelephone," (abstract), *Lab Invest,* 60: 54A.

Martin, E., P. Dusserre, A. Fages, P. Hauri, A. Viellefond, and H. Bastien. 1992. "Telepathology: A New Tool for Pathology? Presentation of a French National Network," *Zentralblatt Pathologie,* 138: 419–23.

Martin, E., P. Dusserre, Cl Got, A. Vieillefond, B. Franc, G. Brugal, and B. Retailliau. 1995. "Telepathology in France: Justifications and Developments," *Archives D'Anatomie et de Cytologie Pathologiques,* 43: 191–95.

Miaoulis, G., E. Protopapa, and G. Delidis. 1990. "Data Administration Concerning Histological Biopsies and Surgical Specimens in a Pathology Laboratory," *Greek Archives of Anatomic Pathology,* 40: 87–94.

Miaoulis, G., E. Protopapa, C. Skourlas, G. Delidis. 1992. "Telepathology in Greece: Experience of the Metaxas Cancer Institute," *Zentralblatt Pathologie,* 138: 425–28.

Murphy, R. L. H., and K. T. Bird. 1974. "Telediagnosis: A New Community Health Resource (Observations on the Feasibility of Telediagnosis Based on 1000 Patient Transactions)," *American Journal of Public Health* 64: 113–19.

Nordrum, I., B. Engum, E. Rinde, A. Finseth, H. Ericsson, M. Kearney, H. Stalsberg, and T. J. Eide. 1991. "Remote Frozen Section Service: A Telepathology Project to Northern Norway," *Human Pathology,* 22: 514–18.

Oberholzer, M., H–R Fischer, H. Christen, S. Gerber, M. Bruhlmann, M. Mihatsch, M. Famos, C. Winkler, P. Fehr, L. Bachthold, and K. Kayser. 1993. "Telepathology with an Integrated Services Digital Network—A New Tool for Image Transfer in Surgical Pathology. A Preliminary Report," *Human Pathology* 24: 1078–85.

Olsson, S., and C. Busch. 1995. "A National Telepathology Trial in Sweden: Feasibility and Assessment," *Archives D'Anatomie et de Cytologie Pathologiques,* 43: 234–41.

Park, B. 1975. "Communication Aspects of Telemedicine," in: Bashshur, R. L., Armstrong, P. A., and Youssef, Z. I. (eds.), *Telemedicine: Explorations in the Use of Telecommunications in Health Care.* Springfield, IL: Charles C Thomas, pp. 54–86.

Phillips, K. L., L. Anderson, T. Gahm, L. B. Needham, M. L. Goldman, B. E. Wray, and T. F. Macri. 1995. "Quantitative DNA Analysis: A Comparison of Conventional DNA Ploidy Analysis and Teleploidy," *Archives D'Anatomie Cytologie Pathologiques,* 43: 288–95.

Riggs, R. S., D. T. Purtillo, D. H. Connor, and J. Kaiser. 1974. "Medical Consultation via Communications," *Journal of the American Medical Association,* 228: 600–02.

Schwartzmann, P. 1992. "Telemicroscopy. Design Considerations for a Key Tool in Telepathology," *Zentralblatt Pathologie,* 138: 183–87.

Schwartzmann, P., J. Schmid, C. Schnörr, G. Sträble, and S. Witte. 1995. "Telemicroscopy Stations for Telepathology Based on Broadband and ISDN Connections," *Archives D'Anatomie Cytologie Pathologiques,* 43: 209–15.

Shimosato, Y., Y. Yagi, K. Yamagishi, K. Mukai, S. Hirohashi, T. Matsumoto, and T. Kodama. 1992. "Experience and Present Status of Telepathology in the National Cancer Center Hospital, Tokyo," *Zentralblatt Pathologie,* 138: 413–17.

Vieillefond, A., F. Staroz, M. Fabre, P. Bedossa, V. Martin-Pop, E. Martin, C. Got, and B. Franc. 1995. "Fíabilté du Diagnostic Anatomo-Pathologique par Transmission d'Images Statiques," *Archives d'Anatomie Cytologie Pathologiques,* 43: 246–50.

Weinberg, D. S., F. A. Allaert, P. Dusserre, F. Drouot, B. Retailliau, W. R. Welch, J. Longtine, G. Brodsky, R. Folkerth, M. Doolittle. 1996. "Telepathology Diagnosis by Means of Digital Still Images: An International Validation Study," *Human Pathology,* 27: 111–18.

Weinstein, R. S. 1986. "Prospects for Telepathology," *Human Pathology,* 17: 433–34, (editorial).

Weinstein, R. S. 1991. "Telepathology Comes of Age in Norway," *Human Pathology,* 22: 511–13, (editorial).

Weinstein, R. S. 1992. "Telepathology: Practicing Pathology in Two Places at Once," *Clinical Laboratory Management Review,* 6: 182–84.

Weinstein, R. S. 1996. "Static Image Telepathology in Perspective," *Human Pathology,* 27: 99–101, (editorial).

Weinstein, R. S., A. Bhattacharyya, B. E. Halliday, Y–P Yu, J. R. Davis, J. M. Byers, A. R. Graham, and R. Martinez. 1995. "Pathology Consultation Services via the Arizona-International Telemedicine Network," *Archives d'Anatomie Cytologie Pathologiques,* 43: 219–26.

Weinstein, R. S., K. J. Bloom, and L. S. Rozek. 1987a. "Telepathology and the Networking of Pathology Diagnostic Services," *Arch Pathol Lab Med,* 111: 646–52.

Weinstein, R. S., K. J. Bloom, and L. S. Rozek. 1987b. "Telepathology: System Design and Specifications," *SPIE Proc Visual Commun Image Process,* 845: 404–07.

Weinstein, R. S., K. J. Bloom, and L. S. Rozek. 1989. "Telepathology: Long Distance Diagnosis," *Am J Clin Pathol,* 91: 39–42.

Weinstein, R. S., K. J. Bloom, and L. S. Rozek. 1990. "Static and Dynamic Imaging in

Pathology," in: Mun, S. K., Greberman, M., Hendee, W. R., and Shannon, R. (eds.), *Image Management and Communications in Patient Care: Implementation and Impact.* Los Alamitos, CA: IEEE Computer Soc. Press, pp. 77–85.

Weinstein, R. S., K. J. Bloom, E. A. Krupinski, and L. S. Rozek. 1992. "Human Performance Studies of the Video Microscopy Component of a Dynamic Tele-pathology System," *Zentralblatt Pathologie,* 138: 399–401.

Wold, L. E., and L. H. Weiland. 1992. "Telepathology at the Mayo," *Clin Lab Manage Rev,* 1: 174–75.

Yu, Y–P, R. Martinez, E. Krupinski, and R. S. Weinstein. 1994. "Analysis of JPEG Compression on Communications in a Telepathology System," *SPIE,* 2165: 283–94.

Chapter 7

TELEMEDICINE AND PRIMARY CARE

James E. Brick

Introduction

This chapter explores the potential contributions which telemedicine can make to the delivery of primary medical care, assesses the status of telemedicine in primary care, and inquires into the constraints currently restricting the growth of telemedicine in the provision of primary care. Certainly, each chapter in this section about the clinical applications of telemedicine (e.g., oncology, psychiatry, and dermatology) addresses to some extent the role of telemedicine as a support system for the remote or isolated primary practitioner dealing with complex or special cases presented in the isolated environment. But in this chapter, issues of telemedicine support for the primary care provider are addressed specifically. Additionally, this chapter focuses on the potential role of telemedicine in enhancing provision of primary care medical services by mid-level health professionals. These discussions are presented against a background which outlines the problems identified and related to the delivery of primary care.

Primary Care

One problem inherent in any discussion of primary care is its definition. A standard medical dictionary defines primary care as "a basic level of health care that includes programs directed at the promotion of health, early diagnosis of disease or disability, and prevention of disease. Primary health care is provided in an ambulatory facility to limited numbers of people, often those living in a particular geographic area. In any episode of illness, it is the first patient contact with the health care system." An earlier dictionary emphasizes primary care as "basic or general health care which emphasizes the point when the patient first seeks assistance from the medical care system and the care of the simpler

211

and more common illnesses." However, when assessing the availability and accessibility of primary medical care, debate, if not outright confusion, centers on just which physicians to include as primary care givers. The general or family practitioner is generally acknowledged as the major purveyor of primary health care. However, some analysts have counted pediatricians and internists as primary care doctors, although many engage in subspecialties—especially internists who focus on cardiology, endocrinology, nephrology, rheumatology, and so on. Some consider obstetricians-gynecologists as providers of primary care, although they obviously deal principally with one organ system in women. In recent assessments, primary physicians include those in general/family practice, internal medicine, obstetrics/gynecology, or pediatrics (Frenzen 1991). The debate and confusion over just whom to include as "primary care givers" derive in part from the issues concerning the most suitable arrangements for primary health care in America, the problem of medical fragmentation/specialization, and who shall be paid and how.

Questions pertaining to the distribution of physicians are complicated by difficulties surrounding the principal measure of physician distribution, namely, physician to population ratios. We are fully aware, of course, that large numbers of patients cross boundaries of variously (usually politically) defined medical areas, thereby negating the conceptual basis for the determination of physician-to-population ratios. Also, to date, no agreement exists about just what constitutes an "appropriate" or optimum number of primary or specialist physicians for any group of people in any given area. In any event, historically, at least since the early part of the twentieth century, using inherently problematical physician to population ratios and some presumed yet unsubstantiated norm, it is the so-called "mal-distribution" of physicians that has attracted attention. A discussion of this problem and its genesis is beyond the scope of this presentation. Suffice it here to say that, throughout most of the twentieth century, both primary and specialist physicians have been more concentrated in urban areas than rural.

In response to this perceived imbalance, the federal government has attempted to redistribute physicians through programs such as the Hospital Survey and Construction Act of 1946 and the National Health Service Corps (NHSC) established by the Emergency Health Personnel Act of 1970. The Area Health Education Centers were initiated in 1972 to address specialty maldistribution, particularly in rural areas. Another strategy to increase the number of physicians in "underserved" areas relied on increasing the production of physicians and relying on market forces to redistribute physicians to smaller towns in accordance with economic location theory. While the relative merits and success of each approach continue to be debated, new federal initiatives such as Medicare reimbursement bonuses for services performed by physicians in designated medically underserved areas and increased funding for the NHSC indicate that the problem remains (Frenzen, 1991). Of course, telemedicine itself must also be included in this package of federal and private initiatives to redress the geographical imbalance of physicians. Obviously, some areas remain unattractive to physicians despite the growth in their numbers and the incentives provided to them by federal programs to locate in designated underserved areas.

Primary Care Physicians

During the 1960s, it was recommended that additional physicians be trained, in part because of a projected decline in the number of general practitioners and a concurrent increase in the number of physicians specializing. As a result, family practice was recognized as a medical specialty; a number of new state-supported medical schools that emphasized primary care training were established; funding increased for primary care training, including physicians, nurses, nurse-practitioners, certified nurse-midwives, and physician assistants; and federal funding was made available for community health centers (Politzer et al., 1991). The total number of primary care physicians grew at an average rate of 3.8 percent per year between 1975 and 1988. During the same period, the overall ratio of primary care physicians per 100,000 persons increased from 61.4 to 85.5 (Frenzen, 1991). However, the ratio was substantially higher at the more urban end of the urban-rural continuum. Nevertheless, by the 1970s nearly every town with 2500 or more residents had at least one physician. And it is generally agreed that the supply of physicians of all types is increasing in rural areas. Yet one estimate places the number of shortage areas at 1900 (HHS, 1990). And, using a threshold ratio of 1 physician per 3500 people, as of 1988 67 percent of the designated primary Health Professional Shortage Areas were in rural areas (U.S. Congress, 1990).

It is apparent that, although the supply of primary care personnel has been expanding, problems of geographic distribution remain. To be sure, earlier a comprehensive study predicted a substantial surplus of primary physicians by the year 1990 (GMENAC 1980). Despite these estimates, states continued to report shortages of primary care physicians and, in fact, subsequent studies again raised the specter of primary personnel shortages (Gamliel et al., 1992). Under the Health Professions Reauthorization Act of 1988, Congress instructed the Secretary of Health and Human Services to survey each state, district, commonwealth, and territory for an assessment of health personnel shortages by discipline. The primary personnel problems indicated by fifty-five responding jurisdictions were: (1) a shortage of primary care practitioners; (2) a shortage of registered nurses; and (3) problems associated with health care provision and inadequate health personnel resources in rural areas (HHS, 1990). Of the respondents, 90 percent said that primary care physicians were in short supply. Further, lack of physicians in rural areas was seen as a serious problem by 82 percent of the jurisdictions responding. A lack of obstetricians/gynecologists was also noted as a particular problem in rural areas. In an effort to provide prenatal care to areas lacking obstetrics/gynecology practitioners, some states have begun to recruit nurse midwives; however, as might be expected, about one-third of all jurisdictions indicated difficulties in locating midwives as well. Also, 90 percent of the jurisdictions responding reported concerns about the availability of registered nurses; 44 percent reported shortages of nurse practitioners; and 40 percent reported shortages of physician assistants, especially in rural and inner city areas.

According to this survey, nearly 75 percent of the jurisdictions' greatest concern had to do with—and the major problem with the delivery of health is in—rural areas. Included among the reasons most frequently reported were an inability of the population to afford care, a lack of a population base sufficient to support the latest technologies and physician specialties, problems recruiting and retaining health care personnel, and financial pressures brought about by lower Medicare/Medicaid reimbursement rates for rural hospitals. Concerns about access to health care services were raised by 80 percent of the respondents, specifically, about the access of rural and inner-city populations, disadvantaged minority groups, the elderly, the medically indigent, and migrant workers. Among the major barriers to providing services to these populations was an inadequate supply of primary care practitioners. One of the major conclusions of the study, once again, was that **states need to implement programs to encourage health care professionals to specialize in primary care fields and locate in areas of greatest need.**

This historic challenge has not gone unnoticed. For example, one medical school has reported success in a selective admissions policy and a special education program, resulting in an improved rate of location and retention of graduating physicians in rural areas and areas with a shortage of physicians (Rabinowitz, 1993). However, this program, developed like many others to address the problem, suffers from a decline in primary care applicants (Colwill, 1992). Statistics indicate a decreased interest in primary care specifically but also in general obstetrics and gynecology, (non-subspecialty) internal medicine, and pediatrics—the core group of primary care providers. The Council on Medical Education has recommended moving to a delivery system with 50 percent of the physicians practicing in "generalist disciplines" (Kindig et al., 1993). Even the most liberal projections have resulted in discouraging time frames, however. For example, if 100 percent of all post-1993 graduates chose a generalist career, a 50 percent supply would be achieved only in the year 2004. A more realistic, but still optimistic, 20 percent generalist graduation rate would mean that by the year 2020 the percentage of general practitioners in the physician supply would be 25 percent and fall to 20 percent by the year 2040 (Kindig et al., 1993). Is telemedicine the missing piece in the primary care shortage/maldistribution problem? To assess the value of telemedicine in addressing these problems, it is necessary first to define the problem more precisely.

Primary Care in Rural Areas

In contrast to urban areas, rural areas have lower population densities, which translates into smaller populations within what may be considered "rational" service areas for the delivery of primary medical services. Therefore, geographically, physicians in rural areas have smaller population bases for their practices. Additionally, per capita income in rural areas has traditionally been lower than in urban areas (Gesler and Ricketts, 1992). Accompanying this is a significantly higher rate of poverty in rural than in urban areas and generally lower percentages of the population with health insurance (Witherspoon et al., 1993). A number of studies also indicate that, when compared to people living in urban areas, rural residents delay seeking health care for similar symptoms. Experience indicates that patients will purchase more medical care at a higher price when they believe in the efficacy of the care, have higher incomes and insurance, and when their physicians suggest that the care is necessary. Thus, characteristics of rural areas and populations indicate that, when compared to urban areas, the demand for health care will be lower.

Rural areas have difficulty in attracting and retaining physicians because of this reduced demand for medical care and consequent lower incomes. The dispersed population, lower income and insurance levels, and lower reimbursement rates among rural populations contribute to making rural areas less attractive to physicians. Other factors include diminished professional status and prestige, professional isolation, general unavailability of continuing medical education, lower levels of continuity of care for patients, and the urban origin of the large majority of physicians (Gorton, 1992). It is into this landscape that telemedicine is being introduced in hopes of redressing the problems induced by these characteristics.

Telemedicine and Primary Care

Proponents of telemedicine in both the private and public sectors advocate telemedicine as a solution to the problem of rural health care delivery. It has the potential to contribute to the resolution of problems through altering the rural landscape, thereby making it more attractive to physicians, and it may simultaneously reduce the need for primary physicians in rural areas.

One way in which telemedicine addresses the primary care shortages in rural areas is by dealing directly with physician concerns about practicing there (McCarthy, 1995). Telemedicine has the potential, given appropriate levels of dispersion, to reduce professional isolation, provide direct and indirect continuing medical education, improve contact with research facilities and up-to-date medical research, increase the continuity of patient care, and increase income of physicians practicing in rural areas.

Depending upon the level of technology employed, telemedicine can reduce professional isolation of the rural primary practitioner in several ways. For example, two-way interactive video consultation with specialists links the isolated practitioner with the specialist community of a large medical center. This virtual support system and contact with professional colleagues should enhance the integration of the rural or otherwise isolated practitioner. However, it must be noted that these contacts are only temporary, will occur only sporadically, and depend on the level of telemedicine technology employed. Therefore, the extent of the integrative possibility of telemedicine remains to be determined. The technology also has the potential to link the primary practitioner with on-line services such as Medline, which provide the opportunity to review the latest medical literature, thereby strengthening links to the professional medical community and improving the quality of care for the rural patient.

Continuing medical education (CME) programs and credits can be obtained via telemedicine technology. Taped programs as well as live-interactive lectures and demonstrations can provide the isolated primary practitioner with ready access to CME. Thus, telemedicine can fill an educational need for rural physicians who cannot readily attend programs offered at urban-based medical centers. While access to this type of education is valuable, perhaps more valuable is the education received by the primary practitioner in directly consulting with a specialist concerning a patient. Anecdotal evidence indicates rural practitioners feel the level and effectiveness of education achieved through telemedicine consultation far surpasses that of the "traditional" continuing medical education programs. In two experiments, the Medical College of Georgia and West Virginia University have established programs that allow rural area physicians to obtain Category 1 CME credit for consultations using telemedicine. One must caution, however, that the extent to which telemedicine can reduce the isolation and increase the education and training of the rural physicians depends upon the dispersion and accessibility of telemedicine technology. The telemedicine systems and networks in place to date by no means offer universal accessibility to the technology for even a small minority of rural practitioners.

While "hard" evidence remains to be obtained, reports indicate that telemedicine substantially increases the percentage of rural patients who can be treated in local community hospitals under the supervision of local physicians. If true, this has several positive implications. Concerns of rural providers regarding continuity of patient care may be significantly reduced. Concomitantly, the income of rural providers as well as of rural community hospitals will be increased. This would add substantially to the attractions of rural practice and the stability of the rural medical care landscape.

One of the earliest experiments in the so-called "first generation" of telemedicine involved the use of two-way interactive video to provide more than consultative support for the delivery of primary medical care by a nurse practitioner. Dr. Kenneth Bird established the link between a "medical station" located at Boston's Logan International Airport and the Massachusetts General Hospital in the late 1960s (Bird, 1975). It was established to serve the airport staff (approximately 6,000) and passengers since there were no medical facilities in the immediate vicinity. The station was staffed by nurse clinicians twenty-four hours a day and physicians attended at peak passenger flow periods each morning and afternoon. One of the primary interests was to determine how well the nurse clinician could be utilized in providing primary medical care under the supervision of a remotely located primary care physician. Microwave transmissions and interpretation of electrocardiograms and roentgenograms were determined to be as accurate as direct visualization. Stethoscopy and microscopy were also accurately performed and interpreted via the system. And satisfactory "telewriting" for record keeping and documentation was used regularly.

Some resistance on the part of patients was noted by one nurse practitioner (Shannon, 1996). She reported that when some patients were informed that they would be seen by a physician over a video network, they "walked out." However, she quickly developed a new strategy which involved ushering the patient into the examination room and telling them "the doctor will be with you shortly." The video system would be turned on and, once in the virtual "presence" of the physician, few patients left.

The important point here is that the value of telemedicine's supportive yet instrumental role in disseminating primary care via mid-level health professionals was recognized from the outset. To date, in the current generation of telemedicine, this potentially important role has not been investigated sufficiently. The place of telemedicine in extending and enhancing the role of the mid-level health professional in delivering care may be viewed as analogous to its role in extending and enhancing the role of the primary care physician. Under supervision of a specialist, the primary care physician can undertake diagnoses and therapies previously precluded by lack of training and/or experience. Similarly, the role of physician "extenders" such as nurse practitioners, physicians' assistants, and mid-wives in the delivery of primary care, under the direct supervision of physicians, can be enhanced and extended. This capacity of telemedicine may be particularly valuable in those more remote areas of the country with chronic shortages of physicians, and, increasingly, in managed care systems.

The nurse practitioner/clinician and physician assistant can be indispensable in the delivery of medical care at any primary ambulatory care site. Currently, they elicit patient histories which assist in defining the patient's problem, and carry out an appropriate physical examination. These tasks are done independently of the physician. In telemedicine, there would be no difference in the use of mid-level health professionals in this capacity. However, with telemedicine linkage, nurse clinicians or physicians' assistants can extend their role. For example, patient history and findings from physical examinations can be discussed with the physician with or without the patient present. If the remote physician feels it necessary, the patient can be examined with the assistance of the nurse or assistant. In these cases the allied health personnel can provide, among other things, visual reinforcement to the physician, describe the auscultatory findings, place a stethoscope, otoscope, or endoscope where the physician requests, relay information regarding palpation, and assist in placing the patient in the proper position for examination. Of course, the use of these personnel in various physician-extender roles must be carefully monitored by the physician. However, telemedicine can permit mid-level health professionals to participate effectively in health care delivery at primary sites. Not only can telemedicine extend the delivery of basic medical care geographically through the use of physician extenders, each patient-extender-physician encounter can also provide additional education to the mid-level professional, thereby increasing the level of expertise and improving the quality of care delivered to the patient.

Telemedicine has the potential to reduce the need for primary care physicians by increasing the pool of possible providers. Telemedicine gives mid-level practitioners, such as nurse practitioners and physician assistants, the ability to act as the primary provider for rural communities because they are supervised and supported by primary care physicians over telemedicine systems. Mid-level practitioners frequently practice in rural (as well as inner-city) areas, and telemedicine would improve their ability to supply more care while assuring quality through supervision. Acknowledging dangers inherent in giving less trained personnel greater responsibility for more seriously ill patients, the access that this level of care would provide could still be a dramatic improvement over the chronic lack of care in many areas. Of course, this same observation applies to the use of telemedicine to extend the delivery of tertiary care to patients through primary care physicians. However, unless telemedicine is proven to be satisfactory for all examinations, it cannot replace the need for physicians in shortage areas by itself.

Thus, telemedicine has the **potential** to address the problem of primary health care delivery in rural areas from several directions. From the supply side, it has the potential to provide and develop a more attractive practice setting for primary physicians. It can accomplish this by reducing professional isolation, increasing access to medical literature, providing for more accessible, relevant, and effective continuing medical education, increasing the level of continuity of care, contributing to the stability and economic viability of rural hospitals, and increasing the income of rural primary providers. Simultaneously, telemedicine could reduce the need for primary care physicians by extending and enhancing the role of nurse practitioners and physician assistants in the direct (supervised) delivery of basic health care in areas faced with chronic shortages of primary care physicians.

Constraints on Telemedicine in Primary Care

There is little doubt that telemedicine has the potential to improve rural primary medical care. However, before telemedicine can be regarded as the answer or as even a partial solution to the problem of chronic shortages in rural areas, a number of questions must be investigated **and** answered. Some have to do with the technology, including such things as availability and quality, while others concern human factors, including acceptance and, ultimately, utilization. And, of course, utilization relates directly to economic viability.

Telecommunications technology has made substantial progress since the inception of the first generation of telemedicine some three decades ago. Indeed, it is this phenomenal progress which forms a large part of the basis for development of the current generation of telemedicine. Nevertheless, proponents of telemedicine must acknowledge, given the current state of the art, that, **ceteris paribus,** in some instances a telemedicine consult is not as effective as that conducted in person. In clinical applications discussed in other chapters, issues such as visual and audio acuity in transmission, and the consulting physicians' lack of tactile and olfactory examination capabilities, are mentioned as limits on clinical effectiveness of the extant technology.

Also related to implementation of telemedicine is cost. Certainly the cost of the equipment has decreased and continues to decrease as more effective means of production are achieved. Nevertheless, costs of installation, maintenance, and transmission remain substantial and often prohibitive for small rural hospitals, not to mention isolated clinics, desirous of linking up with existing telemedicine networks. The most comprehensive units, which are the most desirable, still cost at least $100,000 and recurring monthly transmission costs must be considered as well. Some areas are not yet equipped with the type of telephone lines necessary for the transmission of teleconsult information.

Given this situation, remote telemedicine consult sites are limited in distribution and consequently, accessibility. Generally, only one of several rural hospitals in any given area is selected as a telemedicine network site. To date, no telemedicine systems have reduced costs or been able to place the comprehensive consultation station in the primary physician's office. This is especially true of the more sophisticated telemedicine technology used in several states. It is the primary physician and patient who now must travel to the teleconsult site. Thus, the overall burden of travel is reduced for the patient but increased for the physicians who have less time to spend in this effort. For many of the more remote physicians (and here we are speaking of the office location relative to the teleconsult site) travel distances and scheduling problems greatly diminish or preclude altogether utilization of the network's opportunities.

The issue of reimbursement must also be addressed, since it is related to costs of telemedicine versus direct consultation as well as physician income. Certainly, telemedicine has the potential to change the income of rural practitioners. However, to date, the issue of reimbursement remains unresolved. Coverage under Medicaid is currently not available for many types of clinical telemedicine services. The Georgia Telemedicine Project was the first to receive Medicare and Medicaid reimbursement, but only on an experimental basis. Many other states now provide Medicaid reimbursement, and private insurance and managed care programs pay in some areas as well. Another important consideration is the structure of the consultation itself. As long as the primary physician is required to accompany the patient during a telemedicine consult with a specialist, the cost of the medical care episode to the patient or third-party payer will be greater than the traditional form of consultation. The Health Care Financing Administration (HCFA) is concerned about multiple billings for a single telemedicine consultation as providers at both ends of a video link bill for care. There is an obvious need for guidelines regarding the use and billing for telemedicine services. Unless these are forthcoming, the economic barrier remains mired in the complex questions of financial reimbursement and restrictions on the use of allied health professionals. Currently, HCFA allows reimbursement for mid-level practitioners in rural areas. This policy should be extended to the use of telemedicine services by mid-level practitioners. In this way, small, sparsely populated communities unable to support even small hospitals could still have the benefit provided by supervision of health care delivery by access to primary care physicians through telemedicine (McCarthy, 1995). Again, however, physicians' accessibility to the telemedicine consult site must be considered. Obviously, major concerns remain regarding telemedicine's ability to reduce the net cost of health care.

Despite the advances in technology, the human element is the key to the success of telemedicine's future in primary care delivery. From the outset, these systems/networks must be designed and located with the needs and the abilities of the users—primary physicians, nurse practitioners, physician assistants, and patients—in mind. Technology should never be allowed to drive the process. It is the users on both ends of the teleconsultation who will determine if telemedicine will succeed or fail. In this arena, the question of the ethics of telemedicine must be addressed.

Again, the telecommunications technology must be available and reasonably priced. Unfortunately, the areas most in need of this type of technology are not yet incorporated into the National Information Infrastructure. And, furthermore, the individuals most in need of the technology do not have sufficient funds to purchase the equipment or pay transmission costs. It is likely that the clinical enterprise alone will probably never generate adequate funds to allow the telemedicine systems to be unifunctional. This means that other dimensions of primary health care must be involved including the delivery of health education to the public and practitioners.

Conclusion

Is telemedicine the answer to problems facing the delivery of primary care in rural and other isolated communities? In theory, it holds considerable promise for addressing problems associated with isolated areas, and thus has the potential to increase the number of primary physicians located there. In addition, telemedicine can be used to extend the delivery of primary care to isolated areas via virtually supervised mid-level health care personnel. However, unless and until telemedicine is sufficiently developed for all examinations, it **cannot replace** the need for physicians in shortage areas in and of itself.

Telemedicine, in its current level of development, should be viewed as part of a diversified approach to the problems of providing primary medical care in rural areas. Should telemedicine continue to diffuse, certain legislative and regulatory safeguards could increase efficiency, cut costs, and help assure that medical care delivered via telemedicine networks meets or exceeds the standards of "in-person" medical care. Telemedicine is not a total solution to shortages of primary care in rural areas, but it has demonstrated the potential to address factors that affect physicians' decisions regarding practice in rural areas as well as to extend, geographically, primary care through mid-level health personnel. Once these potential contributions to the resolution of the problem are adequately demonstrated **and** we are satisfied that delivering primary medical care via telemedicine is equal to or better than that delivered in person, telemedicine will surely become an essential and constant factor in the delivery of primary care in the rural United States.

REFERENCES

Bird, K. 1975. "Telemedicine: Concept and Practice," in R. Bashshur, P. Armstrong, and Z. Youssef, eds., *Telemedicine: Explorations in the Use of Telecommunications in Health Care.* Springfield, IL: Charles C Thomas Publisher.

Frenzen, P. 1991. "The Increasing Supply of Physicians in US Urban and Rural Areas, 1975 to 1988," *American Journal of Public Health,* 81: 1141–47.

Colwill, J. 1992. "Where Have All the Primary Care Applicants Gone?" *The New England Journal of Medicine,* 326: 387–93.

Gamliel, S., F. Mullan, R. Politzer, and H. Stambler. 1992. "Availability of Primary Care Health Personnel," *Archives of Internal Medicine,* 152: 268–73.

Gesler, W. and T. Ricketts, eds. 1992. *Health in Rural North America: The Geography of Health Care Services and Delivery.* New Brunswick: Rutgers University Press.

Gordon, R. 1992. "Accounting for Shortages of Rural Physicians: Push and Pull Factors," in W. Gesler and T. Ricketts, eds., *Health in Rural North America: The Geography of Health Care Services and Delivery.* New Brunswick: Rutgers University Press.

Kindig, D. J. Cultice, and F. Mullan. 1993. "The Elusive Generalist Physician," *Journal of the American Medical Association,* 270: 1069–73.

McCarthy, D. 1995. "The Virtual Health Economy: Telemedicine and the Supply of

Primary Care Physicians in Rural America," *American Journal of Law & Medicine,* 21: 111–30.

Politzer, R., D. Harris, M. Gaston, and F. Mullan. 1991. "Primary Care Physician Supply and the Medically Underserved," *Journal of the American Medical Association,* 266: 104–09.

Rabinowitz, H. 1993. "Recruitment, Retention, and Follow-up of Graduates of a Program to Increase the Number of Family Physicians in Rural and Underserved Areas," *The New England Journal of Medicine,* 328: 934–39.

Shannon, G. 1996. Personal communication about conversation with Marie Kerrigan, nurse practitioner at Logan International Airport Medical Station, 1970.

U.S. Department of Health and Human Services. 1990. *States' Assessment of Health Personnel Shortages Issues and Concerns.* Washington, DC: HHS. Publication HRS- -P–OD 90-6.

U.S. Congress. 1990. *Health Care in Rural America Summary 14.* OTA–H-4343 Washington, D.C.: Office of Technology Assistance.

Witherspoon, J., S. Johnstone, and C. Wasem. 1993. *Rural TeleHealth: Telemedicine, Distance Education and Informatics for Rural Health Care.* Boulder, CO: WICHE Publications.

Chapter 8

TELEDERMATOLOGY

Anne E. Burdick and Brian Berman

Introduction

Definition and Description of Clinical Procedure

Teledermatology (dermatology at a distance) is the delivery of dermatologic patient care combining telemedicine technology and specialty expertise. The dermatologist, using telecommunications technology and computer/video equipment, delivers dermatologic care by evaluating clinical and laboratory data, diagnosing and prescribing therapy for patients located at a distance. The goals of teledermatology, as of telemedicine in general, are to provide care to patients in underserved areas, to improve the quality of care, and to decrease the cost of that care by evaluating patients in their local health care setting. Further, telemedicine aims to provide relevant specialty medical education to the referring health care provider.

Patients can clearly recognize when they have skin diseases. Unlike people with internal disease, patients with skin diseases are quite aware of their conditions and do not need a nondermatologist simply to confirm that they have a skin disease. Studies have consistently shown that dermatologists provide the most cost-efficient, highest-quality care for skin diseases when their services are compared to other health professionals. Ramsey and Fox (1981) studied the ability of primary care physicians to diagnose correctly the 20 most common dermatologic diseases and found "primary care physicians deficient in their ability to recognize common dermatoses." Lichen planus was correctly diagnosed by only 16 percent of the nondermatologists, seborrheic keratosis by only 33 percent, seborrheic dermatitis by only 29 percent, atopic dermatitis by only 23 percent, and pityriasis rosea was correctly recognized by only 44 percent of nondermatologists. Malignant basal cell carcinoma was misdiagnosed by 30 percent of nondermatologists.

Pariser and Pariser (1987) studied errors made by nondermatologists prospectively over a 20-month period. There were 319 errors in 260 patients, with 86 percent of the errors in diagnosis. Malignant melanoma is a frequently fatal disease if not recognized and treated in its early states. Cassileth and colleagues (1986) found that, of the nondermatologists studied, "only 12 percent identified at least five of six examples of melanoma as such, and only 11 percent recognized both examples of dysplastic nevi."

Clark and Rietschel (1983) examined quality and cost-effectiveness of dermatologic care. Forty-nine percent of patients (202/410) initially seeing a family practitioner were treated by that physician for the correct disease. The other 51 percent (208/410) was either referred to another physician or treated for the wrong disease. McCarthy and colleagues' study (1991) of the ability of internists appropriately to diagnose, treat, and refer for specific dermatologic disorders revealed that 40 percent of the patients were incorrectly diagnosed.

Health maintenance organizations (HMO) have examined the cost and quality of care when rendered by dermatologists vs. nondermatologists. In 1980, Kaiser Permanente, the largest HMO in America, examined its practice of permitting direct access to dermatologists. At that time Kaiser's cost per visit was $20.90 for internal medicine but only $30.75 for dermatology (Engasser et al., 1981). When Kaiser again examined its policy of direct access for dermatologic care in 1989, it concluded that "it is still true that screening dermatology patients in another department is less cost efficient than permitting them direct access to dermatologists and that direct access means higher quality care in many instances" (Price, 1990). Other HMOs such as the Harvard Community Health Plan, the largest in Massachusetts, and North Carolina's Physicians Health Plan have similarly found that direct access to dermatologists provides the most cost-efficient, highest-quality dermatologic care.

Dermatology is a visual specialty. The American Academy of Dermatology certifying specialty examination includes a section of clinical color slides, from which questions about therapy or mechanisms of disease are asked. Dermatology is particularly well suited for telemedicine.

Dermatologic care includes both medical and surgical interventions. The medical evaluation includes history, visual inspection, and palpation. A dermatology consultant obtains a history relevant to the chief complaint or pertinent to the physical finding, which may not have been appreciated by the referring health care provider. For example, an eczematous eruption on the earlobes may suggest a nickel allergy, which would be substantiated by a history of sensitivity to jewelry or an inability to wear costume jewelry. Additionally, a patient with eyelid dermatitis would be asked about the use of nail polish or aerosolized perfumes or hair products. Aspects of the physical examination will be influenced by the distribution and/or morphology of the lesions. For example, in the evaluation of a patient with lichen planus, a dermatologist will often examine the oral mucosa and nails. For dermatologic evaluations of a widespread eruption, a dermatologist looks globally to determine the extent and distribution of a disease. All areas of the skin are examined, including hair, nails, mucous membranes (i.e., oral lesions), and genitals. Common diseases such as psoriasis or atopic dermatitis are examples of widespread eruptions. Using telemedical systems, dermatologists can evaluate patients by looking at the disease's distribution on the body, then at its regional distribution, and finally at individual skin lesions. The dermatologist carefully evaluates individual skin lesions to determine the type of lesions (pustule, bulla, or nodule), the shape and size (papule or plaque), the surface characteristics (smooth, shiny, scaling, or atrophic), and color.

During direct physical examinations, many dermatologists use a magnifying lens, hand-held or on illuminated stands, to provide 2–10x magnification of lesions to see finer detail which may aid in diagnosis. Many routine diagnostic procedures are conducted during a skin examination. Simple procedures include a hair pull in the evaluation of a patient with alopecia (hair loss) or, in a patient with a history of hair loss, an elicitation of Nicolsky's sign in the evaluation of a patient with a bullous disorder. During an in-person evaluation of a patient with skin lesions with scale on the surface, a dermatologist will often perform a potassium hydroxide (KOH) wet mount microscopic examination using potassium hydroxide either with DMSO or a stain such as Chlorazol fungal stain, which stains fungal hyphae dark green for easy detection to ascertain whether the lesions represent a fungal or yeast infection. Other procedures commonly performed during a skin examination include a Tzanck preparation to confirm a diagnosis of Herpes simplex or Herpes zoster/varicella infection by the presence of multinucleated giant cells; a crush preparation to reveal intracellular inclusions such as molluscum bodies of Molluscum contagiosum; and a Gram stain to detect the presence of bacterial organisms. A Wood's light examination may be conducted to detect certain fungal (*microsporum canis* in tinea capitis) and bacterial infections (*Pityrosporum orbiculare* or *Corynebacterium*) or to assess the extent of pigment loss in depigmented lesions of vitiligo.

Other medical dermatologic procedures include diascopy, capillaroscopy, durometry, and epiluminescence. Diascopy involves pressure applied to the skin either with a magnifying lens or clear glass slide to see if redness of the skin blanches and disappears with pressure, as would happen with blood contained within a blood vessel, or whether blood has extruded from the blood vessel as seen in a vasculitis. Capillaroscopy is examination of the blood vessels of the proximal nail fold with an oil drop and microscope eyepiece of magnification 10x. Tortuosity of blood vessels is commonly seen in patients with connective tissue disease. Durometry is a relatively new technique to establish a numerical assessment of skin induration, and to follow progression of a disease such as scleroderma (Falanga and Bucalo, 1993; Romanelli and Falanga, 1995). A calibrated vertical probe is placed on the skin surface three times for an average measure of induration.

Epiluminescence or dermatoscopy involves the examination of pigmented lesions using a magnifying lens, oil droplet, and a skin microscope or a hand held dermatoscope. Skin microscopes provide magnification ranging from 1x to 10x, while a dermatoscope magnifies 10x. To the dermatologist unfamiliar with dermatoscopy, magnified benign lesions can sometimes appear malignant. Dermatologists are now learning to appreciate the magnified details of color, border, and pigment network patterns provided by the skin dermatoscope through recent articles in the dermatologic literature (Pehamberger et al., 1987; Steiner et al., 1987; Nachbar et al., 1994). However, training is crucial to appreciate these details. In evaluating pigmented lesions during an in-person examination, the "ABCD rule" applies, with the "B" representing border and the "C" representing color. In a direct clinical examination, if the borders of the pigmented lesion are indistinct, or if a "spillage" of pigment is present, and if there is variegation of colors, the lesion is considered suspicious and should be biopsied to rule out melanoma. However, with dermatoscopy, different ABCD rules apply. Using epiluminescence, sharp borders are a worrisome sign, and a variety of color hues is commonly seen in benign pigmented lesions (Pehamberger et al., 1987; Steiner, 1987; Nachbar et al., 1994). In a recent study, it was demonstrated that a dermatoscope enhances the diagnostic ability of dermatologists only if they have been trained in the use of epiluminescence (Binder et al., 1995). In fact, dermatologists not so trained were hindered by epiluminescence in diagnosing pigmented lesions by using a dermatoscope. The need for dermatoscopes in telemedical systems is under debate. Although not helpful to many dermatologists involved in telemedicine, in all likelihood, the use of skin dermatoscopes in telemedicine is similar to in-person evaluations; training dermatologists to benefit from this equipment is probably necessary.

Other medical dermatologic procedures performed during skin exams include patch testing to assess potential contact allergens applied directly to the skin for 48 hours, photo patch testing to test photosensitivity in patients, and cultures for fungus and bacteria.

Another major facet of dermatology is surgical, including biopsies, curettage, and desiccation and surgical excisions, tissue transfer procedures such as surgical flap and grafts, as well as cryosurgery. Also included are such other surgical and therapeutic procedures as incision and drainage, injections, acne surgery (expression of the contents of comedonal lesions), sclerotherapy (injecting superficial veins with saline solutions), chemical peels, dermabrasion, hair transplantation, cryosurgery, and laser surgery. Phototherapy, the therapeutic delivery of ultraviolet light (either UVB or A) to a patient to treat a skin disorder (e.g., psoriasis or atopic dermatitis), is a distinct therapeutic modality offered by dermatologists.

Clinical Application of Telemedicine Within Dermatology

Two telemedicine systems can be used for dermatology: store-and-forward technology, and live interactive technology. Using either one, the medical aspects of dermatology can be achieved for the most part either directly or indirectly, with the aid of the referring health care provider at the remote site. Diagnostic procedures and therapeutic interventions can be recommended to the referring provider. Either system may have a dermatoscope ("skin scope") and a microscope camera as additional features. Store-and-forward technology is convenient to use since the telemedicine consultations are performed asynchronously and do not require the presence of the consultant dermatologist at the same time as the referring physician and patient. The dermatologic images are selected and captured independently by the referring provider and transmitted to the consultant before he or she is involved in the case. The asynchronous mode of consultation in a store-forward modality is time-efficient for the consultant, especially since transmission of an image may take 20 seconds or longer. The time needed to complete the evaluation is often less than that of the live interactive teledermatology consultation, since the social formalities between doctors and between the consultant and patient do not occur. Occasionally, patients give irrelevant history and often want to show the dermatologist skin lesions unrelated to or insignificant for their chief complaint.

There are drawbacks to a store-and-forward system. The dermatologist cannot ask the patient questions and cannot ask for additional views or close-up images to be taken and transmitted. Using live interactive telemedicine systems, dermatologists can interview patients as they would in a face-to-face encounter, and obtain relevant historical information from pertinent questions derived from the physical examination.

An intermediate situation is afforded by the use of the AT&T Picasso Still-Image Phone System, where multiple still images may be transmitted with the benefit of real-time audio.

A limitation of telemedicine for the dermatologist is the inability to palpate skin lesions directly to determine the surface characteristics, e.g., smooth, scaling; or its texture, i.e., whether it is soft, firm, or fluctuant; and the amount of pain or discomfort that light or firm palpation elicits. Often when evaluating a patient who complains of a tender lesion, a dermatologist will touch or palpate the lesion when the patient is unsuspecting. Additionally, the demarcation of the margins of subcutaneous skin lesions, such as lipomas, also cannot be directly appreciated. Although the dermatology consultant cannot directly palpate the skin via telemedicine, the health-care provider in the room with the patient can act as the surrogate for the specialist. That person, whether a physician, nurse, or physician's assistant, can report these aspects of the skin examination to the consultant dermatologist. It is important for the referring health care provider to assess whether a subcutaneous lesion is warm, tender to touch, well demarcated, movable, fluctuant nodule with a central punctum, as characteristic of an infected cyst, or whether the skin is bound-down as in scleroderma. However, sensation is often difficult to verbalize and convey, and the tactile sense is subjective. The temperature of the skin which is increased in cellulitis, and the degree of moisture or dryness of the skin, are also unattainable directly. Direct palpation and surgical intervention via telemedicine may become a reality in the future.[1]

---

[1]A "tactile glove" is being developed to allow doctors to perform direct palpation of patients via

Verbalizing these characteristics and conveying their quality are difficult and require a knowledge of dermatologic terms. In addition, the tactile sense is subjective. Palpation is also important in detecting the temperature of the skin (such as the increase associated with cellulitis) and the amount of moisture or extent of dryness of the skin (as seen in the dry palms of a patient with leprosy).

The dermatologist has to rely on the remote health care provider to perform surgical procedures in some situations. Most primary care providers (general practitioners) can perform punch or shave biopsies. The key is directing the provider to choose a representative or appropriate lesion (an early and late staged lesion, or two different types of lesions, as in pityriasis lichenoides et varioliform acuta) and from an appropriate location, i.e., perilesional skin for bullous disorders. Some primary care providers perform excisions; however, "scoop" shave biopsies and tangential biopsies with razor blades are not generally known or performed by these general practitioners. Training will be required of most health care providers to enable them to perform these basic dermatologic techniques. Nonphysician providers, such as nurse practitioners, and physician assistants can receive training in basic biopsy and diagnostic techniques, and even excisions in some states (Reid, 1996). In the future, it may be possible for consultants to perform surgical procedures with robotics now under development by the U.S. Armed Forces. At the Second Mayo Symposium on Medical Aspects of Telemedicine, suturing by robotics was demonstrated (Satava, 1995). Training and availability, as well as cost of equipment, may limit the use of this advanced technology.

It is important to train referring health care providers in basic dermatologic procedures. Biopsies were recommended by telemedicine dermatology consultants for 33 (29%) of 113 consecutive prison patients examined in an East Carolina University dermatology clinic. Potassium hydroxide preparations were advised for 7 patients (6%). A Wood's lamp examination was recommended for one patient and patch testing for one other patient. Therapeutic interventions included intralesional steroids in five patients (4%) and an Unna boot was advised for another patient (Phillips, 1995). If the health care provider could not perform biopsies at the

---

([1]Continued) telemedicine. Georgia Institute of Technology and the Medical College of Georgia are working on a prototype "tactile image mapper" to detect tactile sensing at the patient's site and provide tactile feedback to the consulting physician (Burrow, 1995, personal communication).

prison, then approximately one-third of the patients would have required travel to the dermatologist.

To expedite telemedicine dermatology consultation, either store-and-forward or live, it is helpful to have the referring provider transmit (fax) the patient's history and physical findings to the consultant. The consultation form used at the Jackson Memorial Hospital/University of Miami Teledermatology Program has been very successful. The referring provider includes the name, age, sex, and race of the patient. Medications, allergies, and a family history of skin cancer or melanoma are listed. The chief complaint is described in terms of duration, location, prior treatment, biopsy reports, and laboratory results with dates. Any significant past medical history is also included. There is space on the consultation form for additional history and physical exam findings obtained by the dermatologist during the live telemedicine consultation, as well as for the assessment and recommended treatment plan. Follow-up with either the primary care provider or the consultant is noted on the consultant form. Both the referring provider and the consultant sign this form, and the consultant faxes the form back once it is completed, to be kept in the patient's chart at the referring clinic. The transmittal form also indicates that the patient has signed the consent form, which includes consent for being videotaped, filmed, or photographed.

The average office visit for a new patient to a dermatologist typically lasts 15 to 30 minutes, depending on the complexity of the skin disorder, and a follow-up visit can range from 5 to 15 minutes. Telemedicine dermatology consultations using live video range from 10 to over 20 minutes. Teledermatology consultations performed at the Medical College of Georgia average 23 minutes in length (Salter, 1996), and at the Eastern Montana Telemedicine Center average 15 minutes (McClosky-Armstrong, 1995). An average of 10 minutes was required at the University of Miami Teledermatology Clinic. It was noted that the time per evaluation shortened as the Miami dermatologist became more accustomed to the technology (Norton et al., 1995). At Tripler Army Medical Center in Hawaii, live interactive dermatology consultations last between 10 and 15 minutes, while store-and-forward consultations take approximately 5 minutes. The University of Texas Medical Branch at Galveston provides dermatologic consultations to the Texas Department of Corrections. While initially live interactive video was used for these consultations, the time required was considered too long for the participating dermatologist, approximately 15–20 minutes per consultation. Since the dermatologist was able to conduct an in-person evaluation in between 5 and 10 minutes, the live interactive video system was changed to a store-and-forward system. The average teledermatology consultation time now averages 6 minutes (Brecht, 1996).

Many dermatologists advocate a complete skin examination during the patient's first office visit (Rigel et al., 1986). Often, the patient is unaware of lesions on the back, or the back of the legs. Thirteen of the 14 melanomas detected in a 2,239-person screening of the Manhattan Melanoma Skin Cancer Detection Screening Program were on anatomical sites normally covered by clothing. Patients having complete skin examinations were 6.4 times more likely to have a melanoma detected than those having partial examinations, in this 1986 study. Practitioners who routinely perform these examinations on all new patients feel that the time spent is short and worthwhile in many ways: the completeness of the examination allows for a higher level of reimbursement, decreases physicians' malpractice risk by detecting lesions early, teaches patients self-examination, and develops a good patient-physician relationship. A live interactive telemedicine consultation is probably required for this complete skin evaluation. It may be possible to have the remote-site practitioner take multiple views of the total body of the patient. The number of images required, however, is large; it takes time to capture the images and to display them on a store-and-forward system. Obtaining and storing these images is also costly. Additionally, the stored images cannot be magnified or be better focused by the consultant, and extra views of the skin from different angles or different areas cannot be obtained easily. Patients would have to return to the referring provider for additional images.

Review of Literature

Early Teledermatology Studies

In 1972, accuracy of dermatologic diagnosis using bidirectional interactive black and white television was assessed (Murphy et al., 1972). Several investigations, reported by Murphy et al., were conducted to assess the diagnostic accuracy obtained with black and white as well as color television. In the initial pilot program, 56 patients with skin lesions located at the Logan International Airport medical station were evaluated by a dermatologist at the Massachusetts General Hospital

(MGH) Telemedicine Center. The consulting dermatologist was able to make the same diagnosis as the direct clinical diagnosis in all but three cases. In a subsequent study, eight patients at the MGH dermatology clinic were examined by a dermatologist located at a distance, using black and white television. The dermatologist using telemedicine could ask the patient his medical history, and could ask the clinic dermatologist who was in the room with the patient about color and palpable alterations of the skin. The diagnoses of 10 of the 12 lesions made by the dermatologist, relying upon the interactive telemedicine system, was "equivalent" to the diagnosis made by clinic dermatologists who had examined the patient in person. It was reported that technical factors of lighting and camera positioning prohibited accurate diagnosis in the remaining two cases. The television system was judged to be cumbersome and the average time to reach a diagnosis was 15 minutes, which is very long to arrive at a dermatologic diagnosis. Murphy also noted that when dermatologists evaluated 75 color slides of dermatologic lesions directly and also by viewing black and white television images of these same color slides. The two dermatologists in the study were 85 percent and 89 percent as accurate with black and white television as they were with direct examination of the slides. Using a color television improved the accuracy only slightly, but was more acceptable to the dermatologists since less time was required to reach a diagnosis.

In 1974, Jackson Memorial Hospital (JMH), the major teaching hospital of the University of Miami School of Medicine, was awarded a grant by the National Science Foundation to determine the feasibility of a two-way microwave video link between hospitals and prisons (Sassmore and Sanders, 1976). The study was funded partly to improve the quality of health care in prisons. For three years JMH medical consultants, including dermatologists, provided consultation via this telemedicine link to prisoners at the Women's Detention Center, the Men's Stockade, and Dade County Jail. In this early project, a nurse at the prison clinic was able to consult with physicians at JMH via two-way video. Many specialists, including dermatologists, participated in this pioneering effort, but no hard data or conclusions were collected or made. It was estimated that over 70 percent of the prison's clinic patients could be successfully evaluated by this system. Even using expensive, bulky, black-and-white equipment, the use of video was reported to be "effective, highly acceptable" to both patients and physicians, and cost-effective.

In 1974, a clinical field trial was conducted providing dermatologic consultations from the Dartmouth-Hitchcock Medical Center (DHMC) in New Hampshire to a clinic in a town thirty miles north, using a microwave interactive closed-circuit television system (Johnson, 1974). A physician's assistant specifically trained in dermatology by the dermatology consultant served as the referring health care provider in this trial. The study included comparisons between simulated remote consultations using the equipment and actual remote consultations using the telemedical system. Five parameters were evaluated: control examinations, lesion clarity, color helpfulness, supervisory completeness by the physician's assistant, and patient reaction. Both black-and-white as well as color monitors were evaluated.

During simulated remote consultations, the dermatologist was located in a TV studio adjacent to but separate from the patient. Twenty-nine consecutive patients were evaluated, using both the color and black-and-white monitors. During the live audiovisual consultations, the dermatologist could ask the physician's assistant to take skin scrapings and palpate a lesion and could also ask for magnification of the lesion. After completing the examination and making a diagnosis, the dermatologist then went into the examining room and determined a diagnosis in person. There were no differences in each of the five parameters evaluated between the simulation and actual remote consultations, indicating that telemedicine dermatologic consultations were both possible and practical.

Forty-one consecutive patients were evaluated via live telemedical consultations from a studio at the DHMC, using both black-and-white as well as color television. The physician's assistant obtained the patient's medical history, and then presented the patient to the dermatologist over the television system. It was determined that the quality of the physician-patient relationship was "not modified" by the television. There was no significant difference in patient attitude between the simulated and remote consultations, with "favorable" attitudes for both consultations. Lesions were seen clearly with the color monitor in all 70 patients. Three diagnoses were impossible to make from the black-and-white images, but possible from the color images. The inability to appreciate depth was partially overcome by manipulating the direction of incident lighting onto the skin, to emphasize the presence of crusts.

Current Range of Application

Dermatology consultations using telemedicine are being conducted at a few medical centers. Since 1991, 52 dermatology consultations have been conducted at the Medical College of Georgia using an interactive telemedicine system (Salter, 1996). At East Carolina School of Medicine (ECU), dermatology consultations have been conducted on 189 prisoners using interactive telemedical technology since the inception of its telemedical program in 1992. Teledermatology consultations make up 50 percent of the 376 ECU telemedicine consults to date (Norton et al., 1995). The Eastern Montana Telemedicine Center reported 808 dermatologic consultations performed out of 1,162 telemedicine visits in fiscal year 1990–1991. In 1994, 21 of the 230 telemedicine consultations were dermatologic (McClosky-Armstrong, 1995). The Texas Tech HealthNet Program has conducted approximately 100 dermatology consults over the past 4 years, using a video link which connects two or three remote sites. Approximately 200 patients with dermatological disease were diagnosed via telemedicine during 1991 (Stenvold and Kvanmen, 1992). By 1993, 200 dermatology teleconsultations had been performed (Holland and Stenvold, 1993) and in 1994, 900 were performed (Petersen, 1995). Teledermatologic consultations will also soon include diagnosis, treatment, recommendations and evaluation plans of patients receiving ultraviolet light treatment at a phototherapy unit in Kirkenes Hospital (Stenvold and Kvanmen, 1992)

The Tripler Medical Center provides telemedical consultations to all federal employees in 52 percent of the earth's surface, approximately 90,000,000 square miles. As of February, 1995, 180 telemedicine consultations have been conducted in interactive sessions held two times a month, with between 8 and 10 patients per session (Norton et al., 1995). Dermatologic consultations have comprised half of these consultations. The AT&T Picasso Still-Image Phone System is also being used with satellites for evaluating personnel on maritime vessels.

The Walter Reed Army Medical Center has conducted telemedical consultations on a large scale in Somalia and Macedonia. Of 170 consultations using digital high-resolution store-and-forward technology, 45 (25%) were dermatologic. In Somalia, dermatologic consultations prevented some evacuations. For example, a soldier had a recurrent rash around both eyes over a two-month period, despite topical medication. An ophthalmologist initially thought it might be due to an autoimmune disease and considered prompt evacuation. The Walter Reed consulting dermatologist reviewed the digital images with the patient's history and diagnosed contact dermatitis secondary to protective rubber goggles and prescribed a stronger topical medication and the patient's eruption resolved (Crowther and Poropatich, 1995). The dermatology consult also facilitated some evacuations. For example, a patient with blisters on the dorsum of his hands taking medications for malaria prophylaxis was thought to have a contact dermatitis unresponsive to a high potency topical steroid. The Walter Reed dermatologist suspected porphyria cutanea tarda (PCT), and inquired about the presence of hypertrichosis, increased facial hair, a feature of PCT and requested a Wood's light fluorescence of his urine, the urine fluorescence was positive and the hypertrichosis was confirmed by the referring physician. The diagnosis of PCT was made and the patient was evacuated home and antimalarials discontinued.

Clinical Effectiveness/Accuracy

Only a few published reports describe dermatologic telemedical consultations; none that rigorously assess their accuracy. In Norway, video conferences have been reported between a university-based dermatologist and general practitioners at a remote hospital. These doctors described good quality images and stated that the system was considered beneficial (Vorland, 1992). Another study reported that video images of 27 patients with dermatologic conditions were transmitted by a two-way telephone and video network to dermatologists at the University Hospital of Tromso (Josendal et al., 1991). These patients, located in a "studio" at a local hospital, received diagnoses and treatments from a dermatologist miles away. All patients felt positively about these remote consultations. Over 130 patients on the U.S. Army base on Kwajalein Atoll in the Marshall Islands have been evaluated with good results by Tripler Army Medical Center in Honolulu (Delaplain et al., 1993). The medically underserved migrant worker population in Homestead, Florida, has received teledermatology consultations from dermatologists at the University of Miami's School of Medicine. During the first six months of operation, 18 patients were seen and both providers and patients reported being "very satisfied" (Burdick, 1995, Norton, et al., 1995, Smith, 1996).

In the autumn of 1989, the first two-way live video teledermatologic consultation occurred at the University Hospital in Tromso (UHT) between a general practitioner and patients at Kirkenes Hospital. During a test phase, diagnoses made by the UHT dermatologist via telemedicine were verified by a second dermatologist present in Kirkenes. It has been reported (Peterson, 1995) that there was "100 percent agreement" in the diagnoses made by the two dermatologists, although the number of patients and types of skin lesions evaluated were not stated. Patients were satisfied with the treatment, and the generalists were satisfied with the training in dermatology that they received.

A preliminary study at the University of Georgia suggests that there is a high degree of reliability in telemedicine dermatology consults. In 22 cases evaluated by telemedicine and by direct examination, there was "100 percent agreement" of the telemedicine diagnoses with the direct examination (Sanders et al., 1995). The details of the cases and the times of the dermatologic diagnoses and format of the telemedical consultations were not provided.

Three studies of the store-and-forward AT&T Picasso Still-Image Phone System have been reported. At Emory University, dermatologic diagnoses made at an actual clinic visit were compared to diagnoses made later on viewing images of the lesions captured at that time and stored by the AT&T Picasso Still-Image Phone System. Seventy images of cutaneous lesions from 38 patients were captured by a video camera and stored. The visual images were reviewed by four dermatologists who were blinded as to the diagnoses made at the actual clinic visit. Diagnoses were made from images alone, without the benefit of the patients' histories. When compared with the original clinical diagnoses, accuracy ranged from 63–88 percent among the four participating dermatologists. The major drawback was difficulty with lighting, which did not always capture the true color or texture of cutaneous eruption. Solitary lesions were more easily recognized than generalized eruptions (Futral et al., 1995).

In another Emory University study, a primary care physician captured images of 30 patients by the AT&T Picasso Still-Image Phone System using a super VHS camcorder and a color monitor. These images were subsequently sent to the consulting dermatologist, who determined diagnoses and recommended treatment plans. The same dermatologist then examined the patients in person and formulated a diagnosis and management plan. Identical diagnoses were formulated using the store-and-forward Picasso System and the live system in 87 percent of the patients. The concordance of the two management plans was excellent in 90 percent of the cases. For 60 percent of the cases, the dermatologist found the utility of the Picasso system to be excellent, and in an additional 33 percent of the cases, the system was rated as adequate to excellent (Fuseltzer, 1995). The author noted the existence of uncontrolled variables and suggested the need for further investigations in this modality. This study, indeed, failed to control for intraobserver concordance rates. It is not clear if the dermatologist had the ability to ask questions directly of the patient or to request additional images of physical finding when the store-and-forward method was used.

A study at the Walker Metherbest Nursing Home in Minneapolis, Minnesota, was undertaken to determine the accuracy of the AT&T Picasso Still Image Phone System in the diagnosis of dermatologic diseases at a nursing home. A designated nurse completed a medical questionnaire on each of the 30 patients requesting a dermatologic consultation, and videotaped cutaneous findings. The nurse selected a representative image using a high-resolution monitor which was then transmitted to the consulting dermatologist via the Picasso System along with a facsimile of the medical questionnaire. A consultation was performed in three ways: the dermatologist made a diagnosis and/or differential diagnosis and a treatment plan initially based upon the sent image alone; then once again after reading the medical questionnaire; and finally after visiting the patient in the nursing home within 24 hours after the image had been sent. Any changes in the diagnosis or treatment plan were documented. Only one original image chosen by the nurse was unacceptable for diagnosis. When compared to diagnoses made at the actual visit, image based diagnoses were greater than 90 percent accurate, whether or not the information on the questionnaire was used. No significant differences in proposed patient management plan resulted from the three types of consultation (Zeilckson, 1995). Unfortunately, this was not a simultaneous study and allowed 24 hours for the dermatologic disease to progress, which may have altered the disease presentation, leading to a different diagnosis and the less than 100 percent diagnostic accuracy. Again, the degree of intraobserver concordance was not established. Also, the bias of the dermatologist was not accounted for; having made a diagnosis via the Picasso system may have biased the dermatologist into rendering the same diagnosis at the in-person visit.

At the University of Miami, the authors have had the opportunity to investigate the accuracy of utilizing the AT&T Picasso Still-Image Phone in teledermatopathology. Seventy-eight sequential specimens submitted for dermatopathology one year prior to our present study were reexamined by the same dermatopathologist using the Picasso Still-Image System, and the diagnoses obtained through teledermatopathology were compared to the dermatopathologist's original diagnoses made through traditional direct microscopy. The dermatopathologist was in real-time audio contact with the microscopist, who was directed by the dermatopathologist and who transmitted the microscopic images. The dermatopathologist intra-observer variation was previously tested and found to be less than 1 percent. Eighty-five percent of the diagnoses made by the dermatopathologist using teledermatopathology correlated correctly with the original diagnoses made by the same dermatopathologist a year earlier using traditional microscopy. Although teledermatopathology was accurate in diagnosing basal cell carcinoma in 100 percent of the cases, there was a tendency to "underdiagnose" squamous cell carcinoma and Bowen's disease (squamous cell carcinoma in situ). No benign lesions were read as malignant using teledermatopathology. Although 85 percent accurate, teledermatopathology using this specific system may have limited usefulness due to the length of time required to transmit two to four images per case, which ranged between four and six minutes per case (Berman, 1995, personal communication).

A recent study by Oregon Health Sciences University demonstrated that there was no statistically significant difference between color slides and digital images in the quality of the clinical information. The study divided 180 slides and digital images into three categories: pigmented lesions, flesh-colored papules, and papulosquamous conditions. Each slide was converted into a digitalized image using a slide scanner and computer. This study used a technique called the multiple-choice, receiver-operating-characteristic analysis technique so as to measure diagnostic capabilities of imaging formats in radiology (Perednia et al., 1995). Common cutaneous diseases with three different levels of diagnostic difficulty were presented and the eight dermatologist viewers were asked to give a diagnosis and also a certainty rating.

A three-year National Library of Medicine study is also underway at Oregon Health Sciences University (Perednia and Brown, 1995). Baseline data on the current management of dermatologic problems in rural health care are being collected. These data will be compared with changes in referrals, patient management, confidence and patient outcomes, and primary care provider knowledge of dermatologic disease once telemedical dermatology consultations become established. Additionally, 1500 photographic images of clinically important skin conditions will be collected and the minimum digital imaging requirements for diagnosing these images will be determined. This study will also attempt to assemble a prototype teledermatology imaging and transmission system that is informative, easy to use, and inexpensive.

The Dermatology Foundation, a private research foundation, has recently funded the authors of this chapter to study the clinical accuracy of telemedicine for dermatologic disease, to be conducted at the University of Miami's Department of Dermatology. In this study, the accuracy of dermatologic diagnosis using an interactive telemedicine system is being tested. An experienced dermatologist evaluates patients in-person at the remote site and another experienced dermatologist, located at the medical center, evaluates patients a few minutes later by interactive telemedicine. The two observers switch their locations on alternate study days. The "gold standard" for diagnostic accuracy is the in-person diagnosis.

The University of Miami study is also comparing the accuracy of the "store-and-forward" telemedicine to the accuracy of interactive telemedicine. The referring health provider captures still video images of the patient and forwards them immediately to the consultant dermatologist at the medical center. The dermatologist then evaluates the patient using the telemedical still images and the written history and physical examination findings from the referring health provider. Once the diagnosis using this "store-and-forward" system is determined, the same dermatologist then examines the patient with the interactive system. During the examination with live video, the dermatologist can examine additional parts of the body, magnify the images, and ask the patient questions.

In addition, the accuracy of using interactive telemedicine for a standard microscopic examination used in dermatology, "telemicroscopy," is being determined in the University of Miami study. The potassium hydroxide wet mount is the test model. Positive and negative preparations using Chlorazol Fungal Stain, which stains hyphae dark green, is evaluated both by direct microscopy and by images transmitted from a video camera connected to a microscope, "telemicroscopy."

The Health Foundation of South Florida has recently supported the authors to determine the accuracy of teledermatologic diagnoses in HIV-infected/AIDS patients using store-and-forward technology. The study will utilize the two dermatologist observer models described above.

Matching the Concept with Reality: Constraints on Developing Telemedicine

Provider Acceptance. While many dermatologists have incorporated computer systems for billing purposes in their offices, many are still unfamiliar with or fear computers. This was demonstrated at the 1995 American Academy of Dermatology annual meeting, where, for the first time, the message center was computerized. While many of the younger clinicians approached the computer keyboard to access or to input messages, quite a few dermatologists were clearly uncomfortable with the technology and opted to hand write their messages and submit them in to an individual who would key them into the computer.

In February, 1995, the Dermatology Telemedicine Special Interest Group of the American Academy of Dermatology held its initial meeting. Participants shared their experiences with telemedical consultations with regard to physician satisfaction, and physicians' ease of diagnosis. Approximately 75 percent of the dermatology consultations have been judged to be "satisfactory" by the dermatologists involved at both ECU and Texas Tech University (Phillips, 1995, Nelder, 1995). Among the problems noted by Nelder were the inability to palpate lesions; the resolution of the HealthNet Program telemedical system; issues of patient privacy, especially with nonmedical personnel in the room; confidentiality of the patient history; and patients' hesitancy to have their genitals filmed. Norton commented that the consultant dermatologist does not need to be 100 percent accurate on the first visit—just better than the primary care physician (Norton, 1995).

The length of time required for a teledermatology consultation is a major factor in provider satisfaction. Dermatologists at the University of Texas Medical Branch at Galveston decided to change from live consultations due to the long (15–20 minutes) consultations to a store-and-forward system which take an average of 6 minutes (Brecht, 1996).

The dermatologist at the University of Tromso indicated "less satisfaction" with the patient-physician interaction using a telemedicine system than with an in-person evaluation. On the other hand, in the Minnesota study of patients in nursing homes undergoing teledermatology evaluations with the Picasso still-image system, the dermatologist found the system to be "user friendly with a short learning curve," and "time efficient" (Zeilckson, 1995). The inability to palpate the skin is a constant drawback in both live and store-and-forward systems. In this Minnesota study, successfully treated patients were presented to the dermatologist by a nurse. However, some dermatologists feel it is necessary for a physician to be at the remote site. Several members of the Special Interest Group on Telemedicine expressed frustration with the time it took to capture an image during live consultations. Some dermatologists have found that papulosquamous lesions are the most difficult to assess, and that evaluating dark-skinned individuals is also difficult. There was general agreement that lighting is a very important issue during telemedicine dermatologic exams.

One serious concern of dermatologists is the lack of reimbursement within the U.S. Telemedical consultations are reimbursed in only a few states, e.g., Montana by Medicaid; Medical College of Georgia by Medicaid, Medicare, and the State Employees Insurance Plan; and Kansas by Blue Shield of Kansas. In three other states, West Virginia, North Carolina, and Iowa, HCFA (Health Care Financing Administration) is reimbursing telemedical consultations as part of a multicenter study of reimbursement and the issue of liability when consulting with patients in states where the dermatologist is not licensed.

Some dermatologists not involved with telemedicine are concerned that local dermatologists will lose patients if academic centers, managed care groups, or HMOs establish teledermatology clinics in those areas. In 1995, the American Academy of Dermatology President, Dr. Peyton Weary, stated that telemedicine may become a potential "turf battle" between academic institutions and community doctors (Weary, 1995). The Medical College of Georgia met this concern by requiring local specialists, if available, to evaluate patients in their community before patients are referred to the University specialist.

Future of Telemedicine Pertaining to Dermatology

Future opportunities for teledermatology include telemedicine for the homebound, chronically ill patient, for example, the individual with leg ulcers, and nursing home patients, e.g., follow-up of decubitus ulcers. Managed care organizations would be interested in telemedical applications particularly in dermatology, which requires "low end equipment," and allows patients to be seen during earlier stages of their diseases.

Conclusion: Utility of Teledermatology

Teledermatology is useful for three types of users of this health care delivery system: the patient, the referring health care provider, and the dermatology consultant. The utility of telemedicine to the patient is probably the most obvious. Access to quality medical care and specialty care is limited in many areas of the country. Maps of the demographic distribution of specialists, including trauma centers, clearly demonstrate the long distances individuals need to travel to receive tertiary and specialty care. Using telemedicine, travel costs for patients are decreased, including transportation expenses and time away from work. The hazard of travel in inclement weather conditions is also reduced. It is also advantageous for patients to receive medical care close to their homes with supportive family and friends nearby, and early in the course of their disease.

Education of primary care providers is one of the most useful aspects of telemedicine. Telemedicine offers the potential for forging professional alliances between general practitioners and specialists which will enhance the generalist's medical knowledge, and keep them abreast of new developments in various fields of medicine. Diagnostic and therapeutic "pearls" from consultants and information about relevant articles can have a great impact on the generalist's patient management. By participating in telemedical consultations, referring practitioners will gain new insights into patient assessment and therapeutic options which can be used in treating patients with similar conditions or presentations in the future. In fact, the interaction between the referring practitioner and the consultant can become a true "mentorship" relationship. Relationships of this kind, based on telemedicine, can decrease the professional isolation felt by practitioners in remote areas. Retention of health providers in isolated communities will improve as a result. In addition, the remote general practitioner may experience an increase in status, once patients learn of the availability of telemedical consultations. This increased status may lead to an increased patient population and higher income.

The utility of telemedicine for patients and referring practitioners, as well as for health insurance companies, lies in establishing an accurate diagnosis rapidly, with a minimum of office visits. The recent studies already described have shown that dermatologists provide the most cost-efficient, high-quality care for skin diseases, when their services are compared to other health providers.

There are many benefits for telemedicine consultants. They get more patients, often due to increased referrals from the general practitioners with whom they have developed professional relationships. Consultants affiliated with medical schools or medical residency training programs find telemedicine useful for educational purposes. The residency housestaff gains exposure to a larger patient population with a more varied case mix. Residency program accreditation approval is based in part on the number of procedures performed or patients examined. With the current emphasis on primary care physician training, and a decreased number of practicing specialists and specialty residency training programs, telemedical consultations will become important in maintaining high-quality patient care and will become a standard part of residency training. Stored dermatologic images can also become part of the teaching syllabus for residents and nondermatologists.

REFERENCES

Berman, B. April 8, 1995. Personal communication.

Binder, M., M. Schwarz, A. Winkler, A. Steiner, A. Kaider, K. Wolff, and H. Pehamberger. 1995. "Epiluminescence Microscopy: A Useful Tool for the Diagnosis of Pigmented Skin Lesions for Formally Trained Dermatologists," *Archives of Dermatology* 131: 286–91.

Brecht, R. February 23, 1996. Personal communication.

Burdick, A. E. 1995. "Response to a Telemedicine Inquiry from the Florida House of Representatives and Text of Model Act to Regulate the Practice of Telemedicine," *Telemedicine Journal,* 1: 309–19.

Burrow, M. March 31, 1995. Personal communication.

Cassileth, B. R., W. H. Clark, E. J. Lusk, B. E. Frederick, C. J. Thompson, and W. P. Walsh. 1986. "How Well Do Physicians Recognize Melanoma and Other Problem Lesions?" *Journal of the American Academy of Dermatology,* 14: 555–80.

Clark, R. A., and R. L. Rietschel. 1983. "The Cost of Initiating Appropriate Therapy for Skin Diseases: A Comparison of Dermatologists and Family Physicians," *Journal of the American Academy of Dermatology,* 9: 787–96.

Crowther, J. B., and R. Poropatich. 1995. "Telemedicine in the U.S. Army: Case Reports from Somalia and Croatia," *Telemedicine Journal,* 1: 73–80.

Delaplain, C. B., C. E. Lindborg, S. A. Norton, and J. E. Hastings. 1993. "Tripler Pioneers Telemedicine Across the Pacific," *Hawaii Medical Journal,* 52 (12): 338–39.

Engasser, P., and R. S. Lyss. 1981. "Direct Access in HMOs," *Journal of the American Academy of Dermatology,* 4: 740–41.

Falanga, V., and B. Bucalo. 1993. "Use of Durometer to Assess Skin Hardness," *Journal of the American Academy of Dermatology,* 29: 47–51.

Futral, M., A. K. Pare, M. Ling, and M. McKay. 1995. Poster: "Tele-Dermatology: Still-Image Transmission from Remote Areas via Telephone," American Academy of Dermatology, Annual Meeting, New Orleans, Louisiana, February 4–9.

Holland, U., and L. A. Stenvold. 1993. "Are Patients Satisfied with Teledermatological Consultations?" Oral presentation, First International Conference on the Medical Aspects of Telemedicine, Tromso, Norway, May 21.

Johnson, M–L. T. 1974. "A Model for Televised Remote Dermatological Consultation," unpublished report.

Johnson, M–L. T., A. B. Burdick, K. G. Johnson, H. D. Klarman, M. Krasner, A. J. McDowell and J. Roberts. 1979. "Prevalence, Morbidity and Cost of Dermatologic Diseases," *Journal of Investigative Dermatology,* 73: 395–401.

Josendal, O., G. Fosse, K. A. Andersen, S. E. Stenvold, and E. S. Falk. 1991. "Long Distance Diagnosis of Skin Diseases," *Tidsskrift For Den Norske Laegeforening,* 111 (1): 20–22.

McCarthy, G. M., C. C. Lamb, T. J. Russell, et al. 1991. "Primary Care Based Dermatology Practice: Internists Need More Training," *Journal of General Internal Medicine,* 6: 52–56.

McClosky-Armstrong, T. September 1, 1995. Personal communication.

Murphy, R. L. H., Jr., R. B. Fitzpatrick, H. A. Haynes, et al. 1972. "Accuracy of Dermatologic Diagnosis by Television," *Archives of Dermatology,* 105: 833–35.

Nachbar, F., W. Stolz, T. Merkle, A. B. Cognetta, T. Vogt, M. Landthaler, P. Bilck, O. Braun-Falco, and G. Plewig. 1994. "The ABCD Rule of Dermatoscopy," *Journal of the American Academy of Dermatology,* 30: 551–59.

Nelder, K. February 8, 1995. Personal communication.

Norton, S. February 8, 1995. Personal communication.

Norton, S., A. E. Burdick, C. M. Phillips, and B. Berman. 1995. "Teledermatology and Underserved Populations," submitted to *Archives of Dermatology.*

Pariser, R. J., and D. M. Pariser. 1987. "Primary Care Physicians' Errors in Handling Cutaneous Disorders," *Journal of the American Academy of Dermatology,* 17: 239–45.

Pehamberger, H., A. Steiner, and K. Wolff. 1987. "*In vivo* Epiluminescence Microscopy of Pigmented Skin Lesions I: Pattern Analysis of Skin Lesions," *Journal of the American Academy of Dermatology,* 17: 571–83.

Perednia, D. A., and N. A. Brown. 1995. "Teledermatology: One Application of Telemedicine," *Bulletin of the Medical Library Association,* 83 (1): 42–47.

Perednia, D. A., J. A. Gaines, and T. W. Butruille. 1995. "Comparison of the Clinical Informativeness of Photographs and Digital Imaging Media With Multiple-Choice Receiver Operating Characteristic Analysis," *Archives of Dermatology,* 131: 292–97.

Peterson, S. 1995. Oral presentation, Second International Conference on the Medical Aspects of Telemedicine and Second Annual Mayo Telemedicine Symposium. Rochester, Minnesota, April 7.

Phillips, C. February 8, 1995. Personal communication.

Phillips, C. February 12, 1996. Personal communication.

Price, V. 1990. "Kaiser Permanente Medical Care Program: Direct Patient Access to Dermatologists in an HMO Practice," San Francisco, California. Unpublished report.

Ramsey, D. L., and A. B. Fox. 1981. "The Ability of Primary Care Physicians to Recognize the Common Dermatoses," *Archives of Dermatology,* 117: 620–22.

Reid, J. 1993. "Telemedicine in Montana, Will it Happen?" Proceedings of the Mayo Telemedicine Symposium, October 1–3, p. 73.

Reid, J. April 9, 1996. Personal communication.

Rigel, D., R. Feldman, A. W. Kopf, R. Weltman, P. G. Prioleau, B. Safai, M. G. Lebwohl, Y. Eliezri, D. P. Torre, R. T. Binford Jr., V. A. Cipollaro, L. Biro, D. Charbonneau, and A. Mosettis. 1986. "Importance of Complete Cutaneous Examination for Detection of Malignant Melanoma," *Journal of the American Academy of Dermatology* 14: 857–60.

Romanelli, M., and V. Falanga. 1995. "Use of a Durometer to Measure the Degree of Skin Induration in Lipodermatosclerosis," *Journal of the American Academy of Dermatology,* 32: 188–91.

Salter, P. April 9, 1996. Personal communication.

Sanders, J., L. Adams, and K. Grigsby. 1995. "Impact of Telemedicine: The Georgia Experience," *Teleconference, The Business Communications Magazine,* 14, (1): 12–14, 27, Telemedicine Issue.

Sassmore, L., and J. Sanders. 1976. "Final Report: An Evaluation of the Impact of Communications Technology and Improved Medical Protocol on Health Care Delivery in Penal Institutions," Volume 1, Executive Summary, NSF Grant #GI 39471, December.

Satava, R., 1995. Oral presentation. Second International Conference on the Medical Aspects of Telemedicine and *Second Annual Mayo Telemedicine Symposium,* Rochester, Minnesota, April 7.

Smith, S., 1996. "Lights, Camera, Diagnosis," Miami Herald, April 7, p. 1.

Steiner, A., H. Pehamberger, and K. Wolff. 1987. "*In vivo* Epiluminescence of Pigmented Skin Lesions II: Diagnosis of Small Pigmented Skin Lesions and Early Detection of Malignant Melanoma," *Journal of the American Academy of Dermatology,* 17: 584–91.

Stenvold, S. E., and B. Kvanmen. 1992. "Teledermatology," *Applied Telemedicine,* Norwegian Telecom Research, Tromso, Norway, pp. 17–19.

Vorland, L. H. 1992. "Teledermatology," *Nordisk Medicin,* 107 (10): 241–43.

Waltzer, J. F. 1995. "The Picasso Still-Image Phone System for Teledermatology," Abstract Book, Second International Conference on the Medical Aspects of Telemedicine and Second Annual Mayo Telemedicine Symposium, Rochester, Minnesota, April 6–9, p. 23.

Weary, P. 1995. Oral presentation. Information Superhighway Course, American Academy of Dermatology, February 7.

Zeilckson, B. D. 1995. "Nursing Home based Teledermatology," Abstract Book, Second International Conference on the Medical Aspects of Telemedicine and *Second Annual Mayo Telemedicine Symposium,* Rochester, Minnesota, April 6–9, 1995, p. 20.

Chapter 9

TELEONCOLOGY

ACE ALLEN AND GARY DOOLITTLE

Introduction

This chapter examines the role of telemedicine in the practice of clinical oncology, and the potential of this "teleoncology" to increase accessibility to cancer care for remote and isolated populations. Teleoncology is defined here as the delivery of clinical oncology consultation using an interactive video telecommunications system to overcome the substantial distance separating the oncology specialist from the patient. The consultation is typically rendered from tertiary medical centers to patients geographically restricted in access to these services, but who may also be institutionalized elderly and prison populations. This chapter reviews the development, scope, and content of clinical oncology and, in particular, the role of the oncology specialist. The capacity of the current generation of telecommunications technology to contribute to the practice of clinical oncology is considered. Studies of patients' and physicians' satisfaction with teleoncology are reviewed. The chapter concludes with speculation about the future of teleoncology.

The Need

Rarely do rural communities have the resources to recruit and retain physician specialists. Most efforts to improve cancer prevention, detection, and treatment have been initiated and implemented at urban cancer centers. Even if the need were identified and financial issues were not of concern, rural areas are sparsely populated and lack the patient base to support a subspecialty practice.

For example, western Kansas, the size of the whole State of New York, has a population of 400,000; New York has 18,000,000. Currently, only two resident medical oncologists serve western Kansas.

Even though we value access to high quality health care for all our citizens, given the demographics of some rural areas, perhaps it is not financially feasible to provide on-site subspecialty care. Unless the delivery of cancer care changes significantly in the very near future, the National Cancer Institute will not be able to improve access in rural areas to most current cancer treatment and prevention measures as called for in its "Objectives for the year 2000."

Clinical Oncology

The care of cancer patients is a combined effort between the primary care physician, nurse practitioner, other support personnel, and specialist. In the traditional practice of medicine, patients diagnosed with a malignancy are usually referred to an oncologist for an initial evaluation. Occasionally, a patient with a suspected malignancy will be referred to a cancer specialist to help with the diagnostic workup. Once the diagnosis is established, care is based upon the specific disease and treatment plan. Routine monitoring may be done by the generalist. However, when disease-specific chemo-/radiotherapy is prescribed, the patient is followed by both the specialist and the primary care physician. In academic centers, oncologists often subspecialize within their field, even limiting their practice to a specific disease (e.g., breast cancer; bone marrow transplantation). Thus, academic oncologists confined to tertiary care centers may receive referrals from generalists or other oncologists.

The practice of clinical oncology is complex and functionally comprehensive. It includes cognitive as well as tactile and other sensory procedures. A large part of clinical oncology consultation concerns elicitation of a patient's medical problems, as well as review of the past medical history, laboratory findings, radiographs, and pathology reports. The discussion of a patient's condition and diagnostic and therapeutic options is equally important. As in all areas of medicine, the physical examination is an important part of the patient evaluation. Components of the exam may include palpation for lymphadenopathy and organomegaly, a neurological assessment, and thorough workup of various pain syndromes. Other elements of the examination procedure may include auscultation and visual examination of the skin, oropharynx, ocular fundus, external auditory canal, and tympanic membrane. While oncology is not a procedural discipline, studies may be ordered at the time of consultation to determine the extent of the malignancy (localized vs. disseminated). This "staging" is of vital importance for understanding prognosis and in the development of a treatment plan. None of these elements of the clinical oncology consultation necessarily requires the simultaneous presence of the patient and practitioner, as long as there are alternate ways of accurately obtaining the information. Telemedicine technology and techniques can provide these alternatives.

To date, telemedicine technology at all levels, from the telephone to two-way interactive televideo, is being used in a great variety of medical applications. Regardless of the level of technology employed, its current utilization is most heavily weighted toward the cognitive dimensions of clinical practice generally and toward the nonprocedural areas such as subspecialties in internal medicine and psychiatry (Perednia and Allen, 1995). This makes sense, since the current generation of telemedicine technology does not support the procedural specialties and "hands-on" dimensions of the clinical examination. Nor is it likely in the near future that telemedicine technology will enable "virtual reality gloves" for tactile examinations and "robotic surgery" procedures outside of rarified and expensive laboratories. Thus, in the coming five years, medical disciplines requiring "hands-on" examination and invasive techniques must be completed with the cooperation of a proxy examiner, who may be the local physician, nurse practitioner, or physician assistant.

Acknowledging existing technological limitations in these areas, telemedicine technology currently lends itself to the practice of clinical oncology, which depends largely on the visual, auditory, and cognitive skills of the oncology specialist. The consultation includes the review of data and communication with the patient and family.

Clinical Teleoncology

The application of telemedicine technologies to clinical oncology is very recent. As a result, published materials on the subject are scarce and limited in scope. The concept of clinical oncology itself was first referred to in the literature in 1990 (Lipsedge et al., 1990). Here the authors, from the Division of Psychiatry of London's United Medical and Dental Schools, speculated about the possible role of digitized interactive video in addressing the unique psychosocial problems of cancer outpatients.

The only published description of clinical teleoncology practice derives from the University of Kansas Telemedicine Program (Fintor, 1993; Allen and Hays, 1995; Allen et al., 1995; Allen et al., 1992). Approximately 200 teleoncology consultations have been conducted there since 1992. Most of these have been conducted between the Hays Medical

Center (Hays, Kansas) and the Kansas University Medical Center (KUMC, Kansas City). These sites are 280 miles apart. The first video linkage occurred in 1991 when interactive video conferencing equipment was installed between KUMC and the Hays Area Health Education Center. In 1992, the first telemedicine consultations were conducted. The most active specialties providing teleconsultations in the first few years of the telemedicine program were pediatric cardiology, child psychiatry, and adult neurology. In the ensuing years, other specialties have used the telemedicine system for consultations including oncology, adult psychiatry, adult cardiology, surgical subspecialties (primarily for post-surgical follow-up), and pediatric subspecialties. At the present time, the telemedicine program is made up of ten rural sites and two urban locations (KUMC—Kansas City and University of Kansas School of Medicine—Wichita).

All telemedicine sites within the project currently use compatible equipment and are linked by a leased fractional T-1 line (the KANS–AN Network) typically running at 384 kbps. All sites are equipped with videoconferencing CODECs, twin 33-inch monitors with high-resolution monitors having far-end camera control, as well as an electronic stethoscope and a document camera/stand transmitting 400 lines of resolution. Video resolution using the proprietary compression algorithm is 368 × 480 lines for motion running at 30 frames per second and 736 × 480 for still graphics.

Until access via telemedicine became available, KUMC oncologists provided on-site care to cancer patients in the Hays catchment area by flying in for biweekly clinics. These were frequently postponed or even canceled due to the erratic weather that typifies western Kansas. In fact, the first teleoncology clinics were arranged on a semiemergent basis as a result of unexpected weather-related cancellation of four fly-in oncology clinics in the winter of 1992–93. Further trials of teleoncology were delayed because the Hays site recruited a full-time oncologist to provide the needed service for cancer patients. When the oncologist from Hays resigned, KUMC teleoncology services were reinstituted in the winter of 1994–95, this time on a regular basis. Currently, teleoncology clinics are scheduled twice weekly with the Hays Medical Center and from three to five patients are seen each week. Teleoncology consults are offered to the other Kansas sites as well. To date, three other sites have arranged for oncology consultation. At a minimum, the telemedicine program has doubled the frequency of the oncology clinics at Hays.

Teleoncology Consultation

A typical teleoncology consultation proceeds as follows. A patient is referred to the teleoncology/hematology clinic by a local practitioner in the Hays area (patients do not self-refer). Prior to the consultation, pertinent patient information, including letters and hospital discharge summaries, and laboratory, radiography, and pathology reports, is transmitted to KUMC via facsimile machine for the consulting oncologist's review. Subsequently, the patient is scheduled for a consultation. A central scheduler from the KUMC telemedicine office arranges the consult. The clinics are blocked for the same time each week, so physician and on-site nursing personnel can plan ahead. In addition, "as needed" consultation may be arranged, if the clinical status warrants an urgent visit. The patient will be accompanied by a nurse practitioner or the referring physician. Typically, a nurse accompanies the patient for routine consults and follow-up visits. If the patient is acutely ill, the referring physician may choose to be present for the telemedicine visit. The decision about which on-site practitioner should be present (physician or nurse) is made at the discretion of the referring physician.

Upon the patient's arrival at the remote consultation site, the nurse introduces him or her to the system, and if he or she agrees to be seen using telemedicine and to have the interaction videotaped, gives him or her a consent form to sign. Only two patients out of over 100 teleoncology consults have declined to use the system, and made this decision prior to the telemedicine clinic. Once the form is signed, the patient is ushered into the telemedicine examination room by a nurse or physician familiar with the operation of the various controls. In the room, the patient is seated before a large television monitor. Above this is a television camera remotely controlled by the consulting oncologist at KUMC. The camera is largely obscured behind a smoked-glass panel. The patient speaks "face-to-face" with the consulting physician, who elicits a medical history in the usual fashion. When required, a physical is conducted with the local nurse or referring physician serving as a proxy examiner. At first, the proxy examination may be somewhat awkward. With time, as with any physician-nurse working relationship, a rhythm is established, resulting in greater comfort.

Under the direction of the KUMC oncologist, a specially adapted electronic stethoscope is used to transmit respiratory and cardiac sounds over the network. In addition, the camera lens can be focused from a distance to assess skin changes, mucositis, petechiae, and the like. The camera can also be positioned to evaluate the patient's gait and movement abnormalities as required. At all times during the examination, "normal," real-time, two-way conversation occurs between the sites. After the physical examination, radiographs (plain films, CTs, or MRIs) are transmitted either from the document stand or from a radiology viewbox, using a 3-CCD camera from the distant site. If needed, other data, such as electrocardiograms and chemotherapy flow sheets, can be transmitted to the consulting oncologist using the document stand. Finally, the results of the consultation may be reviewed with the patient and family at the remote site. Questions and concerns are elicited, and appropriate diagnostic studies and therapeutic interventions are prescribed. Upon completion of the consultation, the patient is offered a copy of the videotape of the clinical interaction.

A new technology is currently being implemented which will facilitate a much higher quality of electronic integration. It will permit the video component of the telemedicine interaction to run as a window within a software shell. The software shell, in turn, will provide software "hooks" to an electronic patient record, digitally stored radiographs and pathology slides, as well as on-line library and other data resources (see below). The new technology will permit easy acquisition of digital audio-video clips which can be bundled with other patient data in a computer "folder" or directory. All this can then be stored as a medical record or forwarded electronically as audio-video e-mail for another medical practitioner to review. Among other things, this capability will obviate the need to use the current awkward system of sending the patient information via facsimile machine prior to the consultation.

Teleoncology: Some Concerns

To date, teleoncology has not been rigorously and systematically assessed evaluating its efficacy. This is true for other telemedicine practice in clinical settings. Until results from studies become available, anecdotal reports provide an indication of the concerns and problems that may affect the acceptance of this technology by people in the medical sector and by the public.

In the Kansas experience, certain problems with the clinics have been identified. For example, the consulting oncologist is highly dependent upon the physical examination skills of the remote health care provider (in many cases a nurse practitioner) at the rural site. Clearly, the on-site practitioner must accurately report important physical findings such as organomegaly, lymphadenopathy, and pain characteristics. On the receiving end, the consulting oncologist must be able accurately to interpret the reported findings. Unless the remote practitioner has had specialized training in the procedures necessary for oncologic evaluation (tumor measurement, grading treatment toxicity, etc.), the consulting specialist cannot be certain of the accuracy of the assessment. For teleoncology to be safely deployed, remote clinical assistants need special and continuing education in the procedures specific to oncology examinations.

Another problem derives from the variation in resolution quality of radiographs transmitted over the current system (480 lines of resolution). This resolution is not generally adequate for diagnostic interpretation of x-rays. For plain films, such as chest x-rays, the resolution for diagnostic interpretation needs to be at least 2,000 lines × 12 bits, according to recently published standards from the American College of Radiology. On the other hand, CT scans, MRIs, and nuclear medicine scans are of lesser resolution and can be reviewed on the compressed-video system with relatively little loss of clarity, making interpretation at a distance reasonable.

It is essential that the technology be user-friendly for physicians and other health practitioners. Physicians' lack of enthusiasm for telemedicine may be related, in addition to the previously mentioned problems, in part, to the complexity of the equipment which must be used. Some physicians learn and adjust more quickly than others to the control panels of the telemedicine system. In our observations thus far, however, in no instance has it taken more than half an hour for a physician to become comfortable using the controls. Additionally, all five participating oncologists have noted that they feel as if they were actually in the same room with the patient within minutes of beginning the teleoncology clinic encounter.

Another important issue is patients' acceptance of the procedure. Patients and the public must be convinced of the efficacy of teleoncology. It must be demonstrated that high-quality care, comparable to that offered by a traditional clinic, can be delivered utilizing the telemedicine system. In our teleoncology experience, only two patients have refused to use the system. Unfortunately, the reasons for refusal were not determined.

Thus, for the practice of teleoncology to become widespread, issues of satisfaction, efficacy, cost-effectiveness, and organizational communications must be addressed. Several pilot studies have been published, looking into patients' and physicians' satisfaction with teleoncology.

Patients' Satisfaction

The objective of this study was to assess levels of satisfaction among rural cancer patients being seen for clinic visits by an oncologist via a telemedicine link to a remote university medical center (Allen and Hayes, 1995). In this study, the satisfaction of 39 cancer patients seen in a teleoncology clinic was assessed, using a 12-item survey. Subsequently, the satisfaction level of 21 of these patients was assessed, using a 9-item survey, after an on-site clinical interaction. In each instance, satisfaction levels were measured on a 5-point Likert scale. The levels of satisfaction with each of the two consultation modes were compared. High levels of patients' satisfaction with the telemedicine encounter were recorded at the time of the initial teleoncology clinic visit, as well as at the follow-up on-site visit. With one exception, for each of the survey items the initial and follow-up mean score was above 3.0, where 3 was the scale midpoint, half way between "very satisfied (5)" and "very dissatisfied (1)." Though only a small pilot study, the results suggest that rural cancer patients are satisfied seeing their oncologist via interactive video, at least occasionally.

Some other observations from the pilot study may be instructive. For example, there were only modest differences in satisfaction levels between the subset of patients meeting the consulting oncologist via interactive video for the first time and that group seen on the system for a follow up visit. Prestudy "wisdom" held that first-time oncology patients would need to be seen in person to establish the confidence obtainable only through direct personal interaction. This supposition may be erroneous.

No age- or gender-related differences in level of patients' satisfaction were found in this pilot study. The level of patients' satisfaction varied, depending on the individual oncologist who conducted the evaluation. This may reflect differences in the level of satisfaction with the physician, independent of the mode of communication (i.e., interactive video versus on-site). However, it may be related to the manner in which the individual physician's personal practice style projects over an interactive video system. If so, this could have very important implications for screening and training of telemedicalists. Some physicians may be less suited than others for interactive video consultation because of their communication style. Others may never become comfortable with the technology. In the traditional practice of medicine, a "good bedside manner" describes the physician with good interpersonal and communication skills. We have to assume that differences in "telemedicine manner" will be identified similarly.

Finally, anecdotally, there was considerable enthusiasm among a sample of patients who took the videotapes of the clinical encounter home with them. In a small, unpublished survey, 13 responding patients reported that they played the videotapes an average of 4 times, that it helped them and their families understand their disease and its treatment, and made them feel more confident in their medical care.

Physicians' Satisfaction

Another pilot study was conducted to assess the levels of physicians' satisfaction with providing outpatient oncology care to rural cancer patients, using teleoncology clinics (Allen et al., 1995). A 9-item survey (Part 1) assessing satisfaction with the patient-physician clinical interaction was administered to three urban-based oncologists immediately after seeing each of 34 patients in rural Hays, Kansas, using interactive video. A 4-item summary survey (Part 2) was completed at the end of each day of teleoncology clinics. Items were scored on a continuous graphical scale from 1 to 100, with the midpoint labeled "neutral." The study results were as follows:

The mean time for the teleoncology consultations was 27 minutes; for comparative on-site consultations the mean time was 22 minutes. The overall mean score for Part 1 was 74, and for Part 2, 66, each well above the "neutral" score of 50, signifying general satisfaction with this modality of providing cancer care. The numbers suggest that physicians were more satisfied with the experience of teleoncology on a patient-to-patient basis, and not as satisfied with the aggregate clinical experience at the end of the day.

It is important to note that evidence supporting even this level of satisfaction is subject to possible bias. For example, the studies were pilots hampered by small numbers and lack of serial follow-up for corroboration and change over time. In addition, there could be potential bias introduced by the researchers who were evaluating their own teleoncology program. Thus, although the levels of satisfaction were reasonably high, even this level could reflect the participants' enthusiasm and subtle pressure from the evaluation process rather than "true" satisfaction. Acknowledging these potential biases, the studies support further evaluation and implementation of teleoncology. Further studies should be directed toward substantiating levels of satisfaction and making appropriate adjustments in the programs to improve it.

Efficacy

Studies of satisfaction with telemedicine are relatively easy to do. At their simplest, useful studies may involve only the investment of intellectual capital, paper for questionnaires, and some computer time for tabulating and analyzing the results. Efficacy studies, on the other hand, are quite difficult. They involve a much greater commitment of the practitioner's valuable time. They involve more resources to set up a control and intervention setting. Finally, they involve the study and comparison of "messy" biological systems (populations of sick human beings) which are extremely complex, hard to control for, and hard to measure. For example, a study comparing the clinical effectiveness of teleoncology vs. on-site oncology for patients with Stage II breast cancer might have to account and statistically adjust for the usual demographic variables including gender (yes, there are men with breast cancer); laboratory, radiography, and pathology findings; past medical history; clinical presentation; physical findings; response to treatment; and other outcomes. Moreover, it would have to account for the extent to which variations in findings and responses were caused by differences in the ability to deliver care utilizing the two modalities (teleoncology vs. on-site) as opposed to differences in physicians' practice styles which are independent of the modality (inter-observer variation). Finally, details of the immediate technological environment would have to be controlled for: the equipment's resolution, motion-handling, color accuracy, audio quality, and so forth.

Though complex, expensive, time-consuming and difficult, it is essential to complete studies of efficacy. As surely as telemedicine won't be used if patients and providers are dissatisfied with it, it *shouldn't* be used if it is not effective for given medical applications. A teleoncology efficacy study at the University of Kansas has accrued its first patients, and should be completed by late 1996.

Cost-Effectiveness

There are no published findings on the financial implications or consequences of teleoncology. In fact, no prospective, hypothesis-driven research on cost-effectiveness has been published since the inception of telemedicine. This is difficult research, but highly important. Compelling evidence suggests that telemedicine programs in North America which were operating in the mid-1970s collapsed, in part, because they were unable to reach financial independence (Park, 1974). Moreover, there is little evidence, with the exception of the prison outreach telemedicine program at the University of Texas Medical Branch at Galveston, that any of the current interactive-video mediated telemedicine programs could justify their existence on the basis of clinical consultations alone (Allen and Allen, 1995). Pursuant to this, most telemedicine programs utilize their teleconferencing infrastructure for medical education as well as for clinical consultations.

Tele-education can be quite specific to oncology. For example, recently, the University of Kansas telemedicine program has conducted a twice-weekly evening training course in techniques of chemotherapy delivery and oncology care for nurses from rural hospitals. As mentioned previously, it is clear that such training is a prerequisite for teleoncology in rural Kansas, since the value of an oncologist's virtual or real presence is decreased in the absence of a trained chemotherapy nurse clinician to serve as the on-site "effector arm." The teleoncology nursing course is financially self-sustaining as students are charged class registration fees.

Organizational Communications

Telemedicine programs operate in an unusual organizational environment. They are composed of distributed systems, usually cooperative ventures linking independent hospitals and clinics. There is often no clear hierarchy of administration, leadership, or authority. Furthermore, payments for participants (physicians and nurses) are often uncertain since reimbursement is not assured, and other benefits from participation in telemedicine consultations are not always clear. Finally, even the language of telemedicine is not always consistent within a given telemedicine program. This can magnify communication and organizational problems. A body of evidence is developing, suggesting that these organizational and communications problems are among the greatest

stumbling blocks to the successful implementation of telemedicine programs (Whitten and Allen, 1995).

Largely anecdotal evidence suggests that nearly every telemedicine
program suffers from the problem of scheduling consultations. This is
especially true for emergency or intermittent consultations. Scheduling
a patient to see a specialist may be difficult even in a traditional hospital
or clinic, but these difficulties are greatly magnified for a telemedicine
consultation. The schedule must satisfy the needs of the rural patient
and practitioner, telemedicine facility (room, technical assistants), the
consulting specialist, and the medical center's telemedicine facility. Also,
because most telemedicine systems are not "dial-up" but are run on
leased networks with a central scheduling node, the telecommunications
lines themselves must be scheduled. Furthermore, there may be conflicts
in the use of the system (which usually has a limited capacity for transmission of teleconferencing traffic) with continuing medical education
programs, administration video conferencing, and demonstrations. These
problems may be addressed by developing a priority of systems use, and
are arguably handled most efficiently by a central administrative office.
This is lacking in most telemedicine programs.

These organizational and communications problems appear to take on
an added significance when imposed on a teleoncology practice. Any
untoward delays or complexities in reaching a physician increase the
stress on cancer patients, already anxious about their condition. This is
also important from a therapeutic perspective, because many chemotherapy regimens are highly time-dependent and must adhere to a rather
rigid schedule of implementation. Within the KUMC teleoncology
program, we have tried to minimize scheduling difficulties by setting
specific time slots for oncology and other specialty clinics. The problems
of scheduling real-time, interactive telemedicine consultations make the
use of store-and-forward technology increasingly attractive for some
telemedicine consultations. While this has worked well in some specialties,
such as dermatology, store-and-forward techniques have not been used
in teleoncology to date.

Current Status

In a recent survey of interactive-video-mediated telemedicine programs in North America, seven of 26 responding programs indicated

that they had done at least one teleoncology consultation in the 1994 calendar year (Allen and Allen, 1995). Unfortunately, the details of the level of oncology activity at each program are not available.

As noted previously, the KUMC teleoncology program now has a regularly scheduled, twice-weekly teleoncology/hematology clinic. The patient volume is 4–6 patients per week. This is augmented by outreach clinics that meet every other week on site at Hays Medical Center. The system is using the new technology noted above, which allows for the simplified integration of audio and video, electronic patient records, radiography data, and the like. It also facilitates the capture and sending of video clips for store-and-forward applications.

Since the establishment of the teleoncology practice between Hays Medical Center and KUMC, an oncologist has been recruited to the Hays community. The traditional and telemedical practices exist in tandem. In addition to direct patient care, the system has been used for second opinions, alleviating the need for patient travel. Also, the teleoncologist has cross-covered for the on-site oncologist during vacations (the on-site oncologist has covered the teleoncologist as well). In the early stages of teleoncology implementation at KUMC, several multispecialty tumor conferences (conferences discussing specific patient cases, involving medical, radiation, and surgical oncologists, as well as radiologists and pathologists, and the referring physician) were held. Eight cases were reviewed. Satisfaction surveys were administered, but the data are too limited for detailed comment. There was enthusiasm for the "tele-tumor conferences" among all participants, with reasonably consistent agreement that the quality of medical data transferred over the telemedicine system was sufficiently acceptable to allow useful discussion of evaluation and management of individual cases.

An underexplored and underexploited area of "low-tech" teleoncology is the use of the Internet and World Wide Web. Tremendous resources are available to any physician or patient with a computer, a modem, and telephone line access (Roelants, 1995; Rice and Allen, 1995; Hancock, 1996). As Internet access improves and drops in price, and as the software interfaces improve, it will become easier to integrate these Internet resources into the daily practice of oncology and teleoncology. Examples of Internet oncology resources for professionals include:

- OncoLink (University of Pennsylvania)
  Internet address: http://cancer.med.upenn.edu

- CancerNet (National Cancer Institute)
  Internet address: gopher://helix.nih.gov/11/clin/cancernet

The Internet can also be an outstanding resource for cancer patients and their families. Examples include:

- Cancer Information for Patients and Families
  Internet address: http://asa.ugl.lib.umich.edu/chdocs/cancer percent3agourhin.htm.

- Cancer-L (Cancer discussion list)
  Internet address: listserv@wvnvm.wvnet.edu

- Breast Cancer Information Clearing House
  Internet address: gopher://nysernet.org:70/11/BCIC

The Future of Teleoncology

Telemedicine techniques appear to be well-suited to the practice of oncology. As we have seen, telemedicine has been used successfully for tumor conferences and nursing education, as well as for direct patient care. It may well be used in the near future for national and international access to top cancer experts, and to new cancer treatments through cooperative group trials. The Internet may develop into a major source of oncology information, especially as access improves and as the functionality of Internet resources—such as the ability to transfer quickly large audio-video files—improves.

The technology of interactive-video mediated teleoncology appears to be moving away from ponderous "rollabout" units, which are essentially teleconferencing units (with some peripheral medical devices such as electronic stethoscopes, attached), to desktop units. These units are fully computer-integrated with immediate access to other electronic resources, such as the computerized patient record, radiographs, laboratories, pathology, library services, and the Internet. Presumably, these new technologies will be configurable for different oncologic specialties. For example, a radiation oncologist might need a higher resolution monitor, with greater gray-scale capability (for checking radiation ports and simulations) than a medical oncologist. Also, the user interface of the new telemedicine units may be tailored to the individual practitioner. Thus, chemotherapy flow sheets might be preferentially graphed out on-screen for one practitioner, and left in tabular form for another.

An interesting implementation strategy for telemedicine that is just becoming available is outreach to the home (Allen, 1995). One company is now offering home health care using televideo transmitted over the local cable television system. Others are using increasingly available digital phone lines (ISDN), or even 56-kbps analog phone lines (Mahmud and Lenz, 1995). Because oncology practice is closely allied with home health issues and hospice care, these telemedicine-to-the-home strategies could have a profound impact on the way cancer care is provided in the future.

Organizational and human factors will have to be addressed forthrightly. Telemedicine disrupts established lines of referral, blurs state and international boundaries, and challenges an organization to develop new ways of doing business. Some medical practitioners may need special training to overcome deficits in their ability to communicate or to project subtle aspects of their personality (warmth, sense of contact, and caring) over a video system.

In the next five years there will be significant changes in telemedicine. Equipment and transmission prices will drop. Functionality will increase. The level of awareness and enthusiasm among practitioners and patients for telemedicine will increase. Medical students will use telemedicine in their training, and will take that knowledge with them to their practices. Research will advance. Somewhere in the interplay of these factors, a critical point may be reached where telemedicine becomes the recognized best answer to a host of problems in access to patients, physicians, and medical data. In the next few years, the practice of oncology may change in ways that our current thinking cannot encompass.

REFERENCES

Allen, A., and D. Allen. 1995. "Telemedicine Programs: 2nd Annual Review Reveals Doubling of Programs in a Year," *Telemedicine Today,* 3:1, 10–14, 18–23.

Allen, A., R. Cox, and C. Thomas. 1992. "Telemedicine in Kansas," *Kansas Medicine,* 93:323–25.

Allen, A., and J. Hayes. 1995. "Patient Satisfaction with Teleoncology: A Pilot Study," *Telemedicine Journal,* 1:41–46.

Allen, A., J. Hayes, R. Sadasivan, S. K. Williamson, and C. Wittman. 1995. "A Pilot Study of the Physician Acceptance of Teleoncology," *Journal of Telemedicine and Telecare,* 1:34–37.

Allen, A. 1995. "Home Health Care via Telemedicine," *Telemedicine Today,* 3 (3): 16–23.

Fintor, L. 1993. "Telemedicine: Scanning the Future of Cancer Control," *Journal of the National Cancer Institute,* 85:18–19.

Hancock, L. 1996. *The Physician's Guide to the Internet.* Philadelphia: Lippincott-Raven Press.

Lipsedge, M., A. B. Summerfield, C. Ball, and J. P. Watson. 1990. "Digitised Video and the Care of Outpatients with Cancer," *European Journal of Cancer,* 26:1025–26.

Mahmud, K., and J. Lenz. 1995. "The Personal Telemedicine System: A New Tool for the Delivery of Health Care," *Journal of Telemedicine and Telecare,* 1:173–77.

Park, B. 1974.

Perednia, D. A., and A. Allen. 1995. "Telemedicine Technology and Clinical Applications," *Journal of the American Medical Association,* 273:483–88.

Rice, R., and A. Allen. 1995. "The Internet as a Health Care Communication Tool," *Telemedicine Today,* 3 (1):6 ff.

Roelants, H. W. M. 1995. "The Worldwide Web: An application of the Health Professional," *Telemedicine Today,* 3 (1):1 ff.

Whitten, P. S., and A. Allen. 1995. "Analysis of Telemedicine from an Organizational Perspective," *Telemedicine Journal,* 1:203–13.

Chapter 10

TELEPSYCHIATRY: APPLICATION OF TELEMEDICINE TO PSYCHIATRY

LEE BAER, PETER CUKOR AND JOSEPH T. COYLE

Introduction

Interest in the problems associated with distance and the use of mental health services may be traced back almost 150 years. During this period, attempts have been made to bring psychiatric services closer to isolated populations (Shannon et al., 1986). Yet the problems persist. Periodically over the past four decades, engineers, physicians, and other health care professionals have explored the use of telecommunication technologies to link specialist physicians with health care providers in isolated areas to provide consultation, diagnosis, treatment, the transfer of medical data, and education. One of the earliest telemedicine experiments involved "long distance" telecommunication for neurological and other consultations from a university department of psychiatry to a state mental hospital. Psychiatry continues to have a strong interest in the potential of telemedicine in general and telepsychiatry in particular for improving the delivery of a wide range of mental health services to isolated providers and populations. This chapter (1) discusses the current situation and problems in mental health care delivery that prompt psychiatry's continued interest; (2) examines the rationale for the use of telecommunications in psychiatry; (3) describes some telepsychiatry projects; and (4) speculates about the future of telepsychiatry. An appendix discusses the specific role of telephone technology in the provision of psychiatric care.

Acknowledgment: The authors wish to acknowledge the help of Craig Burns, Dr. R. Benschoter, John O'Laughlen, and Linda Leahy in preparing this chapter.

Background

In the United States, estimates derived from several National Institute of Mental Health (NIMH) community surveys indicate that, within any one-month period, approximately 15 percent of adults meet the diagnostic criteria for one or more psychiatric disorders (AMA, 1990). Additionally, the current and projected demographic trends for the U.S. population indicate a continuing strong need for the services of general psychiatrists, adolescent and child psychiatrists, and geriatric psychiatrists. But the number of mental health professionals necessary to provide these services is predicted to fall short, while problems are forecast related to geographic distribution, accessibility, and acceptability. Concern about the "mal-distribution" of mental health professionals has emphasized the situation in rural areas. However, other populations are isolated from a comprehensive range of psychiatric services (e.g., in geriatric facilities, prisons, and inner cities).

The Rural Problem

Providers of mental health services are concentrated in urban areas; few rural areas are served adequately (Murray and Keller, 1991). Over the past several decades, both federal and state programs have attempted to address the shortage of mental health professionals in rural communities. In recent years, for example, the National Health Service Corps, established through the Emergency Health Personnel Act of 1970 (P.L. 91-623), has attempted to place psychiatrists in this program in rural mental health centers. At least one state has attempted to familiarize psychiatric residents with the rewards of rural practice (Reed, 1992), to encourage them to enter this type of setting after they graduate.

Nevertheless, psychiatrists in rural mental health centers remain significantly few. Several problems have been identified in attracting psychiatrists to rural areas and keeping them there: a dissonance between the culture of rural and urban areas, reflecting different evaluations and perceptions of mental health and mental health care (Human and Wasem, 1991); difficulties for the psychiatrist in establishing boundaries between professional and private activities in small towns; professional isolation from colleagues and resources; and a dearth of opportunities for continuing medical education (Tucker et al., 1981).

People in rural areas suffer from the long distances they must travel to obtain services, from a lack of public transportation, and from a dearth of accessible mental health outreach services (Human and Wasem, 1991; Bushy, 1994). These problems are not being solved by the provision of federally supported Community Mental Health Centers (CMHCs), partly because the CMHCs are based on an urban model. According to the Community Mental Health Care Act of 1963, CMHCs need a service area population of between 70,000 and 250,000 people. Of course, this number of people can be aggregated most easily in large metropolitan areas. In most rural areas the geographic area required to encompass this number of people is prohibitively extensive. A typical rural mental health CMHC catchment area is 5,000 square miles. The largest such area covers more than 60,000 square miles; its population is distributed across the area in isolated communities and farmsteads (Murray and Keller, 1991). Findings also suggest that a more limited range of services is provided in rural mental health settings when compared to those of urban settings (Sommers, 1989).

Further contributing to the limited range of services generally available within the CMHC is the process of "deprofessionalization" which has occurred. A decrease in the number of mental health professionals employed has a potentially negative impact on the quality and scope of available patient care (Fink and Weinstein, 1979). In part, the decrease has been associated with the use of community residents to deliver services and a concern for the community's social needs rather than the needs of mental patients.

The location of those mental health personnel within rural areas also poses a problem for patient accessibility. It has been noted that 40 percent of the mental health personnel in rural areas are based in hospitals, as opposed to 18 percent for the country as a whole (Bushy, 1994). Consequently, availability of mental health services is dependent upon access to rural hospitals, many of which are in tenuous financial situations and facing closure.

Given the general shortage of mental health professionals and their concentration in urban areas as well as their declining role in CMHCs, it is not surprising to find that the majority of patients do not see psychiatrists or psychologists for mental health concerns. The NIMH estimates that of those persons with one or more mental health visits, only 25 percent were seen by psychiatrists in office practice and an additional 25 percent were seen by psychologists in office practice. Other providers, such as nonpsychiatrist physicians or social workers, saw 40 percent of the patients in office settings. The remainder were seen in organized settings such as hospital outpatient departments, emergency departments and specialty mental health clinics (AMA, 1990).

Cooperation and consultation between psychiatrists, primary care physicians, and other providers have been recommended as essential for the provision of comprehensive mental health care. This is believed to be especially true in rural areas, where primary providers serve as critical gatekeepers for the diagnosis and treatment of psychiatric problems.

The Aging Population

Aging of the U.S. population continues in rural as well as urban areas. Again, it has been projected that the number of geriatric psychiatrists being trained is and will continue to be insufficient to meet the needs of the present and future generations of elderly. Most geriatric psychiatrists currently have academic appointments. Nevertheless, two areas of greatest need are in academic geriatric research and long-term care settings (Siu et al., 1989). Mental disorders in old age are common and have multiple, complex origins. Depression is the most common psychiatric disorder of the elderly. And it is estimated that over one-half of the 1,300,000 elderly in nursing homes suffer from Alzheimer's disease. Treatment programs for the acutely and chronically ill elderly are needed in both institutional and noninstitutional settings.

Other Populations

In 1994, over 1,500,000 men and women were incarcerated in federal, state, and local prisons and jails. This population is expected to increase annually for the foreseeable future. The role of telemedicine in delivering health care to the prisoner population is treated elsewhere in this volume, but it is important to stress here the inclusion of mental health care. The extent of mental illness among prisoners has been documented nationally, but limited statewide surveys indicate that anywhere from 7 to 15 percent of inmates have mental disorders. Further, about one-third of these diagnoses are depression. It is both extremely important and difficult to deliver timely and necessary psychiatric services to this confined population.

Other populations experiencing problems with accessibility to medical care generally, and mental health care specifically, are the increasing numbers of poor and minority people living in most of our largest central cities. In central cities, distances to mental health clinics of only several blocks have been demonstrated to be significant in the utilization of community mental health clinics (Weinstein et al., 1976; Breakey and Kaminsky, 1982). Controlling for race, social "disadvantage," and income, analysis in one large city indicated a significant negative relationship between enrollment in mental health centers and distance from the clinic. The travel distances involved ranged from about 0.2 to about 1.3 miles. It appears that people's willingness to travel for mental health services is particularly sensitive to the distance they have to go.

Given the shortage of psychiatrists and their maldistribution, it appears that telepsychiatry has a potentially important role to play in increasing their accessibility to many now isolated populations. In addition to addressing these problems, an effective and efficient mode of delivering mental health care is needed because the practice of psychiatry is changing rapidly.

Advances in Psychiatry

Within the past quarter of a century, the practice of psychiatry has been radically transformed. Current understanding of the causes of mental disorders and their treatment have little relationship to the models generally used in the mental health field as recently as 15 years ago. The scientific and functional landscape of psychiatry has been radically altered, with enormous implications for the efficient dissemination of new clinical knowledge, diagnostic expertise, and effective treatment methods.

With the demonstration in the 1960s that certain psychotropic agents were unequivocally effective in treating severe depression, schizophrenia, and bipolar disorder, psychiatric researchers increasingly focused on brain mechanisms responsible for major mental disorders. This shift in focus from psychodynamic theory to brain research fortunately coincided with the rapid growth of neuroscience. Postmortem neurochemical studies, brain imaging, and genetic studies provide compelling evidence of the "brain basis" for an expanding number of mental disorders ranging from schizophrenia to obsessive-compulsive disorder (OCD). Furthermore, the psychopharmacologic armamentarium expanded to embrace conditions such as panic disorder, OCD and its variants, and even Alzheimer's disease. Drugs developed on the basis of molecular targets of actions and not serendipity are now regularly introduced into psychiatric treatment.

Psychological and behaviors treatments have not been neglected. These are now subject to empirical studies of specificity and efficacy. In contrast to the recent past when psychodynamic psychotherapy was broadly accepted as a treatment for virtual all psychiatric illnesses, much more selective interventions which require considerable expertise are the norm. Family therapy to reduce expressed emotions in schizophrenia, cognitive therapy in depression, and systematic desensitization in panic disorder represent several specific psychological interventions of proven efficacy.

Finally, psychiatric care has been and continues to be transformed by managed care. The traditional dichotomy between office-based psychotherapy and prolonged inpatient treatment of severe mental disorders is rapidly vanishing. In its place has developed the practice of "brief therapy." The emphasis now is on integrated systems wherein patients move across levels of care differing in intensity and constraint, depending upon clinical improvement. Inpatient stays are now measured in days instead of months with an emphasis on symptom-oriented interventions.

With the possible exception of the innovations of managed behavioral health, the advances in diagnosis, treatment, and understanding of the etiology of mental disorders have occurred at approximately a score of university-affiliated centers for psychiatric research. Thus, the recent transformation of the psychiatric care landscape has been limited. This concentration of expertise is necessary for productive research, but nevertheless it presents a barrier to the transfer of new knowledge to the clinical field. Psychiatric journals represent the major forum for communicating these clinical advances. Yet most research articles are arcane in terminology and design, so much so that the psychiatric literature is becoming increasingly impenetrable for the "front-line" mental health practitioners. It is this widening gap between clinical expertise in tertiary psychiatric centers and the remote community-based provider which telepsychiatry can help to bridge.

Psychiatric Consultation

Psychiatric consultation is typically requested in two clinical situations: (1) to help clarify a difficult diagnosis; and/or (2) for treatment recommendations for a patient unresponsive to standard treatments. Since both situations involve primarily spoken conversation between the consultant and the therapist requesting consultation, the latter often accompanied by the patient, such consultations have been a major focus of telemedicine applications in psychiatry. Other telepsychiatry applications have been in the area of continuing education by lecture or patient interview, which are similar in most respects to courses in other medical specialties. Theoretically, telemedicine supporters argue, the only difference between in-person consultation and interactive video consultation is the distance between the participants. This ignores, of course, any audio or video information lost in transmission. Because diagnostic tests routinely used in psychiatry are psychological tests, there has been little clinical interest to date in psychiatry in using telemedicine technology for the transmission of diagnostic tests or digital images.

Application of Telepsychiatry

The application of telemedicine technology to deliver psychiatric services from a distance has periodically generated great enthusiasm over the past several decades. This is due to its promise to extend needed treatment services to individuals who have been neglected until now due to manpower deficiencies, both in absolute numbers and in unevenness of geographical distribution (Solow et al., 1971). Further, the latest diagnostic and treatment practices in tertiary care centers are available differentially. Thus, given the need for increased accessibility to psychiatric services, the primacy of spoken conversation, and the need for dissemination to the field of recent advances, psychiatry appears to be ideally suited for the incorporation of telemedicine. However, these bursts of enthusiasm and largely anecdotal positive findings from pilot research have not yet resulted in any large-scale development of telepsychiatry. Commenting on this seeming paradox in 1986, Dongier et al. noted in Canada that: "the geographical characteristics of Canada, with its narrow band of populated areas and immense, sparsely-settled territories, should encourage the development of telemedicine and telepsychiatry, both as a partial solution to the well-known problem of medical manpower maldistribution and as a source of support for isolated practitioners. However, more than 20 years after the early trials . . . and in spite of considerable technological progress . . . , telepsychiatry has not spread in any significant fashion in Canada. Its use has been limited by problems with both cost-effectiveness and prejudices of the users" (Dongier et al., 1986).

The Technology

A detailed discussion of telemedicine technology can be found elsewhere in this volume. Suffice it to say here that the quality of the image received is better the less the compression. Unfortunately, cost also increases with less compression. Even images with a modest compression of 128 Kbs cannot be transmitted on ordinary telephone lines but require digital lines. In conjunction with the National Information Infrastructure, these are now being deployed around the country. Still, their cost is somewhat more than the regular analog phone lines. Many experts believe that, optimally, one needs at least 384 Kbs bandwidth to practice telepsychiatry. Most remote settings, however, have access to less bandwidth (128 Kbs). Advances in the diffusion of related technology will eventually resolve these difficulties. In the meantime, choices and compromises have to be made between having no video connection at all and less than excellent video quality.

Telepsychiatry: An Historical Perspective

The history of telepsychiatry via interactive video technology can be traced productively along chronological or geographical dimensions. The chronological approach highlights evolution in technology and applications, while the geographical approach focuses on those few areas successful in obtaining federal funding to purchase the relatively expensive equipment. The review presented here proceeds initially along the chronological dimension and, subsequently, focuses on the few geographic locations involved in major telepsychiatry projects.

As illustrated in Table 10-1, to date, the majority of telepsychiatry applications using interactive video have been in consultation or assessment (N = 8). Others were in continuing education (N = 2) and only one, and it was one of the earliest, was a direct treatment application (Wittson et al., 1961). In the same table the variety of patients populations is illustrated, including adults and children, out- and inpatients, and patients suffering from a variety of problems including mental retardation, as well as psychotic, affective, and anxiety disorders.

Prior to 1993, all studies included for review used closed-circuit television of various forms and, with one exception using one-way video (Wittson and Dutton, 1956), all were two-way interactions without video compression. The two most recent studies assessed lower-cost, *computer-*

based video conferences, using compression and decompression of the video images transmitted over a relatively narrow bandwidth (Ball et al., 1993; Baer et al., 1995). Importantly, these latter studies were also the only ones to assess quantitatively the reliability of the video interview compared to an in-person interview. Both studies used standard psychiatric/psychological tests, whose reliability can also be assessed quantitatively.

At Guy's Hospital in London, sponsored by the European Community, a computer-based, video-link experiment between two psychiatry wards was conducted. Each computer screen displayed a "quarter-screen-sized" window displaying a monochrome real-time image of the person at each end. The sound, of hi-fidelity quality, was generated by an amplifier and speaker connected in parallel with the computer video system. Eleven patients were administered the Mini-Mental Status Exam (MMSE), a widely used cognitive examination in psychiatric interviews. A correlation of 0.89 was obtained between the MMSE administered in the video and the in-person settings. This was identical to the test-retest reliability for the MMSE reported in the original normative sample. It was concluded, in this instance, that a standardized cognitive screening test can be reliably administered using computer-based video consulting in an adult population with acute psychiatric illness.

Baer et al. (1995) used a computer-controlled, digital, compressed video system to test the reliability of several psychiatric rating scales using 26 patients diagnosed with obsessive-compulsive and anxiety disorders. To permit quantitative reliability testing of a wide variety of instruments, a reliability testing paradigm termed the "simultaneous video reliability interview" was developed. This compared the reliability between two "raters" in an interactive video situation with the reliability between the same two raters in a "live" interview setting. Two "Picture Tel 4000 Model 400" units were located at hospitals about 20 miles apart and a connection was established by dialing the other site using the local telephone company's ISDN network. The system contained a digital (ccd) camera with both near and far end controls which provided scanning and zooming capabilities. The primary video output device was a 27-inch monitor. Audio input-output equipment consisted of a high-quality microphone and speaker built into the control keyboard. The study found near-perfect comparative reliability in the video condition (R = 0.99) on the Yale-Brown Obsessive-Compulsive Scale (Goodman et al., 1989), The Hamilton Depression Scale (Hamilton, 1960), and the

Table 10-1.

Telepsychiatric Applications using Audio and Video Communication with Television Equipment

Study	Year	Application	Equipment	Population	N	Locale	Quantitative Assessments?
Wittson et al. [7]	1956	Continuing Education	Closed circuit TV (1-way)	Med students PhD and nursing students	–	Omaha, NE	–
Wittson et al. [8]	1961	Group Psychotherapy	Closed circuit TV (2-way)	Adult Groups	8 groups	Omaha, NE	–
Menolascino et al. [9]	1970	Consultation	Closed circuit TV (2-way)	Mentally Retarded Adults	60	Omaha, NE	–
Solow et al. [4]	1971	Consultation	Closed circuit TV	Psychiatric Inpatients	199	Hanover, NH	–
Dwyer [10]	1973	Consultation	Closed circuit TV (dedicated line)	Psychiatric Outpatients & Prisoners	>150	Boston, MA	–
Straker et al. [11]	1976	Assessment Training	Two-way inter-active cable TV	Inner city Children & Mothers	30 families	New York, NY	–

Dongier et al. [5]	1986	Consultation	Closed circuit TV vs. Face-to-face	Psychiatric Inpatients &	50 35	Quebec, Canada	Pts. no sign. difference Consultee & Consultant: CCTV signif inferior for global assess & diagnosis
Jerome [12]	1986	Assessment	Closed circuit TV (interactive TV link)	Child and Family Out-Patients	—	Ontario, Canada	50% reduction in clinical time for consultant
Fisch et al. [13]	1992	Continuing Education	Closed circuit TV (microwave)	Psychiatrists	—	Boston, MA	—
Ball et al. [14]	1993	Assessment	LCVC (Computer-based videoconf)	Adult Acute Psychiatric Patients	11	London England	r = .89 LCVC to Face-to-face vs. .89 on original face-to-face reliability
Baer et al. [15]	1995	Assessment	Computer-based videoconferencing vs. live-assessment	Adult Out-Patients with OCD	15 12	Boston, MA	intraclass range from .98 to .99 between live and video assessment

Hamilton Anxiety Scale (Hamilton, 1959). The study findings indicated that several semi-structured psychiatric rating scales could be reliably administered using a relatively narrow bandwidth (128 Kbs) in patients with anxiety disorders (Baer et al., 1995).

Although several studies failed to report the number of patients involved, it is estimated that only approximately 600 have participated in telepsychiatric trials to date. Moreover, more than 400 were seen at one of three sites located in Nebraska, New Hampshire, or Massachusetts. These studies are reviewed in some detail here.

The Nebraska Psychiatric Institute Projects

The first description of telepsychiatry published in professional journals reported the pioneering efforts of Wittson and his associates at the Nebraska Psychiatric Institute (NPI). They began in the mid-1950s with one-way closed-circuit television and then proceeded to employ two-way interactive television in many successful applications.

Wittson and Dutton (1956) used closed-circuit television at the NPI as a tool to assist nurses in maximum security wards, to show visitors the facility, and, most frequently, to improve classroom teaching. Teaching sessions were generally loosely scripted to facilitate a coherent demonstration and the "programs" were transmitted to a projector in a large auditorium for students to view. Equipment included a closed-circuit television with mobile camera carts and a 6 × 7-foot projection screen. The entire system was maintained by a full-time electronics technician.

There was no formal evaluation of the system. However, anecdotally, it was reported to have been effective in expanding the number of people exposed to patient interviews and other clinical situations. The underlying premise for the use of the video equipment was based on the question: "Can the student learn the information as well, or better, by hearing (and seeing) a case or by hearing *about* it?" It was observed, for example, that: "A close-up of a patient's face during an interview can teach some things better than any lecture." This initial report was significant in documenting one of the first trials of closed-circuit television for teaching purposes. It was determined to be very effective and recommended for further use and development.

Subsequently, Wittson and his associates (1961) conducted an experiment involving two group psychotherapists, experienced in treating groups of five or six patients. A number of groups were selected for treatment. Half the groups were treated via closed-circuit video and the others were treated conventionally or interpersonally, each for six sessions. Two-way, closed-circuit video was established between rooms at the NPI. The groups were situated in one room and the therapist in the other. In addition, images of both the therapist and groups were shown in a monitoring room. This was to serve as a training room.

Unfortunately, no objective assessments were conducted or, at least, reported. Subjective interpretation of the experience suggested that use of the closed-circuit video link had less influence on the psychotherapeutic outcome than did the therapist and selection of group members. It was further noted that, after the initial adjustment period during the first session, patients felt comfortable using the equipment. Importantly, however, it was also noted that one group, primarily composed of patients with "antisocial" habits, did use the video link to increase resistance and they "whispered to each other without the therapist being able to hear." Further, "no relationship of trust developed in this group" (Wittson et al., 1961). This study was important in providing the first demonstration of the feasibility of two-way interactive video technology in group treatment. Moreover, it suggested some practical ways of how to arrange the group when using the technology and which type of camera lens would minimize distracting camera movements.

Also in Nebraska, somewhat later, Benschoter et al. (1965) summarized findings from their experiment with a 24-hour, closed-circuit, two-way video link between the NPI in Omaha and the state mental hospital located 112 miles away at Norfolk. The experiment was funded by the NIMH for a six-year period. The authors reported that the system was used successfully for a number of purposes, including: (1) consulting with NPI in difficult psychiatric and neurological diagnostic evaluations; (2) evaluating speech disorders and, in some cases, conducting speech therapy; (3) allowing family members in Omaha to schedule appointments to speak with patients, thereby reducing separation and/or loss of interest by family, insuring a continued interest in patients on behalf of the state hospital staff and conducting family therapy; (4) permitting vocational rehabilitation specialists to assist patients in finding jobs in Omaha; (5) discussing collaborative research projects and conducting seminars; (6) transmitting grand rounds from the NPI to the state hospital and, occasionally from the hospital to the NPI; and (7) using interviews with state hospital patients for teaching purposes at the NPI. Again, however, outcomes were not formally evaluated. Subjective assessments indicated the system was useful and ensured privacy. The experiment was significant also in representing the first documented use of two-way video communication between an academic center and a remote psychiatric hospital.

Menolascino and Osborne (1970) reported on an innovative application for the two-way, closed circuit video link (CCTV) used by Benschoter at the Omaha NPI. Sixty mentally retarded adults were transferred from a specialized state institution to the Norfolk State Mental Hospital, which had little experience with this condition. With the video link already in place, the opportunity arose to provide expert psychiatric consultation on mental retardation from the NPI to the resident staff of social workers, nurses, aides, and orderlies at the Norfolk Mental Hospital. The objectives were to train the staff in dealing with the mentally retarded patients and to oversee the implementation of programs to assist them. The link also provided an opportunity for informally assessing the usefulness of the CCTV for this type of consulting and instruction. The link was used one hour a day, three days a week, for 18 months. It was supplemented by monthly one-day visits by NPI psychiatrists to Norfolk.

Initially, the NPI psychiatrists questioned whether the technology would provide sufficiently "close" interpersonal contact to affect long-range treatment. There was also some question whether the Norfolk staff would accept the NPI psychiatrists' advice as valid, considering their separation and distance from the patients. However, over a short period, both psychiatrists and staff adapted to the technology and, reportedly, achieved significant treatment advances. Again, no quantitative assessment was undertaken. However, the descriptive information provided strongly suggests the effectiveness of this intervention. Among the subjective successes were dramatic clinical and quality-of-life improvements in the mentally retarded patients, coupled with a positive effect on the morale of the state hospital staff. The authors were extremely positive in their evaluations of the usefulness of the technology to offset rural manpower shortages, and of its potential as a consulting and training tool for psychiatric residents and other mental health professionals. This experiment is also significant as the earliest documented application of telepsychiatry for reaching remote underserved populations.

Wittson and Benschoter (1972) summarized the experiences of the NPI telepsychiatry projects from their inception in 1955 through 1972. In addition to providing additional information about the various projects, it was the first review to address directly the potential roadblocks to widespread acceptance of two-way interactive video within psychiatry.

Among the problems listed were: (1) initial hesitancy of staff members to use the technology; (2) technical problems such as sound pick-up and camera operations in larger groups; and, most important, (3) the high costs of equipment and transmission time. The then "state of the art" video-transmission equipment cost approximately $18,000. And 45 to 50 hours of transmission time per week amounted to approximately $48,000 per year.

Dr. Benschoter recently reported (personal communication, 1995) that, to her knowledge, there had been no further publications regarding the NPI programs subsequent to 1962. Further, NIMH funding for telepsychiatry at the NPI had expired in 1970, but related activities continued at Veterans' Administration hospitals until the 1980s when this funding ended as well. Apparently, the end of telepsychiatry efforts here did not reflect lack of clinical interest in the system, but the high cost of transmission. Third-party reimbursement for services was never sought because the telepsychiatry services were provided solely to state and VA hospitals. And, since services were provided within Nebraska, there was never any problem with inter-state licensing or credentials. Finally, project directors were careful to have prescriptions written and prescribed by the NPI psychiatrists countersigned by the local "remote" psychiatrist.

Telepsychiatry at Massachusetts General Hospital

In April, 1968, under the direction of Dr. Kenneth T. Bird, Boston's Massachusetts General Hospital (MGH) established a two-way microwave video link with a clinic or medical station at Logan International Airport about three miles away. The project was funded by the Health Services and Mental Health Administration. The station was equipped with cameras which could be controlled from MGH. The station was staffed 16 hours a day by nurse-clinicians and physicians were present for a few hours of peak activity at the airport each day (Dwyer, 1973). Shortly after the system became operational, patients in need of emergency psychiatric help at the station were referred for video consultation to a psychiatrist at MGH. Shortly thereafter, other psychiatrists and psychiatric residents at MGH also provided consultations over the video link. During a period of two-and-one-half years, consultations were provided for about 150 patients; the number of contacts per patient ranged from two to twenty (Dwyer, 1973). The subjective and anecdotal review of the experience concluded that, despite early skepticism, all users responded positively to the experience and were convinced of the potential for both one-to-one and group therapy. Communication with adolescents, children, and some schizophrenic patients was reported to be easier using the technology.

The MGH was next linked via micro-wave, bidirectional video to the Bedford Veterans Administration Hospital some 20 miles away. Funding for this project was provided by the Veterans Administration. This link was used for clinical consultations similar to those conducted at Logan Airport. However, one use of the VA Hospital link was to provide continuing education for foreign-trained psychiatrists (Fisch and Dwyer, 1972). Anecdotal evidence of the greater benefits derived from the two-way, video-audio link versus the unidirectional link previously used was reported. The interactive video was reportedly received positively by all psychiatrists involved as a training tool and interest was retained over the length of the course. Specifically, the following were attributed to the use of interactive video: (1) the participants increased their factual knowledge, (2) increased their ability to access medical literature, (3) increased interest in further learning, and (4) gained greater confidence in their professional abilities. The interactive video improved the attention level and increased responses to questions and other feedback from the students. Apparently, a key factor in the success was the small group experience and the "special relationship that the group members had with the prestigious teaching hospital."

Again, evidence suggests that the telepsychiatric services provided by the MGH to the Logan Airport Medical Station and the VA Hospital at Bedford were terminated with the expiration of funding support.

Telepsychiatry at Dartmouth-Claremont

In 1971, Solow and his associates (1971) reported on experiences with a closed-circuit video link between a Television Consultation Service in the Dartmouth Medical School's Department of Psychiatry and a rural hospital in Claremont, New Hampshire. The link had been operational since its funding by the NIMH in December, 1968. It filled a perceived need for community mental health care in the Claremont area. The local hospitals did not have a staff psychiatrist and the town's mental health clinic was staffed by two psychologists.

The purposes of the Dartmouth–Claremont video link were to: (1)

provide competent and timely mental health consultation without requiring patients to travel from the local environment, (2) enhance the education of family physicians in psychiatry, (3) alleviate the lack of psychiatric care in the area, and (4) evaluate the utility of interactive video as a medium for providing remote physicians readily available expert psychiatric assessment and consultation. The system employed a two-way, closed circuit television transmission with two micro-wave relay stations. Claremont Hospital had a studio and an auxiliary video room, while Dartmouth also had a studio. The system was operated fully by the psychiatrists and physicians. The established protocol allowed a Claremont doctor to call the Dartmouth psychiatrist on-call to schedule a time for a video conference. During the video consultation, the two physicians first discussed the patient. Subsequently, the patient entered the Claremont studio while the accompanying physician observed the interview from the auxiliary video room. Finally, the two physicians again spoke via video, without the patient's presence. The discussion of diagnosis and treatment was recorded.

A subjective assessment of the interaction found no problem with the system. Rapport was reportedly established between the patient and remote psychiatrist, and the psychiatrists reported being able to observe emotional nuances over the video link. Patients accepted the procedure. Importantly, it was believed that even paranoid patients did not suffer increased anxiety, nor did they subsequently incorporate the video interaction and system in their delusions. Patients as well as physicians reported "lost awareness" of the medium. Again, it must be emphasized that these results were only subjective assessments of the process and participants.

During a twelve-month period, 199 interviews were scheduled; only four had to be canceled due to technical difficulties. Of the scheduled interviews, 73 percent involved new patients; the remaining 27 percent were follow-up interviews. Despite initial fears, the system was not used primarily for handling emergency or "unwanted" patients. Rather, it was used for routine clinical purposes. Ninety percent of those patients evaluated via the interactive video system were treated in their local community. Physicians in the Claremont area were reported to have benefited from the experience as they viewed psychiatry as less mysterious. In addition, they reportedly became "more bold in prescribing psychotropic medications" (Solow et al., 1971). Thus, an anecdotal and subjective assessment of the system concluded that it not only provided psychiatric consultations, but also helped teach and demystify psychiatric interviewing and diagnosis for the referring general practitioner.

Despite the positive assessments, by 1976, the use of the Dartmouth-Claremont video link was "reduced to only a rare educational conference (Maxem, 1978). Several factors were cited for the decline from the previous intensive use of the system, including: (1) the presence of a psychiatrist at the Claremont Hospital and, consequently; (2) a decreased demand for consultation; and (3) declining enthusiasm. Although not specifically mentioned, another contributing factor was that the program was not financially self-supporting after the expiration of the NIMH funding.

Summary

The assessments of these three funded telepsychiatry projects are subjective and anecdotal and, therefore, must be viewed cautiously. With some exceptions there appears to have been positive acceptance by psychiatric patients, remote providers, and tertiary providers of psychiatric care and education. However, no project was capable of self-support subsequent to expiration of funding. Some attribute the historical failure of telemedicine systems to prosper beyond the time frame of their funding support to several factors: (1) telemedicine was an information system operating in a period when the knowledge, understanding, and application were in a nascent stage; (2) physicians and other health care providers were concerned primarily with "eliminating" distances and did not exploit other time savings possible through telemedicine; (3) telemedicine was considered merely a communication rather than a medical system; and (4) telemedicine was not organized and managed as a system within the tertiary medical centers or in remote areas and regions (Prestone et al., 1992).

Telephone Technology

The use of telecommunication technology is not new to psychiatry. By the early 1970s, the telephone was already routinely used for all phases of treatment by many psychiatrists (MacKinnon and Michels, 1970; Miller, 1973). In a survey of 58 psychiatrists, 97 percent reported using the telephone for emergencies during treatment, almost one-half used it as a planned adjunct of face-to-face contacts with certain patients, and about one in five used it as a primary or sole mode of treatment.

Table 10-2.

Telepsychiatric Applications Using Audio-Only Telephone Communication

Study	Year	Application	Equipment	Population	N	Locale	Quantitative Assessments?
Mackinnon & Michels [24]	1970	Psychotherapy	Telephone	Adult Outpatients	3	New York, NY	—
Grumet [25]	1979	Psychotherapy	Telephone	Female Outpatient	1	Rochester, NY	—
Wells et al. [26]	1988	Diagnostic Interview (DIS)	Telephone vs. Face to Face	Community Survey	230	Los Angeles, CA	Sens = .71, spec = .89, pred value = .63 for telephone vs. face-to-face Kappa = .57
Baer et al. [27]	1992	Rating Scales	IVR vs. Telephone vs. Paper-and-pencil	Outpatients with OCD	18	Boston, MA	Intraclass r = .99 for telephone, paper-pencil, and talking computer. Intraclass r = 1.00 for CGI
Fenig et al. [28]	1993	Interviewing	Telephone vs. Face to Face	Holocaust Survivors	153	Tel Aviv, Israel	r = .78 live vs. telephone; 6 of 27 items on demoraliz. scale sign. diff

Particularly important and promising are recent innovations using computer-assisted telephone systems incorporating digitized human speech to administer psychiatric rating scales. In one study, patients with obsessive-compulsive disorder were scheduled to complete the scales: (1) through the "talking-computer," (2) over the telephone with a human rater, and (3) using a paper and pencil format (Baer et al., 1992). The scales were completed one hour apart during a single day. The computer was equipped with the Articulate Systems Voice Navigator II and the Magnum TFLX system. The interclass correlation for all three methods of administration for the rating scales ranged from 0.99 to 1.00. The study demonstrated the efficiency and economy of the computer system in data collection and dealing with longitudinal assessments. In addition, patient travel time and clinician time were both reduced through the use of the computer assisted system.

Another recent study demonstrated the equivalency of telephone and face-to-face interviews in conducting a psychiatric epidemiology survey among Holocaust survivors in Israel (Fenig et al., 1993). In the experiment, face-to-face and telephone interviews were conducted among a control group and experimental group totaling 153. In the experimental group, the time interval between the telephone and face-to-face interviews was about 10 days. Both face-to-face and telephone interviews reflected a high level of reliability, with an interclass correlation of 0.78. The reported advantages of the telephone survey were: lower cost, broader sampling framework, capability of reaching people in distant and isolated places as well as in dangerous neighborhoods, and reaching people at odd hours. Among the disadvantages and differences noted were: missing items regarding sensitive issues and more positive responses pertaining to psychiatric symptoms, respectively. Nevertheless, the telephone in its traditional form as well as computer-enhanced telecommunication appear to be useful in clinical and other areas of psychiatry.

Conclusion

The concepts of telepsychiatry and its use for patient diagnosis and treatment as well as staff education have been explored periodically for nearly forty years. Unfortunately, with the exception of only the most recent studies, there has been no significant, rigorous, objective attempt in telepsychiatry to assess "face-to-face" communication vis-à-vis the traditional, direct person-to-person encounter. And, the several early, creative, and ambitious forays into telepsychiatry, as in other medical fields, have not been self-sustaining after external funding expired. In part, this may be attributed to the novelty of the procedure at the time and, in psychiatry, the dissonance between the technology and the absence of technical sophistication in diagnosis and treatment when compared to the practice of psychiatry today.

To sustain telepsychiatry and to enhance its diffusion as an effective and accepted component of psychiatric practice, it must be demonstrated to be a cost-effective and efficient means of improving care of remote and otherwise isolated populations. To this end, careful studies need to be conducted of the parameters limiting the reliability of remote diagnosis via interactive video. A psychiatry oriented to conversant evaluation and intrapsychic conflict, for example, may not suffer from image compression and distortion. However, a psychiatry which utilizes psychotropic medications, considers and relies upon brain imaging results, and is sensitive to neurologic symptoms may not readily tolerate image degradation, distortion, or lack of clarity associated with image compression technology.

Though still limited, information reflecting patient and provider satisfaction with telemedicine appears to be increasing. However, the psychiatric patient may present special issues depending upon diagnosis and setting. Past anecdotal reports are an insufficient basis from which to draw definite and general conclusions. For example, special attention should be paid to the needs of patients with disorders such as paranoid schizophrenia. It would also be useful to determine the extent, if any, to which telepsychiatry confers a sense of confidence in the "remote" consultant and whether compliance is enhanced or reduced.

With the movement towards capitated care, early identification and preventive interventions become much more salient in controlling health care costs. Insofar as current evidence indicates that underdiagnosis of substance abuse and mental disorders contributes substantially to increased health care costs and inappropriate utilization of primary care, cost-effective means of identifying enrollees at risk make sense from both a preventative health and financial perspective. Computer-assisted telephonic screening methods may be an important complementary strategy for addressing these issues. Furthermore, computer-assisted telephonic assessment for symptom change and side-effect profile resulting from the use of psychotropic therapies could prove a cost-effective and convenient way of monitoring treatment response in patients suffering from depression, anxiety disorders, and OCD.

The concentration of resources and expertise of the "new" psychiatry in a relatively few centers requires a delivery system to ensure rapid and geographically comprehensive distribution of information. This is essential to improving diagnosis and care across the United States. However, if these centers cannot "project" knowledge effectively and efficiently, the benefits of the gains will be severely limited. Telepsychiatry's potential, in conjunction with the development of the National Information Infrastructure, to contribute to those benefits is considerable.

REFERENCES

American Medical Association. 1990. "The Future of Psychiatry," *Journal of the American Medical Association,* 264:2542–48.

Ball, C. J., N. Scott, P. M. McLaren, and J. P. Watson. 1993. "Preliminary Evaluation of a Low-Cost Videoconferencing (LCVC) System for Remote Cognitive Testing of Adult Psychiatric Patients," *British Journal of Clinical Psychology,* 32:303–07.

Baer, L., D. G. Jacobs, P. Cukor, J. O'Laughlen, J. T. Coyle, and K. M. Magruder. 1995. "Fully Automated Telephone Screening Survey for Depression," submitted to *Journal of the American Medical Association.*

Baer, L., M. W. Brown-Beasley, J. Sorce, and A. I. Henriques. 1992. "Computer-Assisted Telephone Administration of a Structured Interview for Obsessive-Compulsive Disorder," *American Journal of Psychiatry,* 150:1737–38.

Benschoter, R. A., C. L. Wittson, and C. G. Ingham. 1965. "Teaching and Consultation by Television," *Hospital and Community Psychiatry,* 16:99–100.

Benschoter, R. A. (1995). Personal Communication.

Breakey, W., and M. Kaminsky. 1982. "An Assessment of Jarvis' Law in an Urban Catchment Area," *Hospital and Community Psychiatry,* 33:661–63.

Bushy, A. 1994. "When Your Client Lives in a Rural Area Part I: Rural Health Care Delivery Issues," *Issues in Mental Health Nursing,* 15:253–66.

Dongier, M., R. Tempier, M. Lalinec-Michaud, and D. Meunier. 1986. "Telepsychiatry: Psychiatric Consultation through Two-Way Television," *Canadian Journal of Psychiatry,* 31:32–34.

Dwyer, T. 1973. "Telepsychiatry: Psychiatric Consultation by Interactive Television," *American Journal of Psychiatry,* 130:865–69.

Fenig, S., I. Levay, R. Kohn, and N. Yelin. 1993. "Telephone vs. Face to Face Interviewing in a Community Psychiatric Survey," *American Journal of Public Health,* 83:896–98.

Fink, P., and S. Weinstein. 1979. "Whatever Happened to Psychiatry? The Deprofessionalization of Community Mental Health Centers," *American Journal of Psychiatry,* 136:406–09.

Fisch, A., and T. F. Dwyer. 1972. "Interactive Television in the Continuing Education of Foreign-Trained Psychiatrists," *Journal of Medical Education,* 47:912–14.

Goodman, W. K., L. H. Price, S. A. Rasmussen, C. Mazure, R. L. Fleischmann, C. L. Hill, G. R. Heninger, and D. S. Charney. 1989. "The Yale-Brown Obsessive Compulsive Scale (Y–BOCS). Part I: Development, Use, and Reliability," *Archives of General Psychiatry,* 46:1006–11.

Hamilton, M. 1959. "The Assessment of Anxiety States by Rating," *British Journal of Medical Psychology,* 32:50–59.

Hamilton, M. 1960. "A Rating Scale for Depression," *Journal of Neurology and Neurosurgical Psychiatry,* 23:56–62.

Human, J., and C. Wasem. 1991. "Rural Mental Health in America," *American Psychologist,* 46:232–39.

Jerome, L. 1986. "Telepsychiatry [Letter in Reply to Doninger et al. 1986]," *Canadian Journal of Psychiatry,* 31:489.

MacKinnon, R. A., and R. Michels. 1970. "The Role of the Telephone in the Psychiatric Interview," *Psychiatry,* 33:82–93.

Maxem, J. S. 1978. "Telecommunications in Psychiatry," *American Journal of Psychotherapy,* 32:450–56.

Menolaxcino, F. J., and R. G. Osborne. 1970. "Psychiatric Television Consultation for the Mentally Retarded," *American Journal of Psychiatry,* 127:515–20.

Miller, W. B. 1973. "The Telephone in Outpatient Psychotherapy," *American Journal of Psychotherapy,* 27:15–26.

Murray, J., and P. Keller. 1991. "Psychology and Rural America: Current Status and Future Directions," *American Psychologist,* 46:220–31.

Prestone, J., F. W. Brown, and B. Hartley. 1992. "Using Telemedicine to Improve Health Care in Distant Areas," *Hospital and Community Psychiatry,* 43:25.

Reed, E. 1992. "Adaptation: The Key to Community Psychiatric Practice in the Rural Setting," *Community Mental Health Journal,* 28:141–50.

Shannon, G., R. Bashshur, and J. Lovett. 1986. "Distance and the Use of Mental Health Services," *The Milbank Memorial Quarterly,* 64:302–30.

Siu, A., G. Ke, and J. Beck. 1989. "Geriatric Medicine in the United States: The Current Activities of Former Trainees," *Journal of the American Geriatric Society,* 37:272–76.

Solow, C., R. J. Weiss, B. J. Bergen, and C. J. Sanborn. 1971. "24 Hour Psychiatric Consultation via TV," *American Journal of Psychiatry,* 127:120–23.

Sommers, I. 1989. "Geographic Location and Mental Health Services Utilization among the Chronically Mentally Ill," *Community Mental Health Journal,* 25:132–44.

Straker, N., P. Mostyn, and C. Marshall. 1976. "The Use of Two-Way Interactive Television in Bringing Mental Health Services to the Inner City," *American Journal of Psychiatry,* 133:1202–05.

Tucker, G., J. Turner, and R. Chapman. 1981. "Problems in Attracting and Retaining Psychiatrists in Rural Areas," *Hospital & Community Psychiatry,* 32:118–20.

Weinstein, A., A. Hanley, L. Scott, and R. Stronde. 1976. "Services to the Mentally Disabled of Metropolitan Community Mental Health Center Catchment Area," NIMH Pub. No. B10-ADM-76-373. Washington, D.C.: U.S. Department of Health, Education, and Welfare.

Wittson, C. L., and R. Dutton. 1956. "A New Tool in Psychiatric Education," *Mental Hospitals,* 7:11–14.

Wittson, C. L., D. C. Afflect, and V. Johnson. 1961. "Two-Way Television in Group Therapy," *Mental Hospitals,* 12:22–23.

Wittson, C. L., and R. A. Benschoter. 1972. "Two-Way Television: Helping Medical Center Reach-Out," *American Journal of Psychiatry,* 129:624–27.

SECTION IV
TELEMEDICINE SYSTEMS

Chapter 11

TELEMEDICINE IN THE UNITED STATES

JIM GRIGSBY

Introduction

Telemedicine utilizes an array of technologies including facsimile, medical data transmission, audio-only format (telephone and radio), still visual images, and full-motion video. In addition, robotics (Minsky, 1979; Satava, 1992) and virtual reality (Kelly 1994) have been introduced into some experimental applications. Information is transmitted via terrestrial lines (copper telephone wires, coaxial cable, and optical fibers), terrestrial microwave, radio, and satellites.

During the past three decades, interest in telemedicine has waxed and waned. Recent improvements in communications and computer technology have made possible more accurate transmission of medical and other data. In conjunction with advances in the digitization and compression of data, high speed-computing, software innovations (e.g., data compression, image enhancement, graphical user interfaces), and the decreasing costs of technology, the availability of an improved transmission medium has moved telemedicine closer to wide-scale implementation (Grigsby et al., 1994).

Telemedicine has been adapted by a wide range of applications and specialties. Indeed, it is difficult to imagine a medical specialty or health care discipline that could not use telemedicine for at least some applications. This chapter discusses the current status of these activities in the United States. It begins with a brief survey of the scientific literature, followed by a detailed discussion of current telemedicine activities throughout the country. Research in radiology, pathology, and other clinical services and education has proceeded on relatively independent tracks, and the literature review is divided accordingly.

Current Status of Medicare Coverage and Payment Policy

From a policy perspective, teleradiology and telepathology stand apart from all other clinical applications. Traditionally, the Health Care Financing Administration (HCFA) has established a policy (the "face-to-face" rule) that denies coverage for Medicare beneficiaries of medical consultations rendered on the telephone out of concerns about excessive utilization, potential abuses, and uncertain quality. Hence, reimbursement for telemedicine services requires face-to-face contact if these services are ordinarily provided in person. Since diagnostic radiology and pathology do not ordinarily require face-to-face contact between provider and patient and specialists, these two services are exempt from the face-to-face rule. Consequently, the move from radiology to teleradiology and from pathology to telepathology was almost insignificant for a payer standpoint. Physicians continued to bill for these services as they had previously, and the same billing codes were used for teleradiology as radiology.

Other telemedicine applications approved for reimbursement by HCFA include the transmission and interpretation of electrocardiograms, and analysis of both single and dual chamber pacemakers (Physician Payment Review Commission, 1995). HCFA's coverage policy has been a contributing factor to the independent and rapid development of teleradiology, while inhibiting, to a certain extent, the growth of clinical telemedicine.

Teleradiology

Like pathology, diagnostic radiology deals primarily with the analysis of visual images. These two specialties are therefore the most easily adapted to telemedicine. One of the earliest papers on the feasibility of teleradiology was an editorial by Andrus and Bird in 1972, in which the authors described technical aspects of the use of interactive television for displaying images. Murphy et al. (1970) had previously shown that microwave transmission of radiological images was feasible, and Webber et al. (1973) developed an inexpensive system for radiology, using an amateur band television transmitter of low power. The latter system was found to be "less than satisfactory" for transmission of radiographs. But the authors suggested that technological advances would improve its efficacy.

By the 1980s, the technology had improved significantly. Mun and collaborators (1988) dealt with the general outline and advantages of, and approach to, technology assessment for such a system, and the same was done more recently by Ho et al. (1995). Templeton et al. (1988) discussed the technical parameters and equipment requirements for a comprehensive teleradiology system, and Huang et al. (1990) described image communication methods, focusing their attention especially on communication networks.

Systems have been developed in a variety of settings. One system was intended for use within a large general hospital (Hickey et al., 1990); another provided radiological consultation for a teaching-hospital-based family medicine center (Franken et al., 1989); and a military application involved field units in the Persian Gulf war (Cawthon et al., 1991). It is estimated that there are currently several thousand telemedicine consults in the United States (Allen and Allen, 1994; Franken et al., 1995). Most of these (perhaps as many as 7,000 or more) transmit images within an institution (intrafacility programs), or from a hospital to the homes and offices of radiologists who work in that hospital. There are, however, several regional telemedicine programs that provide services over intra- or interstate networks.

Allen and Allen (1994) attempted to identify teleradiology programs that provide services between facilities. Acknowledging that their survey may have missed some sites, they estimated that approximately 22,000 interfacility teleradiology cases were handled by about 160 radiologists from 12 programs in the first six months of 1994. To put these figures in perspective, the authors noted that approximately 257 million diagnostic radiological procedures were done in 1990. Hence, this is about .0085 percent of the total. The practice of teleradiology is likely to continue to expand rapidly. It is already becoming integrated into the routine practice of medicine differently from most of clinical telemedicine, and the American College of Radiology has now published standards for the use of teleradiology (ACR, 1994).

The effectiveness of teleradiology has improved significantly over the years. Early studies (e.g., Jelaso et al., 1978) found a relatively high rate of both false negatives and false positives, with overall accuracy unacceptably low (82% in the above-cited study). The Jelaso study utilized ordinary telephone lines for communication, with a video camera recording radiographic images from films displayed on a view box. The TV equipment had inadequate spatial resolution, and when the zoom capability of the camera was used, the signal-to-noise ratio deteriorated.

Recent research on the electronic transmission of radiological images has demonstrated higher levels of effectiveness. However, when used across the spectrum of diagnostic requirements, in a primary medical care setting (Paakkala et al., 1991), and in the emergency room (Kagetsu et al., 1987), video radiological images were associated with more incorrect diagnoses than conventional films, even though digital teleradiology was more accurate than in older studies. But the technology used in these studies was substandard in comparison to available equipment.

Teleradiology has been shown to be effective for consultation, although some reservations have been expressed about its use for primary diagnosis (Grigsby et al., 1994). For the detection of certain kinds of pathology, digital teleradiology, using either films or interactive display, is equal to conventional radiographs (Cox et al., 1990). It appears that even when digital methods show a statistically significant inferiority to analog display, the magnitude of the difference is not large. Effectiveness for some applications, however, has been somewhat problematic. This includes the interpretation of chest films, on which interstitial infiltrates and pneumothorax have been somewhat difficult to detect (Cox et al., 1990; Goodman, 1986; Kundel, 1986). Digital radiology also has been found to make the identification of subtle, nondisplaced fractures somewhat more difficult (Scott et al., 1993). Accurate mammography, difficult even using conventional methods, may be rather problematic when the data are digitized (Karssemeijer et al., 1993). The accuracy of excretory urography was reported to be somewhat limited (DiSantis et al., 1987), although significant improvements in technology since the time of that study probably would produce more favorable results today.

Telepathology

As is the case for much of telemedicine, many of the papers published on telepathology simply describe existing or proposed systems, providing no data on their effectiveness. Schwarzmann (1992), for example, described a telerobotic microscope system with which the pathologist is able to examine tissue specimens prepared by an on-site technician. Most work in pathology has involved the viewing of intra-operative frozen-tissue sections, but in several cases (e.g., Kayser and Drlicek, 1992) the methods and results were not clearly described, and the value of the research is difficult to determine. Shimosato et al. (1992) reported diagnostic accuracy of at least 90.5 percent with a fiberoptic high-definition television (HDTV) system using a real time moving image, and 69 percent with a static image over an ISDN network with high-definition TV monitors. The findings suggest that moving images permit greater accuracy than static ones, but the reader is left in the dark about several aspects of design, and cannot put much confidence in the results.

In recent years, the effectiveness of telepathology has increased significantly. The use of telecommunications networks to provide remote pathological services has been evaluated in a number of studies (Bhattacharyya et al., 1995; Eide and Nordrum, 1992; Eide et al., 1992; Nordrum et al., 1991; Weinstein et al., 1987; Weinstein et al., 1992), and the use of the technology has spread rapidly for both diagnosis and consultation. According to Perednia and Allen (1995), approximately 60 dedicated telepathology systems were in operation in the United States by the end of 1994.

Despite its growth and proliferation, the effectiveness of telepathology has not been as firmly established as that of teleradiology. Research has been conducted with a limited range of tissue types, and has focused primarily on frozen-section studies. Nonetheless, the feasibility of telepathology has been established, and those pathologists who use it are obviously persuaded of its accuracy despite the limited support of empirical research.

Telerobotically controlled microscope systems, which many pathologists consider necessary, are expensive. The greatest need for telepathology in rural areas is likely to be for intraoperative frozen-section studies; most other pathological studies do not require immediate analysis and can wait for a visiting pathologist, or they may be sent elsewhere by overnight delivery. Because the number of frozen-section studies required may be small, the growth of telepathology is likely to be considerably slower than that of teleradiology.

Clinical Telemedicine

Feasibility of Telemedicine

The bulk of the published literature supports the feasibility of telemedicine, but does little to illuminate its effectiveness (Grigsby et al., 1995; Perednia and Allen, 1995). Papers concerning feasibility generally fall into one of four categories: case studies (e.g., Siderfin, 1995), descriptions of telemedicine projects (e.g., et al., 1992; Delaplain et al., 1993), descriptions of specific applications (Hubble, 1992), and discussions of the use of telemedicine by specific specialties or subspecialties (Perednia and Brown, 1995).

The use of telemedicine to provide remote consultation can be accomplished readily using current technology. Projects in a wide range of venues and clinical contexts have demonstrated the feasibility of telemedicine beyond any doubt. More specifically, the feasibility of a number of specialties and subspecialties has been demonstrated, including anesthesiology (Gravenstein et al., 1975), cardiology (Murphy et al., 1973; Mattioli et al., 1992), critical care medicine (Grundy et al., 1982), dermatology (Murphy, et al., 1972; Perednia and Brown, 1995), neurology (Chaves-Carballo, 1992; Hubble, 1992), otorhinolaryngology (Pedersen et al., 1994; Rinde et al., 1993), pediatrics (Cunningham et al., 1978), and psychiatry (Brown, 1995; Preston et al., 1992).

It also has been demonstrated that telemedicine is workable in a broad range of environments. NASA, for example, was one of the earliest users of telemedicine as a means of providing care for astronauts (Pool et al., 1975). More recently, NASA has used the technology to establish links with Armenia and Russia for purposes of disaster relief (Houtchens et al., 1993; Nicogossian, 1989). The Department of Defense attempted telemedicine in the context of desert warfare (Cawthon et al., 1991), in conjunction with mobile medical units, demonstrating its integrative capability under difficult geographic and climatologic circumstances. Finally, telemedicine has shown its utility in the more prosaic settings of the home, clinic, hospital, and long-term care facility.

Effectiveness of Clinical Telemedicine

Obviously telemedicine is clinically feasible, but the degree to which it is medically effective in all clinical areas is still subject to investigation. Several clinical studies have been conducted; most of these were narrowly focused and used very small samples. The results, therefore, are not fully generalizable, and their reliability is not always certain. For example, interactive video was compared with in-person consultation for the examination and rating of patients with Parkinson's Disease (Hubble, 1992; Hubble et al., 1993). The Hoehn and Yahr score (Hoehn and Yahr, 1967) was used for staging Parkinson's Disease (PD), and the Unified Parkinson's Disease Rating Scale, or UPDRS (Fahn et al., 1987). These are two standardized measures of functioning for persons with PD. The authors reported a correlation of 0.91 between in-person and interactive television (IATV) ratings, which suggests good reliability. A close examination of the table containing individual patients' scores, however, reveals a significant amount of variability for some patients. One patient, for example, obtained a UPDRS motor score of 31 during face-to-face consultation, and 22 over IATV; another obtained scores of 8 and 14, respectively.

These differences may be a function of a number of factors, including the normal variability of PD patients who are being treated with dopaminergic medications, as well as interrater reliability of the UPDRS (Martínez-Martín and Bermejo-Pareja, 1988). Also, they may reflect real differences between IATV and face-to-face examination. Parkinsonian tremor may be difficult to detect with compressed video, and the assessment of dystonia depends on the skill of the examiner. While these authors discussed some of these issues, given the small sample size (n = 9) and the lack of covariate data, it is not possible to interpret the observed differences.

Cardiology is probably the best studied specialty in clinical telemedicine. Murphy et al. (1973) examined the use of an electronic stethoscope for auscultation, examining 50 persons, 27 of whom were patients whose diagnoses were already known. All systolic murmurs heard by direct auscultation were also detected by the electronic stethoscope. There were neither false positives nor false negatives. There was a slight difference in the grading of murmurs by the two methods in five of 24 cases, but no consistent direction to the discrepancy. The findings were somewhat more mixed with respect to diastolic murmurs. There was 100 percent agreement on diastolic murmurs heard at the left sternal border, but only 75 percent (six of eight) on those at the apex, in favor of direct auscultation. The two missed murmurs were described as "faint (Gr 1/6), rumbling murmurs." In three cases there were differences in grading.

The study by Murphy et al. is over 20 years old, and the equipment has undoubtedly improved over time, but it appears that only two other studies have examined the issue of the accuracy of auscultation using a remote electronic stethoscope. Mattioli and associates (1992) used an electronic stethoscope for remote auscultation with a sample of seven pediatric patients. The authors reported generally good sensitivity and specificity when the data were compared with the results of conventional auscultation. They noted that the remote equipment demonstrated 100 percent sensitivity for the presence of cardiac disease and for the need for follow-up. In a later study of acoustic and electronic stethoscopes among 78 conventional pediatric cardiology outpatients and 38 telecardiology patients, Belmont et al. (1995) found telestethoscopy useful for screening (with 89.5% accuracy), but less accurate than conventional methods. The authors discussed a number of reasons for the sometimes unsatisfactory interrater reliabilities in auscultation (with either type of stethoscope). The results of these studies of auscultation suggest that the electronic stethoscope may be adequate, but further research is clearly indicated.

Sobczyk et al. (1993) studied the accuracy of pediatric echocardiography, transmitting data via modem over standard telephone lines. In a series of 47 patients (24 of whom were normal), 83 percent "were thought to give accurate diagnostic impressions compared with videotape review." Of the incorrect diagnoses, the authors asserted that only one resulted from problems with the transmission of information. The others, they argued, resulted from "the selection and transmission of an image without sufficient information to allow a definite diagnosis." They did not describe the data from which they reached this conclusion.

Transmission of electrocardiograms has been studied by at least two groups who reported that an off-the-shelf fax machine (Bertrand et al., 1994) and a digitizing flatbed scanner with transmission modem (Ong et al., 1995), produced images that receiving cardiologists rated as either "good" or "excellent" in quality in every case. EKG transmission is relatively common in the United States; analysis of Medicare claims data found that approximately 200,000 such transmissions took place for Medicare beneficiaries in 1993 (Physician Payment Review Commission, 1995).

Murphy et al. (1972) were the first to study the accuracy of dermatologic diagnosis, using a set of 75 color slides projected onto a screen to produce an image measuring 3 × 2.5 feet. The image was then photographed by a television camera and displayed on both black and white and color television monitors. Physicians made diagnoses from either direct viewing of the slides or from the televised images. The actual diagnoses were known for each slide, serving as a "gold standard" against which accuracy could be compared. The authors reported that diagnostic accuracy was slightly lower for the televised images than for direct examination of the slides, due in large part to the equipment used. Current video monitors provide much better resolution than those of 23 years ago. Color images yielded slightly greater accuracy than did black and white. Perednia and Brown (1995) described a teledermatology research project in progress, but apparently there have been no other studies of the effectiveness of teledermatology.

The effectiveness of psychiatry is notoriously difficult to demonstrate. Outcomes are difficult to define, and concrete, measurable indicators often fail to capture the salient phenomena. Diagnostic accuracy would be a poor measure of telepsychiatry's effectiveness, since diagnostic accuracy and reliability for conventional psychiatry are poor. Nevertheless, there have been attempts to evaluate the effectiveness of telepsychiatry. Dongier et al. (1986) reported preliminary data from a study of 50 patients selected for IATV interviewing and 35 controls who were interviewed face-to-face. They reported no significant differences in level of satisfaction with the two approaches on the part of patients. In contrast, both consultants and primary care providers rated the IATV interviews as inferior to in-person interviews.

It appears that the only other attempt at a controlled study of telepsychiatry was conducted by Ball et al. (1995). They examined four modes of interaction: face-to-face, telephone, hands-free telephone, and a desktop computer-based video-conference system. The research design was not clearly described, and the sample was very small (only six patients) and heavily weighted toward severe psychopathology. Three patients were schizophrenics, one had a paranoid disorder, one had depression, and one had a "mixed neurotic disorder." All were well-known to their providers. The authors reported that, among these seriously ill patients, the video system "induced the greatest frustration, the least sense of having been understood, and the most disappointment with the consultation" (p. 24). On the other hand, there were no differences among the six physicians with the four modes. The sample was too small and unrepresentative, and the findings are inconclusive. These results were the opposite of those obtained by Dongier et al., who had found providers to be more critical than patients.

Pedersen and his associates (1994) conducted a study of the effectiveness of telemedicine for otorhinolaryngology, finding that an ENT (ear, nose and throat) specialist was able to make diagnoses using the telemedicine system. In one experimental condition, after having examined the patient by interactive television from another room, the specialist joined the patient and conducted a face-to-face examination. There was complete concordance for the specialist's diagnoses on all 17 patients. These examinations were not, however, validated by independent observers; the same physician appears to have done both examinations.

In a study of fetal ultrasound, Fisk et al. (1995, p. 41) reported that physicians found ultrasound images sent via telemedicine indistinguishable from standard ultrasound, "with almost no perceptible loss of picture quality or frame rate at the receiving end."

Clinicians and investigators alike have expressed considerable interest in trying to understand how telemedicine affects patient management. Very few studies on this issue have been published. A report by Houtchens et al. (1993) on NASA's Spacebridge projects to the former Soviet republics of Russia and Armenia provided some useful information. In these two projects, consults were conducted for a total of 209 patients. In 54 of these cases, diagnoses were changed as a result of the consults. In addition, the interpretation of 27 diagnostic studies was altered, and the diagnostic process and treatment plans were both changed in 47 cases. Nearly half of the diagnoses were changed in surgical cases. In large part, this reflects the differences in diagnosis and treatment between American and Russian medicine. Also, the data suggest that patient management might be significantly altered when telemedicine is used internationally. When used domestically, these effects are likely to be less striking.

Fit Between Technology and Clinical Functions

An important issue in telemedicine is the level of technology necessary for specific clinical tasks. On the one hand, it may be argued that telemedicine should be organized as comprehensive systems of care, and that telemedicine networks should invest in the technology that permit the provision of a wide range of clinical services. But it seems likely that there will always be areas that are too remote, too poor, or otherwise too disadvantaged to support the infrastructure and equipment necessary for such a comprehensive system. The question then becomes: What can be accomplished with different levels of technology?

It has been demonstrated, albeit in only a handful of research projects, that the telephone alone may be effective in providing a good deal of medical care. For example, in one VA Medical Center substitution of clinician-initiated telephone calls for some clinic visits reduced medical care utilization without adversely affecting patient health status (Wasson et al., 1992). Patients randomly assigned to a group receiving in-person care at twice the usual interval until clinic follow-up, augmented by scheduled telephone consultation between clinic visits, had fewer total clinic visits, less medication use, fewer days of hospitalization, and fewer ICU days than patients assigned to the in-person group. Similar findings have been reported in other studies (e.g., Bertera and Bertera, 1981).

Few studies have compared the effectiveness of different levels of technology in telemedicine consultation, but those that have yielded some interesting findings. Moore et al. (1975) examined the use of telemedicine in three clinics staffed by nurses able to consult with physicians at a different site by telephone. In this study, in accordance with a predefined schedule, nurses who needed a physician consult used either the telephone or an interactive television system. During the study, 354 patients were seen for whom the nurses required physician consultation (out of a total of 1,408 visits). Telephone consults were twice as likely to result in a patient traveling immediately to the hospital for a visit. With television, when follow-up was required, it could more frequently be done in a neighborhood outpatient clinic. Patients described themselves as generally satisfied with both modes of delivery. Providers tended to prefer whichever mode they had just used.

An experimental study compared the effectiveness of conventional medical care with four modes of telemedicine: color television, black and white television, still black and white images, and hands-free telephone (Conrath et al., 1977). Each of 1,015 patients attending rural clinics in Ontario was randomly assigned to one of these four groups while waiting to be examined face-to-face by another physician. Thus, every patient had two successive physician appointments, one by telemedicine and one in person.

In comparing all telemedicine diagnoses with those obtained in person, concordance was 61 percent, which compared favorably to what these researchers had previously found for two attending physicians, each of whom conducted face-to-face examinations with the patient. Accuracy of diagnosis did not differ for the groups assigned to different levels of technology, with the exception of dermatology, where the ability to see skin lesions was very important. Despite the equivalence of diagnostic accuracy among the four types of technology, physicians expressed greater confidence in diagnoses made using the interactive television systems. There were no significant differences in patient management among the four modes, or between these modes and face-to-face consultation. Similarly, there were no striking differences in patient satisfaction among the different modes, except for a slight preference for color IATV.

The quality of video equipment has improved significantly since this study was conducted. It is possible that, if this study were replicated with current equipment, the findings would be different, with the results favoring color IATV. The color television of the mid-1970s lacked the fidelity and resolution of the equipment now available. Nevertheless, it appears likely that the use of audio-only systems might be clinically effective for a variety of health problems.

Extrapolating from the findings of the Moore et al. and Conrath et al. studies, it was speculated that a larger percentage of the care rendered by telemedicine may be completed with audio information only. The addition of static visual images would provide an incremental improvement in diagnostic accuracy, especially for such specialties as dermatology. The use of short, full-motion video clips might be beneficial in another percentage of cases, especially for the observation of such phenomena as gait, tremor, and range of motion. These could all be handled by telephone, radio, or store-and-forward video technology (in which visual, verbal, or other data are acquired and transmitted to another site for later review). Hence, given current capabilities and costs of existing technology, full-motion real-time interactive video may be necessary in only a small percentage of cases coming to the attention of the remote provider. However, significant changes that might occur in any or all of these variables (the technologic capability, the cost of equipment and the specific uses and clinical applications) will probably alter this assessment.

Acceptability of Clinical Telemedicine to Patients

Anecdotal data suggest that most patients are satisfied with the services they receive via telemedicine, but little research has been published on patient satisfaction. Bashshur (1978) examined both community attitudes toward telemedicine, and the effects of experience on telemedicine on those attitudes. The community survey found that, among persons not yet exposed to telemedicine, large majorities believed that telemedicine would be less satisfactory than face-to-face medical care.

In the same study, Bashshur asked a sample of 72 telemedicine patients from rural Maine to complete attitudinal instruments prior to and after their first telemedicine encounters. Before telemedicine, the sample was almost evenly divided among those who thought telemedicine would be the same as an in-person visit ($n = 25$), those who thought it would create problems ($n = 23$), and those who said they didn't know ($n = 24$). Following the telemedicine session, 67 percent thought it had been about the same as in-person care and only 17 percent thought it was less satisfactory than a face-to-face visit. The remainder were unsure, and no one thought telemedicine was superior. As Bashshur noted, "familiarity did breed comfort" (p. 36).

Allen and Hayes (1995) administered a short survey to 39 cancer patients following their first tele-oncology session, and followed up with a similar survey with 21 of these patients, after a subsequent in-person visit to the same physician who had conducted the telemedicine consult. Patients were generally satisfied with telemedicine, although they found it more difficult to be candid over the video system. After the in-person visit, they described themselves as less inclined to use the IATV system again. Their reluctance to use the system again is in contrast with the findings of Pedersen and Holland (1995) of the University Hospital of Tromsø in Norway, who surveyed 24 ear, nose, and throat patients after they had undergone tele-endoscopy. Only one patient expressed dissatisfaction; 18 were "very satisfied" with the examination. Twenty-one of the 24 said they would prefer to undergo tele-endoscopy again on another occasion rather than having a specialist travel 250 kilometers from a tertiary care facility or traveling that distance themselves.

Based on the available data, whether anecdotal or obtained systematically, most patients tend to be satisfied with the services they receive. Even among psychotic patients utilizing telepsychiatry, informal reports of negative reactions were rare. Clearly, many people are satisfied with telemedicine, but more important are questions concerning why some patients don't like telemedicine, how these attitudes may change over time with more telemedicine experience, and what changes can be made in the process of care delivery in telemedicine to produce more favorable responses.

In evaluating patient satisfaction with telemedicine, it should be appreciated that the base rate for satisfaction with conventional medical care is generally quite high. People are usually satisfied with the care they receive. However, in view of the limited experience with telemedicine and self-selection among telemedicine patients, it is difficult to assess stable attitudes and satisfaction with telemedicine. Finally, just as few patients have used this technology, few physicians have done so. Those providers currently using telemedicine are generally interested in it; want to promote it; and are thus invested in making their patients feel comfortable. As more providers become involved in telemedicine, patients' perceptions may change.

Provider Attitudes Toward Telemedicine

In view of providers' role as key gatekeepers in the care process, their satisfaction with the technology may be more critical than patients' satisfaction. Higgins et al. (1984) studied provider acceptance of telemedicine in the Sioux Lookout program in Ontario. The sample was too small to draw reliable conclusions, and the method was not described in detail. Nevertheless, it appeared that the 34 nurses surveyed were more positive about telemedicine than were the four physicians. Two physicians described themselves as "positive" about the system, while two were "neutral." The authors concluded that "provider acceptance of telemedicine is extremely difficult to measure" (p. 288).

Fuchs (1979) studied the satisfaction with telemedicine of providers in the STARPAHC project. He interviewed 47 physician and nonphysician providers and administrators in 89 interviews. Concerns were frequently expressed about the reliability of the equipment (which failed "only 2 to 5% of the time"), and about the time required for teleconsultation.

More recently, Allen and his associates (1995) examined medical oncologist satisfaction with IATV for an initial patient visit. Three oncologists completed a questionnaire that inquired about their contacts with each of 34 patients over a period of four days. They also completed a second, brief questionnaire at the end of each teleoncology clinic day. A variant of the first questionnaire was completed after each patient visit on a day when seven patients were seen face-to-face in clinic. As the authors noted, the sample was too small to reach any conclusions, and there was no variability in the responses to the survey regarding in-person care (all three physicians gave the maximum possible score for face-to-face consultation). Nevertheless, ratings of telemedicine were generally favorable. Specific responses reflected frustration with use of the equipment, concern about the amount of relevant information transmitted, and difficulty asking intimate questions.

In addition to concerns about the effectiveness of telemedicine, there are several other reasons for physicians' reluctance to become involved in telemedicine. First, physicians in rural areas tend to react to the introduction of telemedicine in much the same way that most people respond to anything new: they think about its likely effects on them. In remote areas with few resources, isolated providers are likely to see the advantages of telemedicine in providing them and their patients ready access to clinical expertise. They may perceive other benefits as well, such as computerized databases, literature searches, opportunities for interaction with colleagues, and continuing medical education.

In rural areas not severely underserved, the reaction may be quite different. For instance, when a community has a 100-bed hospital and 10–20 physicians, practice patterns are generally well established, and a referral network is in place. The introduction of a telemedicine program into that type of community may be perceived as a threat to the *status quo*, including the potential loss of patients and income. Telemedicine sites have been established in some rural communities that were rarely or never used, in large part because local physicians perceived them as a predatory marketing scheme.

In academic medical centers, physicians are generally on salary and the income they generate from patient care goes to the practice plan. In these settings, there may be little incentive to assume what appears to be additional work especially if it has no effect on tenure or promotion, if it increases clinical responsibilities without offsetting savings elsewhere, and if the telemedicine clinic is inconveniently located.

Inertial resistance to telemedicine may best be overcome through actual use of the technology. Resistance due to a reluctance to change one's habitual mode of practice, to the extent it exists, is likely to dissipate with experience. Telemedicine may be considered a novelty, even a fad by some physicians. For telemedicine to become an integral part of the health care systems, the technology may have to become a part of medical training. This holds true for nurses, physician assistants, medical students, residents, fellows, and therapists (e.g., physical therapists, speech and language pathologists).

Current Telemedicine Activities in the United States

Scope

In 1990, only four telemedicine programs in all of North America used real-time videoconferencing for patient care. By the beginning of 1993, there were perhaps 15 operational telemedicine programs in the United States, excluding teleradiology (Grigsby, 1995). As of spring, 1995, as many as 30 or 35 programs may be actively involved in patient care, and another 40 to 50, in various stages of development. Perednia and Allen (1995) reported that at the end of 1994 there were telemedicine programs in at least 40 states, although many of these were probably not being used for regular patient care.

Patient volumes have been quite low, and even many well-established, well-funded telemedicine programs see as few as two to four patients a week. In an informal survey of the most active programs in North America, it was estimated that 2,250 patients were seen through IATV consultation in all of North America (Perednia and Allen, 1995). A substantial percentage of those patients were served through Canadian programs, and about 1,000 were renal dialysis patients seen for 3- to 5-minute nephrology consults through the Texas Telemedicine Project. In the first six months of 1994, approximately 500 IATV mental health interactions occurred between patients and providers in the United States (Allen, 1994).

The small numbers of patients receiving services through telemedicine should not be viewed as a reflection of intrinsic limitations in the concept or merit of telemedicine. Rather, many factors are included, including the relative newness of telemedicine, the attitudes of health care providers, the lack of insurance coverage for telemedicine services, and others. With time, and with the growth of telemedicine into such niches as managed care and home health care, volumes will increase significantly.

Technology

Transmission Media. Telemedicine data transmission has employed terrestrial lines (copper telephone wires and optical fibers), terrestrial microwave, radio, and satellites. The *bandwidth* or *bit rate* of the transmission medium (these terms refer to the number of bits that may be sent per second) has limited the type of telemedicine system to be used. Narrow bandwidth systems, such as ordinary telephone lines, are relatively inexpensive to operate, but because of their limited capacity they are restricted to the transmission of voice and digitized data which do not need to be sent in real-time.

Very broad bandwidth networks, including fiber optic cable and many satellite systems, are capable of carrying sufficient data to allow transmission of interactive, full-motion video. Lines designated as *T1* have a capacity of 1.544 megabits per second (mbps), and T3 lines carry 44.736 mbps. By using data compression algorithms, interactive television (IATV) may be used with bandwidths as narrow as 384 kilobits per second (kbps), or approximately $1/4$ of the capacity of a T1 line, but the images thus transmitted frequently appear jerky. The use of broad band networks is expensive, since bit rate is directly related to line charges.

In the last ten years, fiber optic cable and T1 twisted copper lines have become available even in non-metropolitan areas of many states. North Carolina is developing a statewide T3 network, and Iowa has an extensive fiber optic telecommunications network. Millions of businesses and private residences are now attached to cable networks for receiving television, but these connections have the potential to allow two-way communication, and could be used for telemedicine. Currently, most telemedicine programs rely on terrestrial land lines, which are generally quite expensive, but which are less costly than satellite transmission.

Technological advances over the next ten years may make the use of communications satellites affordable. For now, their cost is a significant barrier to their utilization for telemedicine. Nevertheless, for certain very remote areas, satellites may be the only feasible option. This is not a problem for an agency like NASA, which has access to its own broadband satellites for international disaster telemedicine. Smaller operations not as well funded, however, may be able to afford only older, narrow bandwidth satellites. This is true for SatelLife, a program that provides limited medical consultation services to a number of Third World countries.

Equipment. Early telemedicine programs in the United States primarily utilized IATV, which was employed in the service of real-time patient examination. The Memorial University of Newfoundland (MUN) experimented with a one-way video, two-way audio link between the medical center and several other provincial hospitals, but the cost of live television was prohibitive, and MUN moved to two-way audio conferencing, supplemented by digitizing tablets. MUN does have, however, the capacity to use full-motion video and transmission over T1 or fractional T1 lines.

Most telemedicine projects developed over the past five years have also used IATV. In 1993, telemedicine was almost synonymous with real-time videoconferencing. This has begun to change since that time. Although costs for equipment have been falling (they may now be at one-quarter to one-third of their level at the beginning of 1993), standard videoconferencing IATV may not be as cost-effective as desktop computer-based systems (which are themselves one-quarter to one-third as costly as studio video-conferencing equipment). The larger systems are therefore falling somewhat out of favor. Moreover, the great expense of line access charges, especially for the bandwidth necessary for real-time video transmission, has made store-and-forward systems very attractive. It has also encouraged the development of techniques for using the Internet in telemedicine.

Organizational Characteristics of Telemedicine Programs

In the early days of telemedicine, a relatively wide variety of institutions established telemedicine projects. These included NASA, the Indian Health Service, general hospitals, a state hospital, at least one county government, group practices, nursing homes, prisons, neighborhood health centers, and academic medical centers. Early funding was provided by the Department of Health and Human Services (formerly Health, Education and Welfare), the National Science Foundation, the Office of Economic Opportunity, the Department of Veterans Affairs, and NASA.

In its recent resurgence, telemedicine has been fostered primarily within academic medical centers, although the level of commitment and support in this environment has varied. In few cases, strong administrative support for telemedicine has been provided throughout the institution. For most programs, support has been lacking, central coordination has been unavailable, and telemedicine has not developed as rapidly as it might. The Mayo Clinic began using telemedicine in the late 1980s, and, gradually, has been developing a national and international network. Rural Options for Development and Educational Opportunities, a mental health program in Eastern Oregon (RODEO NET), the Eastern Montana Telemedicine Project in Billings, and the Texas Telemedicine Project were among the few programs that developed outside of academic institutions.

In many cases, telemedicine networks have been initiated by academic medical centers, and rural communities and providers played a secondary role in planning and implementation. In a few cases, the failure to design such networks on the basis of the needs and preferences of the local communities and providers has resulted in limited support for them. As word of telemedicine spread, the administrators of many rural hospitals began to view the technology as a possible boon to their facilities, or even as a key to survival. Rural hospitals have subsequently taken a more active role in joining and planning these partnerships.

Recently, telemedicine has moved out of the academic environment. Through 1994 and 1995, telemedicine programs have been developing in a number of different kinds of institutions. The Department of Veterans Affairs established a Task Force on Telemedicine in March 1995, and charged it with identifying and evaluating the best means of maximizing the benefits of telemedicine. A relatively large number of teleradiology and telemedicine systems is either already in operation or in development throughout the VA nationwide, but these systems have been established primarily by individual VA facilities, and have not had any central direction. The Department of Defense has been exploring the use of telemedicine for CHAMPUS beneficiaries, and for its personnel, in many areas of the country.

Some private corporations and investor groups have been lured by the potential for earnings. Private urban hospitals often view telemedicine as a means of obtaining referrals from rural markets, and many plan to use the technology to secure their positions in an expanding health care network. Many domestic enterprises have turned their attention overseas to such foreign markets, including China, Japan, Korea, and the Middle East.

With a few exceptions, managed care organizations have been slow to start telemedicine projects of their own. It seems likely, however, within closed health care systems such as HMOs, the VA, prisons, and the military, that telemedicine is most likely to prove a cost-effective vehicle for delivery of health services. For fully capitated HMOs, telemedicine could permit the efficient penetration of smaller markets, with optimal use of local nonphysician providers (nurse practitioners and physician assistants), and utilization of specialists in urban areas.

Home health care has grown dramatically since 1970, accompanied by a striking increase in the provision of Medicare-covered home health services (Kemper, 1992; Shaughnessy et al., 1994). Telemedicine has made some preliminary forays into this market, with programs for monitoring the pulmonary function of persons who have undergone lung or heart-lung transplantation (Finkelstein et al., 1993), and for cardiac rehabilitation (Sparks et al., 1993). Several telemedicine-home health care projects are in the planning stages, although the costs for equipment remain somewhat high. The Medical College of Georgia, jointly with the Georgia Institute of Technology and the U.S. Army, hopes to take advantage of the presence of cable installed into individual residences for home television for its Electronic House Call project. The use of telemedicine also is being evaluated in two programs in Kansas. Long-term care facilities represent another largely untapped market.

Populations Served

Bashshur and his associates have discussed the wide range of settings and populations served by the first generation of telemedicine programs (Bashshur, 1979; Bashshur and Armstrong, 1976; Bashshur et al., 1975; Lovett and Bashshur, 1979). Since that time, telemedicine has changed significantly, irrespective of the technological improvements.

Early telemedicine was established in both urban and rural areas, and in a variety of places such as nursing homes, jails, airports, Indian reservations, and state psychiatric hospitals. In recent years, the emphasis has been on providing medical care in rural areas. In part, this may reflect the rural focus of the federal agencies providing funding for research and demonstration projects (e.g., the Rural Utilities Service [formerly Rural Electrification Administration], and the Federal Office of Rural Health Policy). Nearly all of the 10–15 operational telemedicine programs in the United States that started in 1993 served rural populations.

Generally, the primary rationale for the development of telemedicine has been to serve populations with limited access to traditional medical services. Rural residents have received considerable attention. Depending on definition, between one-sixth to one-quarter of the U.S. population lives in rural or non-metropolitan counties (Office of Technology Assessment [OTA], 1990). The Western and Great Plains States, in particular, have large, sparsely populated rural areas. For instance, in Montana in 1991, fewer than 200,000 persons of the state's population lived in metropolitan areas. The entire state has a population of about 800,000 persons, less than the population in and around Birmingham, Alabama. But Montana is nearly three times as large as the entire state of Alabama (U.S. Bureau of the Census, 1993).

Overall, residents of rural areas do not differ significantly from urban dwellers with respect to the incidence of acute health problems. Many chronic medical conditions such as arthritis, hypertension, diabetes, and heart disease are more prevalent in rural than urban areas (Norton and McManus, 1989). There are also more elderly persons in rural areas who usually have greater need for health services than younger persons (Meade, 1992). Despite higher prevalence of chronic illness among rural residents, they are somewhat less likely than their urban counterparts to utilize outpatient health services (OTA, 1990), especially specialty and subspecialty care.

Other medically underserved segments of the population targeted by telemedicine include American Indians (Bashshur, 1979; Justice and Decker, 1979), residents of islands that are former U.S. Trust Territories in the Pacific (Delaplain et al., 1993), U.S. military personnel, and inmates of federal, state, and local correctional facilities (Sanders and Sasmor, 1973).

One of the early telemedicine programs was Space Technology Applied to Rural Papago Advanced Health Care (STARPAHC). This project was funded by NASA, equipped by Lockheed, and implemented by the Indian Health Service on the Papago (Tohono Odham) Indian reservation in Arizona in cooperation with the Tohono Odham. STARPAHC utilized telecommunications technology in conjunction with a fixed service unit and a mobile health unit—a bus outfitted as a small clinic, demonstrating the feasibility of using advanced technology to bring medical services to remote areas (Bashshur, 1979), in this case in conjunction with mobile health services. According to Justice and Decker (1979), the STARPAHC program was as clinically effective in providing health care as was the case for the regular clinics staffed by physicians within the Indian Health Service system.

Prisons are good candidates for telemedicine systems for several reasons. Prisons frequently are located in rural areas. Often, when prison inmates become ill, they must be transported from the prison to a secure medical facility elsewhere. This poses an escape risk, and the prisoner ordinarily must be accompanied by two or three guards. The cost of obtaining care for inmates is thus inflated by the expense of transportation and security. This can be obviated with telemedicine.

Moreover, it is not uncommon for prisoners to feign illness to get a break from the monotony of prison life, or to escape temporarily from a difficult environment. By reducing the incentive to malinger, telemedicine has the potential to reduce frivolous utilization of medical services. Finally, prisoners have a proclivity to allege cruel and unusual punishment in civil litigation, sometimes as a consequence to delayed or inappropriate medical care. Telemedicine may permit more timely attention to medical problems, and the videotaping of telemedicine consults can provide the documentation necessary for assessing the quality of care delivered. Anecdotal data from North Carolina suggest a decrease in such litigation following the establishment of a telemedicine link with East Carolina University.

The first telemedicine link with a correctional facility was established between Jackson Memorial Hospital and several county institutions in

Florida in the 1960s. More recently, East Carolina University began
providing services to North Carolina's Central Prison in Raleigh, a
maximum security penitentiary. The University of Texas Medical Branch
at Galveston has deployed a very large prison telemedicine program;
755 prisoners were examined using telemedicine over the first six months
of the program (Brecht et al., 1996). The HealthNet program at Texas
Tech University Health Sciences Center, in Lubbock, provides telemedicine services to prisons in West Texas. Patients are typically presented to
a consultant over the IATV link by a physician assistant. The Georgia
Statewide Telemedicine Program (Adams and Grigsby, 1995) has provided consultation for prisons, and reported that as many as 90 percent
of prisoners who previously would have required transportation elsewhere for medical care subsequently were seen exclusively in the con-
fines of the prison.

Telemedicine Applications

In view of the variety of telemedicine applications, a conceptual
scheme for classifying them is useful. Telemedicine is not a single,
homogeneous system but a variety of uses and processes utilizing a broad
spectrum of technologies. Hence it makes little sense to speak globally of
the effectiveness or cost-effectiveness of telemedicine. Instead, the empirical and conceptual issues become more manageable when we consider
the effectiveness of specific telemedicine applications or processes. There
are many approaches to classifying telemedicine activities, and the choice
is largely a function of what one wishes to study or to analyze. It may be
useful to study the effectiveness of telemedicine in the diagnosis or
management of individual diseases. Nondisplaced fractures, for example,
may be difficult to discern on radiographs (Scott et al., 1993), whereas
other orthopedic problems may be less problematic. The problem with
this approach, however, is that if one wishes to study effectiveness of
telemedicine even within a single specialty, one ends up with a separate
cell for each disorder and few cases in each cell.

A second approach is to order telemedicine activities by specialty, or
by classes of service in specialty or subspecialty areas, since different
specialties may require various kinds (e.g., different sensory modalities)
of information. Dermatology and radiology, for example, rely heavily
on visual information, while cardiac auscultation requires high-quality
audio data. Consequently, dermatology might be provided less effectively
with an audio-only telemedicine system than would cardiology (Conrath

et al., 1977). Similarly, teleneurology depends more on the transmission of moving images, while teledermatology requires only static ones.

A third approach, developed especially for use in health services research, is to organize telemedicine services into classes of clinical problems. One such conceptual framework (Grigsby et al., 1995) describes nine such classes, including:

1. Initial emergency evaluation, triage, and pretransfer arrangements
2. Medical and surgical follow-up
3. Routine consultations and second opinions
4. Extended diagnostic work-ups or short-term management of self-limited conditions
5. Management of chronic diseases and conditions
6. Supervision and consultation for primary care encounters when a physician is not available
7. Transmission of diagnostic images
8. Transmission of other medical data (e.g., fax of an EEG or EKG, labs, chart data)
9. Preventive medicine, public health, and patient education.

This approach assumes that the management of chronic diseases, for example, has certain processes in common across specialties, especially in the mix and intensity of services provided. Emergency consultation, regardless of specialty, would involve a unique set of processes of care. Clearly this approach to classification is arbitrary, and any of these nine categories may overlap with others, or could be subdivided further.

A fourth approach to classification of applications that also may be useful is by general technological category. Different levels of technology vary with respect to the amount and type of information they convey. Fax transmission provides a graphic presentation of either visual or verbal material. Audio-only methods (radio and telephone) typically are used to convey only verbal information, creating an image in the mind of the listener. The addition of visual imagery simplifies this process and permits the transmission of much more complex information than could be conveyed verbally. Visual information may consist of still images (e.g., of a dermatologic lesion), short video sequences (e.g., of a patient's festinating gait or intention tremor), or full-motion video. A simpler way to classify applications is by real-time transmission (in which case patient and consultant actually interact with one another, as in telepsychiatry) or store-and-forward methods.

Network Architecture

The terminology used to describe telemedicine networks is sometimes confusing. Tertiary care facilities that provide consultants are perhaps most frequently referred to as "hubs," a term derived from the "hub and spoke" model. In this configuration, the spokes are the smaller institutions—community hospitals and rural clinics—whose patients are referred for consultation. These smaller institutions are sometimes referred to as "remote sites," "rural sites," "spokes," and in a few cases, "satellite facilities." Furthermore, when these smaller institutions function both as sources of consultants and as sources of referrals, they are called "sub-hubs."

To simplify terminology, Perednia (1995) has suggested the use of terms reflecting the location of consultants and patients. Thus, a tertiary care medical center that provides consultation for others would be called a *consulting site*. A rural community hospital that refers patients for consultation to the tertiary care hospital would be called a *referring site*. This works well until we have to deal with those intermediate institutions that provide both patients and consultants. They might be designated in general as *referring/consulting sites*, while in any given situation they are *either* a referring or consulting site. The terminology problem becomes even more complex in a distributed (or "cloud") network.

In general, there are at least four basic configurations of telemedicine networks. The first three are predominantly hierarchical in structure. In the first, a *simple hub and spoke network*, a single tertiary care facility provides medical expertise for one or more referring sites. Most current telemedicine programs take this form. East Carolina University, for example, has a single consulting site and several referring sites. In the second model, a *multiple hub and spoke network*, two or more consulting sites provide medical expertise for one or (usually) more referring facilities. This multiple-consulting-site model is also common, and is typified by the High Plains Rural Health Network in northeast Colorado, southwest Nebraska, and northwest Kansas. In this network, two institutions function as sources of consultants to many rural facilities around the state. The third model is a less strictly hierarchical scheme comprised of one or more consulting sites, one or more referring/consulting sites, and one or more referring facilities. This might be considered a *complex hub-and-spoke* arrangement. The Georgia Statewide Telemedicine Program, for example, has facilities that serve both as referring and consulting sites.

The fourth type of network, of which there are currently no examples, is likely to develop on a large scale as desktop-computer-based telemedicine systems become more common. This is a *partially* or *fully distributed* network (referred to some as a "cloud"), in which the different sites are nodes in a network that is, to a large extent, interconnected. In a fully distributed network, the arrangement is not at all hierarchical. That is, there is no central consulting facility that coordinates the entire network. Instead, the nodes are interconnected with one another, and they may seek or provide consultation from and to several or all facilities in the network. In a partially distributed network, the arrangement is more hierarchical, but the referring and referring/consulting sites have some degree of autonomy. A system of this sort could be established, for example, linking the offices of a number of private practitioners, both specialists and consultants, by means of a personal computer-based system.

Schematic diagrams of these kinds of networks are presented in Figures 11-1, 11-2 and 11-3. This scheme for classifying networks represents an idealized model. Some networks do not fit any of these models, and some may have both hierarchical and distributed characteristics, but these represent the predominant types.

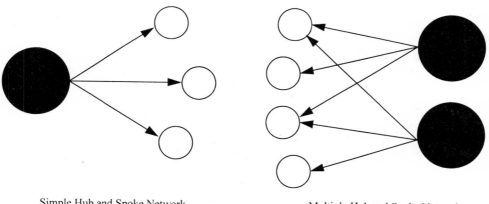

Simple Hub and Spoke Network   Multiple Hub and Spoke Network

Figure 11-1. Schematic diagram of *Simple* and *Multiple Hub and Spoke Networks*. In this and subsequent figures, the arrows point from the consulting sites to the referring sites.

Providers

At present, much of telemedicine is conducted using real-time IATV consultation between a specialist or subspecialist at a tertiary care facility and a primary care physician with the patient at a rural facility. This

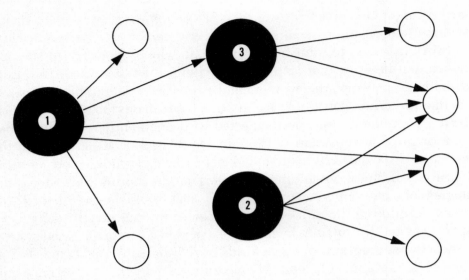

Figure 11-2. Schematic diagram of a *complex hub and spoke network.* This particular network consists of two consulting sites (1 and 2), a single referring/consulting site (3), and six referring sites.

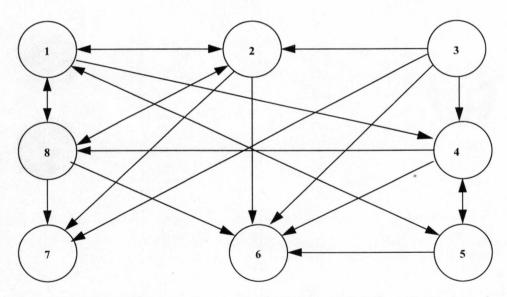

Figure 11-3. Schematic diagram of a *partially distributed network.* This particular network contains two sites that only make referrals (6 and 7), one site that only provides consultation (3), and five sites that both make referrals and provide consultation.

arrangement involves the inconvenience of coordinating two physicians' schedules, and may not be cost-effective, since there presumably is little need for the primary care physician to be present during such an encounter. In regions with severe shortages of medical personnel, nurse practitioners and physician assistants are sometimes utilized in lieu of the primary care physician. Mental health professionals—psychiatrists, psychologists, nurses, and social workers—have used telemedicine for assessment and intervention, although treatment planning and patient disposition, in the patient's absence, are more common. Services like speech and language therapy or physical therapy do not seem to have been provided over these networks as yet.

Future Directions for Telemedicine

Early telemedicine projects did not survive. Most depended on federal funds for their operation, and when the grants ran out, the programs folded (Bashshur, 1995). Equipment was expensive, bulky, immobile, and lacked the sophistication of current technology. The transmission of video images was problematic. Video equipment was not always capable of faithful representation of color, and its resolution was inferior. Telecommunications equipment has improved significantly, as has the infrastructure for transmission of information. In the 1970s, for example, many rural areas still had telephone lines, but not telemedicine. Computing was done on large mainframe machines which lacked the speed and flexibility of desktop computers of the 1990s. Although fax technology was available twenty-five years ago, it was expensive and not widely used. There was nothing like the current capacity for digitization and compression of data.

The health-services sector of the economy has changed drastically since that time as well. The average in-patient length of stay has dropped significantly since the introduction of DRGs and managed care. A significant amount of care that used to be offered only on an in-patient basis has moved into the out-patient clinic and the home. Individual hospitals have merged into vertically integrated health care systems, and large national hospital corporations have swallowed up locally owned hospitals and medical practices. Medicare and Medicaid, established in the 1960s, have grown into dominant forces in health care financing. During the decade between 1980 and 1990, Medicaid expenditures grew by 287 percent, and Medicare by 352 percent (Koch, 1993). The Health Care Financing Administration, which administers these programs, now plays a major role in formulating national policies on payment and coverage.

Whereas in the 1970s managed care consisted of Kaiser, Group Health Cooperative of Puget Sound, and a fledgling federal HMO program, it now has become widespread, with market share increasing by a factor of greater than 3 over the past 15 years. Physician payments are no longer handled primarily on an unregulated fee-for-service basis.

The environment of telemedicine today is vastly different from what it was in the early days. Public and private investment today is occurring at a scale not seen in the 1970s. Yet the early work established the clinical feasibility of telemedicine, and it produced research models that can be utilized today (e.g., Bashshur, 1978; Conrath et al., 1977).

Telemedicine technology is likely to become dominated by desktop computer systems. These are relatively inexpensive, and are capable of integrating with audio/video information with electronic medical records and picture archiving communication systems (PACS). Both home health and long-term care facilities are likely to adopt telemedicine, especially as private provider groups establish corporations for this purpose. Early reports from pilot home health programs (Allen, 1995b; Lindberg, 1996) suggest that telemedicine may have a major impact on the delivery of home care. Although the question of Medicare coverage may remain unresolved in the short term, Medicaid, commercial insurers, and private pay patients are capable of supporting the development of such programs.

Two other likely areas for significant growth are managed care organizations and regional telemedicine centers. Telemedicine using nonphysician providers linked electronically with physicians is ideally suited for managed care, because it permits the distribution of specialty and subspecialty service available at central sites to an entire network of geographically dispersed clinics. Telemedicine would also enable managed care providers to provide rural health care at lower cost, while avoiding the need to rotate specialists in rural sites or require patients to travel long distances. Regional telemedicine centers are already in development, especially in radiology, where two or three corporations are capturing a significant share of the telemedicine market. Multispecialty clinics could provide services over a large geographic area, and could establish major tertiary care facilities in different regions.

Although it is rapidly evolving, we can expect telemedicine in some form to become a wide-spread and permanent fixture of the medical care landscape. It already has gained a secure beachhead in radiology, and while a number of factors continue to impede the full implementation of clinical telemedicine (Puskin, 1995; Sanders and Bashshur, 1995), declining costs and changes in the health care sector are creating more favorable conditions for telemedicine than existed twenty years ago.

REFERENCES

Adams, L. N., and R. K. Grigsby. 1995. "The Georgia State Telemedicine Program: Initiation, Design, and Plans," *Telemedicine Journal,* 1: 227–35.

Allen, A. 1994. "Telemental Health Services Today," *Telemedicine Today,* 2 (Spring): 12–15, 24.

Allen, A. 1995a. "UTMB–Galveston: In a League of its Own," *Telemedicine Today,* 3 (Spring): 15–16.

Allen, A. 1995b. "Home Health Care via Telemedicine," *Telemedicine Today,* 3 (Spring): 26–27.

Allen, A., R. Cox, and C. Thomas. 1992. "Telemedicine in Kansas," *Kansas Medicine,* 93: 323–25.

Allen, A., J. Hayes, R. Sadasivan, S. K. Williamson, and C. Wittman. 1995. "A Pilot Study of the Physician Acceptance of Tele-oncology," *Journal of Telemedicine and Telecare,* 1: 34–37.

Allen, D., and A. Allen. 1994. "Teleradiology 1994," *Telemedicine Today,* 2 (Fall): 1, 21–23.

American College of Radiology. 1994. *ACR Standards for Teleradiology.* Reston, VA: The College.

Andrus, W. S., and K. T. Bird. 1972. "Teleradiology: Evolution through Bias to Reality," *Chest,* 62: 655–57.

Ball, C. J., P. M. McLaren, A. B. Summerfield, M. S. Lipsedge, and J. P. Watson. 1995. "A Comparison of Communication Modes in Adult Psychiatry," *Journal of Telemedicine and Telecare,* 1: 22–26.

Bashshur, R. 1978. "Public Acceptance of Telemedicine in a Rural Community," *Biosciences Communications,* 4: 17–38.

Bashshur, R. L. 1995. "Health Policy and Telemedicine," *Telemedicine Journal,* 1: 81–83.

Bashshur, R. 1979. *Technology Serves the People: The Story of a Co-Operative Telemedicine Project by NASA, The Indian Health Service, and the Papago People."* Tucson, AZ: The Indian Health Service.

Bashshur, R. L., and P. A. Armstrong. 1976. "Telemedicine: A New Mode for the Delivery of Health Care," *Inquiry,* 13: 233–44.

Bashshur, R. L., P. A. Armstrong, and Z. I. Youssef. 1975. *Telemedicine: Explorations in the Use of Telecommunications in Health Care.* Springfield, IL: Charles C Thomas.

Belmont, J. M., L. F. Mattioli, K. K. Goertz, R. H. Ardinger, and C. M. Thomas. 1995. "Evaluation of Remote Stethoscopy for Pediatric Telecardiology," *Telemedicine Journal,* 1: 133–49.

Bertera, E. M., and R. L. Bertera. 1981. "The Cost-Effectiveness of Telephone vs.

Clinic Counseling for Hypertensive Patients: A Pilot Study," *American Journal of Public Health,* 71: 626–29.

Bertrand, C. A., R. L. Benda, A. D. Mercando, M. M. Taddeo, and K. E. Bailey. 1994. "Effectiveness of the Fax Electrocardiogram," *American Journal of Cardiology,* 74 (3): 294–95.

Bhattacharyya, K., J. R. Davis, B. E. Halliday, A. R. Graham, A. Leavitt, R. Martinez, R. A. Rivas, and R. S. Weinstein. 1995. "Case Triage Model for the Practice of Telepathology," *Telemedicine Journal,* 1: 9–17.

Bird, K. T. 1972. "Cardiopulmonary Frontiers: Quality Health Care via Interactive Television," *Chest,* 61: 204–05.

Brecht, R. M., C. L. Gray, C. Peterson, and B. Youngblood. 1996. "The University of Texas Medical Branch—Texas Department of Criminal Justice Telemedicine Project: Findings from the First Year of Operation." *Telemedicine Journal,* 2 (1): 25–35.

Brown, F. W. 1995. "A Survey of Telepsychiatry in the USA," *Journal of Telemedicine and Telecare,* 1: 19–21.

Cawthon, M. A., F. Goeringer, R. J. Telepak, B. S. Burton, S. H. Pupa, C. E. Willis, M. F. Hansen. 1991. "Preliminary Assessment of Computed Tomography and Satellite Teleradiology from Operation Desert Storm," *Investigative Radiology,* 26: 854–57.

Chaves-Carballo, E. 1992. "Diagnosis of Childhood Migraine by Compressed Interactive Video." *Kansas Medicine,* 93: 353.

Conrath, D. W., E. V. Dunn, W. G. Bloor, and B. Tranquada. 1977. "A Clinical Evaluation of Four Alternative Telemedicine Systems," *Behavioral Science,* 22: 12–21.

Cox, G. G., L. T. Cook, J. H. McMillan, S. J. Rosenthal, S. J. Dwyer III. 1990. "Chest Radiography: Comparison of High-Resolution Digital Displays with Conventional and Digital Film," *Radiology,* 176: 771–76.

Cunningham, N., C. Marshall, and E. Glazer. 1978. "Telemedicine in Pediatric Primary Care: Favorable Experience in Nurse-Staffed Inner-City Clinic," *Journal of the American Medical Association,* 240: 2749–51.

Delaplain, C. B., C. E. Lindborg, S. A. Norton, and J. E. Hastings. 1993. "Tripler Pioneers Telemedicine across the Pacific," *Hawaii Medical Journal,* 52: 338–39.

DiSantis, D. J., M. S. Cramer, and J. C. Scatarige. 1987. "Excretory Urography in the Emergency Department: Utility of Teleradiology," *Radiology,* 164: 363–64.

Dongier, M., R. Tempier, M. Lalinec-Michaud, and D. Meunier. 1986. "Telepsychiatry: Psychiatric Consultation through Two-Way Television. A Controlled Study." *Canadian Journal of Psychiatry,* 31: 32–34.

Eide, T. J., and I. Nordrum. 1992. "Frozen Section Service via the Telenetwork in Northern Norway," *Zentralblatt Pathologie,* 138: 409–12.

Eide, T. J., I. Nordrum, and H. Stalsberg. 1992. "The Validity of Frozen Section Diagnosis Based on Video-Microscopy," *Zentralblatt Pathologie,* 138: 405–07.

Fahn, S., R. L. Elton, and members of the UPDRS Development Committee. 1987. "Unified Parkinson's Disease Rating Scale," in Fahn, S., Marsden, C. D., Calne,

D. B. and Goldstein, M. (eds.), *Recent developments in Parkinson's Disease,* vol. 2. Florham Park, NJ: Macmillan Health Care Information, pp. 153–64.

Finkelstein, S. M., B. Lindgren, B. Prasad, M. Snyder, C. Edin, C. Wielinski, and M. Hertz. 1993. "Reliability and Validity of Spirometry Measurements in a Paperless Home Monitoring Diary Program for Lung Transplantation," *Heart and Lung,* 22: 523–33.

Fisk, N. M., S. Bower, W. Sepulveda, P. Garner, K. Cameron, M. Mathews, D. Ridley, K. Drysdale, and R. Wooton. 1995. "Fetal Telemedicine: Interactive Transfer of Realtime Ultrasound and Video via ISDN for Remote Consultation," *Journal of Telemedicine and Telecare,* 1: 38–44.

Franken, E. A., A. Allen, C. Budig, and D. Allen. 1995. "Telemedicine and Teleradiology: A Tale of Two Cultures," *Telemedicine Journal,* 1: 5–7.

Franken, E. A., C. E. Driscoll, K. S. Berbaum, W. L. Smith, Y. Sato, L. W. Steinkraus, and S. C. S. Kao. 1989. "Teleradiology for a Family Practice Center," *Journal of the American Medical Association,* 261: 3014–15.

Fuchs, M. 1979. "Provider Attitudes toward STARPAHC: A Telemedicine Project on the Papago Reservation," *Medical Care,* 17: 59–68.

Goodman, L. R., W. D. Foley, C. R. Wilson, A. A. Rimm, and T. L. Lawson. 1986. "Digital and Conventional Chest Images: Observer Performance with Film Digital Radiography System," *Radiology,* 158: 27–33.

Gravenstein, J. S., L. Berzina-Moettus, A. Regan, and Y-H Pao. 1975. "Laser mediated Telemedicine in Anesthesia," *Anesthesia and Analgesia,* 53: 605–09.

Grigsby, J., M. M. Kaehny, E. J. Sandberg, R. E. Schlenker, and P. W. Shaughnessy. 1995. "Effects and Effectiveness of Telemedicine," *Health Care Financing Review,* 17: 115–31.

Grigsby, J., E. J. Sandberg, M. M. Kaehny, A. M. Kramer, R. E. Schlenker, and P. W. Shaughnessy. 1994. *Telemedicine case studies and current status of telemedicine.* Denver, CO: Center for Health Policy Research.

Grigsby, J., R. E. Schlenker, M. M. Kaehny, P. W. Shaughnessy, and E. J. Sandberg. 1995. "Analytic Framework for Evaluation of Telemedicine," *Telemedicine Journal,* 1: 31–39.

Grundy, B. L., P. K. Jones, and A. Lovitt. 1982. "Telemedicine in Critical Care: Problems in Design, Implementation, and Assessment," *Critical Care Medicine,* 10: 471–75.

Hickey, N. M., J. G. Robertson, and M. Coristine. 1990. "Integrated Radiological Information System: A Radiology Multimedia Communication System," *Journal of Thoracic Imaging,* 5: 77–84.

Higgins, C. A., D. Conrath, and E. V. Dunn. 1984. "Provider Acceptance of Telemedicine Systems in Remote Areas of Ontario," *Journal of Family Practice,* 18: 285–89.

Ho, B. K. T., R. K. Taira, R. J. Steckel, and H. Kangarloo. 1995. "Technical Considerations in Planning a Distributed Teleradiology System," *Telemedicine Journal,* 1: 53–65.

Hoehn, M. M., and M. D. Yahr. 1967. "Parkinsonism: Onset, Progression, and Mortality," *Neurology,* 17 (5): 427–42.

Houtchens, B. A., T. P. Clemmer, and H. C. Holloway, A. A. Kiselev, J. S. Logan, R. C. Merrell, A. E. Nicogossian, H. A. Nikogossian, R. B. Rayman, A. E. Sarkisian, and J. H. Siegel. 1993. "Telemedicine and International Disaster Response: Medical Consultation to Armenia and Russia via a Telemedicine Spacebridge," *Prehospital and Disaster Medicine*, 8: 57–66.

Huang, H. K., S. L. Lou, P. S. Cho, D. J. Valentino, A. W. K. Wong, K. K. Chan, and B. K. Stewart. 1990. "Radiological Image Communication Methods," *American Journal of Radiology*, 155, 183–86.

Hubble, J. P. 1992. "Interactive Video Conferencing and Parkinson's Disease," *Kansas Medicine*, 93: 351–52.

Hubble, J. P., R. Pahwa, D. K. Michalek, C. Thomas, and W. C. Kollar. 1993. "Interactive Video Conferencing a Means of Providing Interim Care to Parkinson's Disease Patients," *Mov. Discord*, 8: 380–82.

Jelaso, D. V., G. Southworth, and L. H. Purcell. 1978. "Telephone Transmission of Radiographic Images," *Radiology*, 127: 147–49.

Justice, J. W., and P. G. Decker. 1979. "Telemedicine in a Rural Health Delivery System," *Advances in Biomedical Engineering*, 7: 101–71.

Kagetsu, N. J., D. R. P. Zulauf, and R. C. Ablow. 1987. "Clinical Trial of Digital Teleradiology in the Practice of Emergency Room Medicine," *Radiology*, 165: 551–54.

Kayser, K., and M. Drlicek. 1992. "Visual Telecommunications for Expert Consultation of Intraoperative Sections," *Zentralblatt Pathologie*, 138: 395–98.

Karssemeijer, N., J. T. Frieling, and J. H. Hendriks. 1993. "Spatial Resolution in Digital Mammography," *Invest Radiol*, 28: 413–19.

Kelly, P. J. 1994. "Quantitative Virtual Reality Surgical Simulation, Minimally Invasive Stereotactic Neurosurgery and Frameless Stereotactic Technologies," presented at the conference on Medicine Meets Virtual Reality II: Interactive Technology and Healthcare. San Diego, CA: January, 1994.

Kemper, P. 1992. "The Use of Formal and Informal Home Care by the Disabled Elderly," *Health Services Research*, 27: 421–51.

Koch, A. L. 1993. "Financing Health Services," in Williams, S. J., and Torrens, P. R. (eds.), *Introduction to Health Services*, 4th edition, Albany, NY: Delmar, pp. 299–331.

Kundel, H. L. 1986. "Digital Projection Radiography of the Chest," *Radiology*, 158 (1): 274–76.

Lindberg, C. 1996. Personal communication.

Lovett, J. E., and R. L. Bashshur. 1979. "Telemedicine in the USA: An Overview," *Telecommunications Policy*, 3 (1): 3–14.

Martínez-Martín, P., and F. Bermejo-Pareja. 1988. "Rating Scales in Parkinson's Disease," in Jankovic, J., and Tolosa, E. (eds.), *Parkinson's Disease and Movement Disorders*. Baltimore-Munich: Urban and Schwarzenberg, pp. 235–42.

Mattioli, L., K. Goertz, R. Ardinger, J. Belmont, R. Cox, and C. Thomas. 1992. "Pediatric Cardiology: Auscultation from 280 Miles Away," *Kansas Medicine*, 93: 326–50.

Meade, M. S. 1992. "Implications of Changing Demographic Structures for Rural

Health Services," in Gesler, W. M., and Ricketts, T. C. (eds.), *Health in Rural North America.* New Brunswick, NJ: Rutgers University Press, pp. 69–85.

Minsky, M. 1979. *Toward a Remotely-Manned Energy and Production Economy.* Cambridge, MA: Massachusetts Institute of Technology Artificial Intelligence Laboratory, A.I. Memo No. 544.

Moore, G. T., T. R. Willemain, R. Bonanno, W. D. Clark, A. R. Martin, and R. P. Mogielnicki. 1975. "Comparison of Television and Telephone for Remote Medical Consultation," *New England Journal of Medicine,* 292: 729–32.

Mun, S. K., H. R. Benson, B. Lo, B. Levine, R. Braudes, L. P. Elliott, T. Gore, and M. L. Mallon-Ingeholm. 1988. "Development and Technology Assessment of a Comprehensive Image Management and Communication Network," *Medical Informatics,* 13: 315–22.

Murphy, R. L. H., D. Barber, A. Broadhurst, et al. 1970. "Microwave Transmission of Chest Roentgenograms," *American Review of Respiratory Disease,* 102: 771–77.

Murphy, R. L. H., T. B. Fitzpatrick, H. A. Haynes, K. T. Bird, and T. B. Sheridan. 1972. "Accuracy of Dermatologic Diagnosis by Television," *Archives of Dermatology,* 105: 833–35.

Murphy, R. L. H., P. Block, K. T. Bird, and P. Yurchak. 1973. "Accuracy of Cardiac Auscultation by Microwave," *Chest,* 63: 578–81.

Nicogossian, A. E. 1989. "Final Project Report. U.S.–U.S.S.R. Telemedicine Consultation Spacebridge to Armenia and Ufa," presented at the Third U.S.–U.S.S.R. Joint Working Group on Space Biology and Medicine, December, 1989, Moscow and Koslovodsk, U.S.S.R.

Nordrum, I., B. Engum, E. Rinde, A. Finseth, H. Ericsson, M. Kearney, H. Stalsberg, and T. J. Eide. 1991. "Remote Frozen Section Service: A Telepathology Project in Northern Norway," *Human Pathology,* 22: 514–18.

Norton, C. H., and M. A. McManus. 1989. "Background Tables on Demographic Characteristics, Health Status, and Health Services Utilization," *Health Services Research,* 23: 725–56.

Office of Technology Assessment, U.S. Congress. 1990. *Health Care in Rural America,* OTA–H–434. Washington, DC: Government Printing Office.

Ong, K., P. Chia, W. L. Ng, and M. Choo. 1995. "A Telemedicine System for High Quality Transmission of Paper Electrocardiographic Reports," *Journal of Telemedicine and Telecare,* 1: 27–33.

Paakkala, T., J. Aalto, V. Kähärä, and S. Seppänen. 1991. "Diagnostic Performance of a Teleradiology System in Primary Health Care," *Computer Methods and Programs in Biomedicine,* 36: 157–60.

Pedersen, S., G. Hartviksen, and D. Haga. 1994. "Teleconsultation of Patients with Otorhinolaryngologic Conditions," *Archives of Otolaryngology and Head and Neck Surgery,* 120: 133–36.

Pederson, S., and U. Holland. 1995. "Tele-Endoscopic Otorhinolaryngological Examination: Preliminary Study of Patient Satisfaction," *Telemedicine Journal,* 1: 47–52.

Perednia, D. A. 1995. *Definition of Terms: Surveyor Telemedicine Evaluation Software Manual.* Portland, OR: Telemedicine Research Center.

Perednia, D. A., and A. Allen. 1995. "Telemedicine Technology and Clinical Applications," *Journal of the American Medical Association,* 273: 483–88.

Perednia, D. A., M. L. S. Brown. 1995. "Teledermatology: One Application of Telemedicine," *Bulletin of the Medical Library Association,* 83: 42–47.

Physician Payment Review Commission. 1995. *Annual report to Congress.* Washington, DC: PPRC.

Pool, S. L., J. C. Stonsifer, and N. Belasco. 1975. "Application of Telemedicine Systems in Future Manned Space Flight," paper presented at Second Telemedicine Workshop, Tucson, December, 1975.

Preston, J., F. W. Brown, and B. Hartley. 1992. "Using Telemedicine to Improve Health Care in Distant Areas," *Hospital and Community Psychiatry,* 43: 25–32.

Puskin, D. S. 1995. "Opportunities and Challenges to Telemedicine in Rural America," *Journal of Medical Systems,* 19: 59–67.

Rinde, E., I. Nordrum, and B. J. Nymo. 1993. "Telemedicine in Rural Norway," *World Health Forum,* 14: 71–77.

Sanders, J. H., and L. Sasmor. 1973. "Telecommunications in Health Care Delivery," National Science Foundation, *Proceedings of the First Symposium on Research Applied to National Needs* (RANN), 167–69.

Sanders, J. H., and R. L. Bashshur. 1995. "Perspective: Challenges to the Implementation of Telemedicine," *Telemedicine Journal,* 1: 115–23.

Satava, R. M. 1992. "Robotics, Telepresence and Virtual Reality: A Critical Analysis of the Future of Surgery," *Minimally Invasive Therapy,* 1: 357–63.

Schwarzmann, P. 1992. "Telemicroscopy: Design Considerations for a Key Tool in Telepathology," *Zentralblatt Pathologie,* 138: 183–87.

Scott, W. W., Jr., J. E. Rosenbaum, S. J. Ackerman, R. L. Reichle, D. Magid, J. C. Weller, and J. N. Gitlin. 1993. "Subtle Orthopedic Fractures: Teleradiology Workstation versus Film Interpretation," *Radiology,* 187: 811–15.

Shaughnessy, P. W., R. E. Schlenker, K. S. Crisler, M. C. Powell, D. F. Hittle, A. M. Kramer, M. J. Spencer, S. K. Beale, S. G. Bostrom, J. M. Beaudry, P. A. DeVore, V. Chandramouli, W. V. Grant, A. G. Arnold, M. K. Bauman, and J. Jenkins. 1994. *Measuring Outcomes of Home Health Care,* Final Report from "A Study to Develop Outcome-Based Quality Measures for Home Health Services," funded by the Health Care Financing Administration. Denver, CO: Centers for Health Policy and Services Research.

Shimosato, Y., Y. Yagi, K. Yamagishi, K. Mukai, S. Hirohashi, T. Matsumoto, and T. Kodama. 1992. "Experience and Present Status of Telepathology in the National Cancer Center Hospital, Tokyo," *Zentralblatt Pathologie,* 138: 413–17.

Siderfin, C. D. 1995. "Low-Technology Telemedicine in Antarctica," *Journal of Telemedicine and Telecare,* 1: 54–60.

Sobczyk, W. L., R. E. Solinger, and A. H. Rees. 1993. "Transtelephonic Echocardiography: Successful Use in a Tertiary Pediatric Referral Center," *Journal of Pediatrics,* 122: S84–88.

Sparks, K. E., D. K. Shaw, D. Eddy, P. Hanigosky, and J. Vantrese. 1993. "Alternatives for Cardiac Rehabilitation Patients Unable to Return to a Hospital-Based Program," *Heart and Lung,* 22: 298–303.

Templeton, A. W., G. G. Cox, and S. J. Dwyer 3rd. 1988. "Digital Image Management Networks: Current Status," *Radiology,* 169: 193–99.

U.S. Bureau of the Census: 1993. *Statistical abstract of the United States: 1993.* (113th edition). Washington, DC: Bureau of the Census.

Wasson, J., C. Gaudette, F. Whaley, A. Sauvigne, P. Baribeau, and G. Welch. 1992. Telephone Care as a Substitute for Routine Clinic Follow-Up." *Journal of the American Medical Association,* 267: 1788–93.

Webber, M. M., S. Wilk, R. Pirruccello, and J. Aiken. 1973. "Telecommunication of Images in the Practice of Diagnostic Radiology," *Radiology,* 109: 71–74.

Weinstein, R. S., K. J. Bloom, E. A. Krupinski, and L. S. Rozek. 1992. "Human Performance Studies of the Video Microscopy Component of a Dynamic Telepathology System," *Zentralblatt Pathologie,* 138: 399–401.

Weinstein, R. S., K. J. Bloom, and L. S. Rozek. 1987. "Telepathology and the Networking of Pathology Diagnostic Services," *Archives of Pathology and Laboratory Medicine,* 111: 646–52.

Chapter 12

TELEMEDICINE AND THE MILITARY

Jesse C. Edwards, Jr. and Camille A. Motta

Introduction

The purpose of this chapter is to describe the development of telemedicine within the context of the military sector and to explore the potential linkages and implications, from the lessons learned here, to the development of telemedicine in the civilian community. Specifically, recent developments in military telemedicine are described which focus on clinically relevant applications of the technology.

The military environment provides a unique setting for development of telemedicine. First, it is not encumbered by the same organizational or regulatory restraints applicable to medical care delivery in the civilian sector. Second, military personnel represent a population quite often isolated from medical care and frequently in need of rapid access to specialty care. Telemedicine has the potential to increase accessibility to a wide range of medical care, and the application of telemedicine in the military arena can serve to illustrate the advantages and potential problems with similar applications in the civilian sector.

The role of military medicine is to support the forces in combat and to conserve the fighting strength in peace and war. In the past, it has made contributions to the practice of medicine in the civilian sector as a derivative of the continuing need to develop new approaches to a wide variety of changing circumstances. Examples include improved sanitation techniques, medical triage, and helicopter aeromedical evacuation techniques which have spread beyond the military structure to the civilian community. It can be expected that application of telemedicine in the military will continue this trend as it strives to support military forces at home and abroad, during peacetime as well as under battlefield conditions.

Medical Care in the Military

The presence of a staff model, wherein physicians are key leaders and "employees" of the health delivery system, may be one of the military's greatest advantages in planning and executing telemedicine projects. This close organizational relationship may be the best to support telemedicine initiatives because the agendas and incentives are largely the same, regardless of the professional discipline of those involved. Additionally, the state licensure issue, one of the most obvious regulatory obstacles facing civilian delivery systems, does not affect physicians in the military. To circumvent this issue, some argue that the distant patient is "transported" to the site of the physician's practice. However, if legally challenged, the argument may not stand in court, and indeed some states are proposing restrictive legislation requiring physicians practicing interstate telemedicine to be licensed in each state. Physicians interested in providing services across state lines are compelled, therefore, to obtain a license in each state where they render their services. The military is exempt from interstate licensure restrictions.

Telemedicine in the Military

The United States military formally launched a coordinated telemedicine program in 1994, with the establishment of the Department of Defense Telemedicine "Test Bed." The ultimate purpose of the Test Bed is to investigate clinically relevant applications of rapidly emerging technology. The military's efforts in telemedicine are part of a broad attempt to "re-engineer" the delivery of health services. The reengineering is premised on several principles, including: (1) the need to reduce geographic and temporal barriers to care, (2) the reorganization of the medical establishment into a more fluid and continuous horizontal structure, and (3) the introduction of parallel and asynchronous communications based on the increased role of information/communication as a basic component of medical care. Thus, telemedicine is viewed as a "holistic system wherein each individual and relevant provider has total access to information, knowledge, and wisdom enabling enlightened self-care or expert intervention, any time, anywhere" (Telemedicine Test Bed CEO Support Group Meeting, 1995).

The central goals of such a reengineered health delivery system are the increased productivity for providers and accessibility for patients. That increase is to come largely from eliminating unnecessary travel and reducing the need for all communication related to medical care episodes to require the synchronization of two or more people. The phrase "time and distance independence" is often used to characterize an integrated health delivery system using telemedicine. Similarly, much of the same level of care obtained at major medical centers would be delivered to remote stations and battlefield locations via telemedicine. Underlying the vision of military telemedicine is an information-based theory of both the practice of medicine and the business of health delivery operating in a free marketplace, with the free flow of information. In this marketplace, information is the currency and telecommunications is the transfer medium.

Military in the Third Wave of Civilization

Alvin Toffler (1980), the futurist, describes a "Third Wave of Civilization," bringing with it extraordinary change. This wave is to embrace the Space Age, the Information Age, the Electronic Era, and the Global Village. Its inception may be dated from 1956, the first year when there were more white-collar than blue-collar workers in the United States. Instead of accelerating production, the emphasis of the future will be on continual individualization of the output. This Wave will be marked by significant social reorganization and, necessarily, creative restructuring. Because the central factor in the information age is knowledge, the key is to have "the right knowledge in the right head at the right time and in the right place" (French, 1995).

Dubik and Sullivan (1994) speak of the changes having an impact in the military arena. The concept of war is changing and expanding to include not only the armies of one or several nations/states but also religious groups, terrorist organizations, tribes, guerrilla bands, clan leaders, and others. "The net result is a blurring of the distinction between war and operations other than war." Military foes will arise unexpectedly, and the conditions for decisive victory will differ with each use of military force. Victory will entail not only sufficient destruction of the enemy's armed forces, but also dominance over the enemy's information system. This changing military and conflict environment requires changes in the military structure. In fact, the new construct needs to be one in which the "Information Age Army" continuously reinvents itself.

Challenges to Military Medical Planners

A profile of battlefield casualties has been created from data on past wars (Zajtchuk and Sullivan, 1995). It includes the assessment of combat trauma in terms of mortality and morbidity. Combat trauma is usually assessed in terms of mortality, that is, the percentage of the casualty population that is fatally wounded. Combat mortality is measured by two normalized statistics: the percentage killed in action (KIA) and the percentage who die of wounds received in action after having reached a medical facility (DOW). The definition of "killed in action" is a battle casualty who is killed outright or who dies as a result of wounds or other injuries before reaching a medical treatment facility. Combat trauma is also assessed in terms of morbidity, which refers to the total number of personnel-days lost, or the average number of days that a combat casualty is not effective for medical reasons.

Data from seven major wars during the last 150 years revealed that about 20 percent of those who sustain combat injuries are killed in action. Substantial progress has been made in reducing the mortality of those who survived to receive treatment, that is, the number of DOWs has decreased. In part, this decline may be due to the development of rapid and safe evacuation procedures as well as to the achievements of modern surgery.

Data concerning combat injuries during the Vietnam conflict shows about 50 percent of U.S. soldiers killed in Vietnam bled to death on the battlefield. Simple first aid could have saved up to 20 percent of these casualties (Bellamy, 1994). Hence, **swift intervention** is a key element in battlefield medicine since most battlefield deaths occur rapidly. In Vietnam, for example, about 67 percent of those killed in action died within ten minutes of wounding (see Figure 12-1). Of the remaining 33 percent, most died between 10 minutes and 1 hour after wounding (Pruitt, 1993). Many casualties did not receive timely treatment because the battlefield medic could not locate them, or because battlefield conditions prevented the medic from reaching them (Zajtchuk and Sullivan, 1995, p. 2).

Other data sources show that about 15 to 25 percent of casualties are severe. This underlines the importance of the medic as "first responder" and the need to have him perform whatever lifesaving procedures are possible on the battlefield. Although the principles of Advanced Trauma Life Support (ATLS), the standard basic trauma resuscitative skills, are

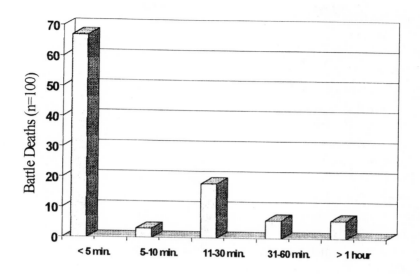

Estimated Time, Wounding until Death

Figure 12-1. During the years 1967, 1968, and 1969 of the Vietnam conflict, the Wound Data and Munitions Effectiveness Team accumulated detailed tactical and medical information on 7,682 U.S. Army and U.S. Marine casualties, including the only data available on how rapidly a casualty dies following war trauma. Of 100 consecutive casualties (for whom data exist) in the WDMET database, who were killed in action, 67 died within 10 minutes and 33 died after 10 minutes. Reprinted with permission, *Military Medicine*, Jan. 1996.

taught to most military physicians, no formal ATLS training is required for medics. They get no formal training beyond the didactics in their individual skill development course. Each medic should receive this level of training. But, even after such training, the capacity of the medic to perform resuscitative duties on or adjacent to the battlefield would be enhanced by direct **mentoring** from a physician.

Combat casualty care begins with the medics expected to render resuscitative care to patients with life-threatening injuries. Edward Jeffer (1994) described the central component of military medicine during the Cold War as the concept of **medical triage.** The philosophical and operational concept of medical triage was to ensure the survival of the maximum number of those who could be saved, that is, the greatest good for the greatest number of wounded. Those most severely wounded would be labeled "expectant" (expected to die) and set aside to be treated only after it was determined that all those individuals who had a greater likelihood of surviving were fully stabilized. The principle in operation here was to minimize the use of precious and scarce resources (surgeons) in futile surgery.

Data from past military battles show that about 15 to 25 percent of casualties will be classified as **severe.** Koehler, Smith, and Bacaner (1994) state that "there has been no fundamental change in the way the military medicine system trains for and practices combat trauma care since the last major advance came during the Vietnam Conflict with the advent of rapid aeromedical evacuation from the front line, a practice first utilized during the Korean Conflict."

Reliance on air evacuation at the front line, as in the Vietnam conflict, may never again be possible due to the growing sophistication and availability of shoulder-held surface-to-air weapons. Thus, ground evacuation needs to be far more combat-capable while allowing ATLS support en route. Present military transportation equipment does not allow for the initiation or continuation of ATLS care of casualties during transportation. This is true for all modes of transportation up to the C-130 cargo plane, rarely used to transport casualties from the primary point of aid to the second or third echelon of care.

Medical Care in the Persian Gulf

The Persian Gulf War highlighted the practical need to restructure military medical capabilities to meet the demands of short, simultaneous, and intense campaigns. The military medical leadership has taken careful heed of a report (U.S. General Accounting Office, 1992) critical of the Army's medical preparedness during the Persian Gulf War. The GAO concluded that "Due to problems with equipping and supplying units in theater, the Army was not adequately prepared to provide medical care prior to the start of the ground campaign, and the Army's ability to provide adequate care, had the war lasted longer or had the predicted number of casualties occurred, would have been questionable."

Among the problems cited by GAO were doctors and nurses who had not been trained during peacetime to perform their assigned tasks, hospital units lacking sufficient mobility to perform their mission, due to the speed of the battle and weight and configuration of the hospitals, and the inefficiency of the Professional Officer Filler System (PROFIS) which is designed to identify and reassign active Army physicians and nurses from hospitals and clinics to deploying medical units. The system did not function as planned: information used to identify physicians and nurses was outdated and unreliable, resulting in a massive "scramble" to configure fully units deploying from the United States and Europe.

A principal finding of GAO concerned the effective use of ambulances and the evacuation of casualties. Ground ambulances could not be used as much as planned due to the rugged terrain, lack of navigational equipment, communication difficulties, and the long distance between field hospitals and the front lines. Air evacuation units were taxed by the distances from pickup points to the hospitals. The long distances required refueling and crews had trouble locating fuel points.

Nearly 15,000 personnel were medically evacuated during this brief war, of whom fewer than 3 percent of these were combat casualties. This provided an opportunity for military commanders to assess the actual and potential shortcomings of the aeromedical evacuation system. Some of these shortcomings occurred in the medical regulating system, the U.S. military services' system for finding the medical facility that will provide the best care for the patient at the lowest possible cost. During Operation Desert Storm, more than 60 percent of the patients arrived at the wrong European aerial port of debarkation, and more than 50 percent arrived in the wrong country. In response, it has been suggested that a single seamless system should be created, capable of tracking patients from point of injury through definitive care, meeting changing situations, and financially accountable (Gunby, 1994).

Additional medical care shortcomings related specifically to communications and technology were identified (Nguyen, 1994). These included: (1) Largely outdated and only marginally reliable radio communications. Division ambulances had only nonsecure communication and corps ambulances did not have radios. Organizations controlling medical evacuations and personnel cross leveling were frequently out of range. Often, there were no direct communication links with the next higher medical treatment facility, resulting in many difficulties because messages were received through secondary channels. With the outdated and limited amounts of communication equipment, ambulance drivers often got lost in the vast desert. Computerized land navigation systems were not distributed to all medical vehicles. Precious evacuation time was wasted as a result. (2) Field hospitals were not very mobile and could not be prepared for transport quickly. A more mobile hospital was needed. Radiology equipment capable of meeting clinical needs and withstanding austere field environments was not available.

Addressing the Problems

Medical military planners met in 1992 at the Seventh Conference on Military Medicine, hosted by the Uniformed Services University of the Health Sciences, in Bethesda, Maryland. They developed a list of challenges confronting military medicine. The following summarizes the consensus of the conferees with specific emphasis on communications and mobility which military medicine must strive to achieve (Gunby, 1992). These include better communication at all levels, better design of medical facilities and equipment to meet the challenges of severe climates and rapid deployment, more mobile medical facilities to keep pace with rapidly advancing forces, and planning flexibility for future military medical needs.

A wide array of efforts is underway to address the challenges confronting military medicine. Among these is the coordination of these efforts so that they can continue to develop along parallel tracks and yet be mutually supportive over the coming decade. The next section describes many of the most critical initiatives underway.

The Use of Telemedicine

Telemedicine has the potential to address the significant problems identified as facing modern military medicine. If properly developed, it will facilitate swift intervention and triage, provision of on-site access to specialty care for trauma and other severe medical episodes, and increased mobility of and flexibility in the delivery of medical care in remote areas under a variety of field conditions. At the same time, it will facilitate the development of a seamless medical care delivery system for military personnel located at installations at home and abroad.

Telemedicine and Preventive Medicine

As mentioned earlier, a key element of the military telemedicine vision is the aim of enlightened self-care or expert intervention, any time anywhere. This part of the telemedicine vision requires a major shift in health services delivery. The existing illness-based, acute care, episodic treatment mode offered to mostly passive patients must give way to a new mode. The new focus will be on disease prevention and health promotion, offering continuous care to active patients who will be encouraged and enabled to assume responsibility for their own health.

Preventive medicine before deployment minimizes the number of persons unable to deploy for medical reasons and also minimizes the average time to deploy the total force. During the deployment process, information technologies can reduce mobilization and deployment cycle times and prevent a mismatch between theater requirements and deployed forces and equipment. In an active theater of operations, the early identification of disease vectors can minimize the incidence of disease among deployed forces and allies, as well as civilian populations.

Telemedicine initiatives are currently translating to improved care for the nondeployed military medical beneficiary. For example, linkages such as those between Brooke Army Medical Center in San Antonio and Darnell Army Community Hospital at Fort Hood, Texas, initially covered four clinical areas that benefit from the early involvement of appropriate specialists, including those with expertise in high-risk obstetrics, pediatric asthma, hypertension, and diabetes mellitus. Real-time, freeze-frame, recorded video and digitized images are transmitted from Darnell to Brooke while communication is supported by interactive, two-way video.

The closure of military bases, along with their hospitals, and the rise of health care costs have helped to motivate the search for better ways to provide health services consistent with the goals of preventive medicine. American military installations are closing in the United States and around the world, mostly in Europe and the Pacific basin. The effect of these closings can be reduced by the application of telemedicine technology. For example, telemedicine linkages from Tripler Army Medical Center in Hawaii to areas across the Pacific and services from Landstuhl, Germany to field clinics, will fill many of the gaps left by the closings. The expertise of medical specialists stationed in the United States can be transmitted to facilities located abroad, including primary care institutions, thereby maximizing the use of available physicians. Even in relatively peaceful 1994, more than 63,000 aeromedical evacuations were made in the Defense Department at a cost of more than $300 million. Effective use of telemedicine can minimize the number of evacuations for these nonbattle injuries and diseases (Berry, 1995).

TRICARE, the new health care plan for military beneficiaries, is a program of managed care for families and retirees under 65. In TRICARE, families and retirees will be treated at military facilities or will have lead agents contract with civilian doctors and hospitals near military facilities to provide necessary care through regional networks. For the Army, this has meant the creation of a "Gateway to Care" program, allowing local medical commanders to expand access to primary care, to win back patients now using the CHAMPUS program, and to reduce unnecessary bureaucracy. The Gateway to Care program is also creating local networks of physicians to meet medical needs. In fiscal 1994, Army military treatment facilities had 209 Gateway to Care initiatives.

In July, 1993, the Defense Department established 12 regions for TRICARE, varying from dense clusters of military installations to situations where beneficiaries live far from military hospitals. In each region, a commander of a large military medical center was designated as the "Lead Agent" in putting into effect the department's managed care approach to medicine. All military medicine resources, regardless of branch of service affiliation, are coordinated by the Lead Agent. Each region will function as follows.

 I. All the medical treatment facilities in the region will jointly develop a plan based on local and regional needs to deliver the necessary care. The lead agent will coordinate the development of that plan.

 II. The regions will award managed care support contracts to help expand a network to deliver care. The contractor will provide finders for health care, enroll people in the network, keep the necessary records, and put marketing and educational plans into effect.

 III. The lead agent will determine, on a site-by-site basis, whether enrollees look mainly to the military hospital or to the contractor for care. The services, the contractor, and the lead agent will cooperate to ascertain where beneficiaries are now going for medical care and how best to bring them back to military facilities or contractors. TRICARE will be phased in and the lead agent will serve as a hub to coordinate services (Anonymous, 1995).

Telemedicine will help to meet many of the challenges associated with access, cost, and quality within TRICARE. Employing managed care principles, TRICARE is creating a new peacetime environment for military medicine by organizing all health-service resources in a geographical area through the coordination of a single Lead Agent.

In Region 1 (the Northeast), consolidation of costly inpatient services is taking place. In the past, the Army, Air Force, and Navy all operated a full range of adult, child/adolescent, and substance abuse services at their own large hospitals. Under the new system, inpatient services have been consolidated into centers of excellence. Walter Reed Army Medical Center is now the center for adult inpatient; Malcolm Grow Air Force Medical Center is now focused largely on substance abuse; and the National Naval Medical Center houses the area's only neonatal intensive care unit.

The organization of child and adolescent inpatient services is being decided as this is being written. The location of the pediatric inpatient unit will drive the integration of the general surgery and specialty surgical service, planned to take place concomitantly when the inpatient units are centralized (Dineen and Larson, 1995).

These consolidations have created situations where the inpatients are located in one institution while certain subspecialists are located in another. This, in turn, creates an opportunity for telemedicine to serve as the mode of clinical consultation between these units and the subspecialists. A clinical trial in one of the affected environments is now underway. Neonatologists in the Neonatal Intensive Care Unit at the National Naval Medical Center in Bethesda, Maryland, are being linked via ISDN Lines to the pediatric subspecialists (mainly pediatric surgeons) located at Walter Reed in the District of Columbia. Clinical consultations will take place using state-of-the-art desk-top telemedicine workstations. In a randomized, controlled trial, clinical outcomes will be measured, the technology will be assessed, and attending physicians' and consultants' satisfaction will be gauged.

Region 3 (the Southeast) was designated as the Center for Total Access. Dwight David Eisenhower Army Medical Center is Region 3's Lead Agent. Regional resources are being pulled together collaboratively to work on the development of sound telemedicine approaches. In June, 1995, a research contract was awarded by the U.S. Army Medical Research and Materiel Command to a consortium lead by the Medical College of Georgia (MCG) in Augusta, including the Georgia Institute of Technology in Atlanta. These funds were matched by state funding to enable military beneficiaries to access the statewide MCG telemedicine network. The MCG network of over 60 telemedicine consult sites was begun in 1991.

The purpose of the collaborative research effort is to test tertiary care among rural telemedicine nodes, to extend the distribution of Georgia's statewide telemedicine system by adding PC-based teleconferencing systems and still-image phone systems, and to deliver medical care to the homes of selected patients in Augusta and Fort Gordon, Georgia.

The ultimate telemedicine system for home use will consist of desktop platforms linked, via cable with multipoint capabilities, to patients' homes and nursing homes. Participating homes will be equipped with a reverse amplifier and an internal "bridge" or digital telephone circuitry. The system will combine various components in a multimedia platform that integrates a variety of diagnostic devices, including electronic stethoscope, EKG, digital blood pressure and pulse monitor, Doppler, pulse oximeter, spriometer, and blood chemistry analysis (Medical College of Georgia, Telemedicine Center, 1995).

The Center for Total Access also takes advantage of the co-location of the U.S. Army Signal Center and School. These two organizations work effectively together at Fort Gordon to jointly develop an understanding of the roles and bandwidth requirements that telemedicine will demand of battlefield systems.

Region 11 (the Northwest), coordinated by Madigan Army Medical Center, is the first DoD region to award the TRICARE support contract which embodies a "new way of doing business" for all of DoD. Here, applications targeted for telemedicine include distance learning for providers; distance learning for patients; distance administration; modifications of existing processes (e.g., scheduling); and clinical care to include services delivered through interactive video and through advanced radiology image management systems.

The Telemedicine and Radiology Image Management piece of the project is referred to as Project Seahawk. The primary objectives of Seahawk in Washington State are to use commercially available telemedicine and networking technologies to improve the quality of health care services provided to patients, improve access to care, support emergency care (after hours and on weekends), and reduce the costs of delivering health care services for patients at participating sites.

Individual site surveys were undertaken with the expectation that remote sites would draw expertise from the larger sites. It was found, however, that each site had some capability that could be exported via telemedicine to the others where it might be needed. For example, McChord Air Force Base was able to export its flight medicine expertise. In turn, it received help with dermatology and radiology consults. Recognizing that Seahawk was not capable of meeting all the telemedicine needs of the participating locations, the project was planned for a gradual, phased-in implementation (Haines, 1994).

The DoD Telemedicine Test Bed

The DoD Telemedicine Test Bed was established to manage the rapid medical application of emerging digital technologies. The purpose of the project is to integrate telemedicine technologies into the Military

Health Services System (MHSS). The Test Bed is an integrated joint initiative of the Army, Navy, and Air Force elements of the MHSS. Each of the military services may pursue Test Bed initiatives based on mission differences, resource capabilities, and needs. A joint review and coordination of initiatives will ensure consistency within the emerging architecture. The Test Bed vision is to ensure rapid development of a prototype program focused on reengineering the delivery of military health services through aggressive exploitation of evolving technology and other improvements. This vision is supported by an interconnected diagnostic and therapeutic telecomputational system.

The DoD Executive Agent, with the guidance of a Board of Directors, coordinates the tri-service telemedicine initiatives through the U.S. Army Medical Research and Material Command. The Medical Advanced Technology Management Office (MATMO) was formed to provide a multidisciplinary team of physicians, clinical engineers, systems managers, health care administrators, and support staff to oversee, consult on, and execute specified telemedicine projects.

The Test Bed recognizes that telemedicine research and development are comprised of complex initiatives using several different technologies. The clinical, technical, and managerial impacts of these technologies on a number of distinct subparts of the health services system are being explored. A largely decentralized management approach is employed to meet the clinical requirements of a wide range of technical applications.

The organizational structure of the Telemedicine Test Bed embodies a shift in power from bureaucracy to "ad-hocracy," wherein task forces, *ad hoc* groups, committees, and other ransient teams come together to solve a specific problem and then separate (Medical Advanced Technology Management Office, U.S. Army Research and Material Command, 1994).

The following were adopted as guiding principles of MHSS telemedicine activities and the Test Bed:

(1) Readiness Orientation—Focusing on the evolving nature of military battlespace and the military medical environment to support assigned or implied operational missions.
(2) Patient Focus—Developing total access for difficult-to-serve, expensive-to-serve, and under-served patients who rely on military medicine for prevention, diagnosis, and treatment.
(3) Rapid Prototyping and Process Re-engineering—Using rapid prototyping

and process re-engineering for development of new medical practice and advanced technology applications.

(4) Outcome-Based Metrics—Testbed projects must have potential for measurable, reproducible improvement of outcomes. Outcomes evaluation will combine peer review, clinical trials, economic analysis, and statistical techniques to insure quality, access, and cost benefit.

(5) Open and Integrated System Architecture—Testbed projects must be designed to achieve interoperability, scaleability, and upgradeability. Moreover, they must be totally integrated with existing DoD medical systems to provide patient- and provider-friendly systems.

(6) Leveraged Development—Mandating active outreach and collaboration with academia, industry, and other governmental agencies to leverage development. Commercial and/or Government Off-the-Shelf Technology use will be maximized.

(7) Sound Business Practice—Testbed project management will utilize a management approach that follows sound business practices (Department of Defense Draft Telemedicine Test Bed [TTB] Management Plan, 1994).

In January, 1995, the Advanced Research Projects Agency (ARPA) announced the initiation of an effort entitled the Advanced Biomedical Technology Program, with the focus on medical support in the battlefield. Of particular interest were those technologies which have dual use (i.e., both commercial and military utility), including advanced diagnostics, advanced trauma care, health care information structure, telemedicine, and medical simulation.

Each of these technologies pertains directly to the quality and accessibility of medical care that reaches the individual soldier in the combat zone. The key to their success is the use of advanced information systems to detect casualties instantly, to assess the degree of care needed, to project physicians' skills and experience on the battlefield, and to bring all resources to bear to assist physicians who are treating difficult cases within a theater of war or a civilian disaster area. These efforts are detailed in another chapter in this volume.

Other Projects

Region 12 (the Pacific Basin), coordinated by Tripler Army Medical Center, initiated a five-year effort in 1993 named Project Akamai. The project goal is to improve the timeliness and accessibility of health services throughout the Pacific Rim region for federal beneficiaries during peacetime or contingency and humanitarian relief operations. The service area includes the Republic of Korea, where nearly 36,000 US troops are stationed, the islands of Japan and Hawaii, Alaska, and the Trust Territories of the Pacific Islands (e.g., the Federated States of Micronesia, American Samoa, the Marshall Islands, and Johnston Island). The ultimate goal is to create a virtual health services system for patient care, for medical education, research, and the shared subspecialty use. Health care providers will have access to all patient records and perform consultations from any location through a single desktop platform configuration. The Medical Diagnostic Imaging Support (MDIS) will create digital pathways and platforms throughout the Pacific (Telepresence: the Akamai Project business plan, 1995).

Medical Diagnostic Imaging Support (MDIS)

The MDIS system is a large Picture Archiving and Communications System (PACS) and teleradiology project initially undertaken collaboratively by the U.S. Air Force and Army. The system acquires, moves, stores, transmits, and displays radiology images within a digitized network, thereby allowing the use of most images without ever needing to print film. The joint project with Loral and Siemens began in late 1991. The MDIS project was the seed from which the present day MATMO grew. The hospital-wide system has undergone phased implementation at Madigan Army Medical Center at Fort Lewis, Washington, Brooke Army Medical Center at Fort Sam Houston, Texas; and the hospital at Wright Patterson Air Force Base in Dayton, Ohio.

The MDIS system is a "filmless" radiology system using digitized radiography. The overall system provides images equal in diagnostic quality to current film-based images while dramatically improving the productivity of the hospital staff. It allows instant access to radiological images simultaneously by multiple users in multiple locations in an on-demand environment. The MDIS system currently uses an optical disk "jukebox" to store images. The system also allows the "rescue" of an underexposed radiograph by adjusting brightness and contrast. It can also selectively accentuate features through magnification or highlight radiologically translucent areas, such as adipose tissues. It can stack serial images, such as those from a computer tomography study, and allow a physician to follow a structure, such as a blood vessel, through several actions. Film retrieval rates exceed 99.1 percent. As the computing power for the hardware and software increases, three-dimensional reconstruction of the body and pattern-recognition computer-aided diagnosis will be possible.

Since its initial implementations, MDIS has come on line at several new locations in the United States and the Republic of Korea, with more implementations planned. Additionally, installations of commercial equivalents have been completed at the Veterans' Administration Medical Center in Baltimore, Maryland and the Samsung Hospital in Seoul, Korea. The Air Force has exploited the teleradiology component of MDIS at numerous installations and is planning the largest hospital-wide implementation in the world at Wilford Hall Medical Center in San Antonio, Texas. Currently, the system uses T1 telephone lines to transfer images between facilities. Project Seahawk will extend the reach of the MDIS system by using the DS-3 SONET ring network around the Puget Sound to connect approximately eight medical facilities. Overseas, MDIS projects include the Akamai in Hawaii which links U.S. military medical facilities in the Republic of Korea.

As in fixed facilities, the deployed medical structure suffers from the slow delivery of radiological information. Delays in the interpretation of roentgenographic images and delays in transcription and report approval often lead to delays in diagnosis with subsequent delayed or inappropriate treatment and increased costs. These problems are compounded in a rugged field environment due to the sensitivity of much of the equipment and the tremendous logistical burden of providing film and chemicals into forward deployed areas. Through the enabling technologies of MDIS, on-line standard reports will result in timely radiological diagnosis and communication while logistical support is greatly simplified. Based on personal statements made by several physicians at Madigan Army Medical Center, combined with current commercial pricing documents, it may be concluded that digital radiology networks will be widely available in the civilian sector considerably earlier than otherwise possible.

Other Technology Applications

Global telemedicine was tested during the U.N. peace-keeping mission in Somalia in late February, 1993. A Remote Clinical Communication System (RCCS) was deployed. The chief components of the RCCS–FlyAway field unit were a high-resolution digital color camera (Kodak DCS 200ci, Eastman Kodak Company); a laptop computer (Macintosh Powerbook 180, Apple Computer); image acquisition software (Adobe PhotoShop, Adobe Systems, Inc.); and a portable International Maritime Satellite (INMARSAT) transceiver (MX 2020P MAGNAPhone, Magnavox Electronic Systems). Readers should refer to the article by Crowther and Poropatich (1995) for a complete description of the RCCS–Systems.

In this system, images taken with the digital camera were transferred to the laptop computer via a SCSI interface. Adobe Photoshop software was used to import, display, compress, and store the images. A computerized consult sheet was developed using Claris FileMaker Pro. Images and associated data were transmitted by modem to the RCCS–Base unit at Walter Reed, using either dial-up telephone lines or satellite communication.

For a thirteen-month period beginning in February, 1993, 74 cases consisting of 248 images were transmitted from Somalia. Several "case studies" demonstrated the value of telemedicine and telecommunications technology. For example, high-resolution digital imaging was used to examine skin rash around the eyes of a U.S. soldier. It was determined that the images were of diagnostic quality. In another instance, an electronically transmitted history, EKG, and digital image of an echocardiogram of a shrapnel wound to the chest of a Kenyan soldier permitted the consultants at Walter Reed to arrive at a definitive recommendation. And, finally, high-resolution detailed images of blood vessels around the cornea of a French soldier permitted successful on-site treatment of the injury and prevented the need for evacuation (Crowther and Poropatich, 1995).

Advanced Teleconsultation

Advanced teleconsultation allows a physician in a forward medical unit to consult with a specialist located elsewhere. Currently, the staff at Walter Reed Army Medical Center, Washington, D.C., is able to support medics in Germany, Somalia, Croatia, and Macedonia by means of two-way interactive video, still images, and textual data. The consultations may include multiple, simultaneous "windows" on the video monitors, which permit the physician at the medical center to interact and collaborate with the remote provider by simultaneously viewing the wounded soldier's medical record; graphic displays of the soldier's vital signs and ECG; a transmitted, digitized x-ray film; and real-time interactive video.

The virtual consultation is effective enough to reduce the number of medical and logistical support personnel in some deployments while providing expert medical support far forward (Zajtchuk and Sullivan, 1995).

A formal evaluation of the Army's telemedicine experience in Somalia, Macedonia, Croatia, and Haiti has been conducted on the basis of a retrospective case study of 114 completed consults (out of 171 consults completed from February 1993 to March 1995). The study revealed a significant change in diagnosis in 30 percent of the cases, a significant change in treatment in 32.5 percent, and a moderate change in treatment in 25 percent of the cases. Telemedicine consultation prevented air evacuation in 11 patients, or about 10 percent.

Real-Time Combat Medic Telementoring

As discussed previously, it is imperative that the medical care provided by the medics be as comprehensive as possible to resuscitate and stabilize the soldier on the battlefield, thereby reducing casualties. Telementoring is the provision of expert advice to a combat medic on the battlefield via telecommunications, providing the medic with access to patient data and early decision support in the so-called "golden hour" during which the patient's chances of survival are greatest. Telementoring can offer the capability to provide high-tech "opportunity training" during noncombat situations. In addition, when combat soldiers are equipped with geosynchronous injury alert systems, telementoring can also support more efficient patient acquisition.

The hardware that will make this possible is one-way video, integrated into the medic's protective goggles, and two-way voice communication, provided by a throat microphone and a small earphone, both attached to the goggles. Data will be transmitted by means of a palmtop computer. This equipment set can be linked via Broadband Code Division Multiple Access to the mentor. The physician will be able to observe the first aid being administered on the field and will be available to coach the medic during the process. The coaching role may be particularly useful during a mass casualty situation when appropriate triage is especially critical. The health care system then has an opportunity to better prepare for further treatment upon evacuation (Zajtchuk and Sullivan, 1995).

Portable Emergency Medical Information System

As an alternative and/or supplement to interactive telementoring, some military people are advocating the further development of a wearable computer system. A prototype has been developed which provides access to medical information and supports remote primary care by offering remote diagnostic consultation and treatment advice.

The system consists of a small 4 × 4 × 3 -inch chassis that houses the CRT, interface boards, battery mount and a panel of standard I/0 ports. The miniature VGA CRT, mounted on a band on the waist, displays text, graphics, or information relayed from the medical information system or from the medical consultant. The wearable computer system would interface with the medical information center via standard military communications equipment, such as SINCGARS radio, connected to the computer via the RS232 communications port. The addition of a miniature digital camera would allow real-time images of the patient's condition, or of the wound or injury, to be relayed to the diagnostician as indicated.

This computer system would also allow transmission of text and graphics (e.g., critical care guidelines, anatomical diagrams, etc.). It would allow the medic or non-expert attendant to perform diagnosis and treatment without the physical presence of a medical expert, but with access to the best on-line medical information and expertise available. On-line communication capability would be activated by voice commands, and a voice communication link would be established. Since all commands given to the computer would be verbal, the medical attendant would have his/her hands free to continue medical treatment (Altieri and Ishman, 1995).

Advanced Trauma Care—The Critical Care Pod (LSTAT)

Patients must be stabilized prior to being transported. The goal of advanced trauma care (stabilization) is to preserve critical organ function at minimal physiologic status while controlling hemorrhage, reversing systemic shock, and preventing hypoxia by using auto-controlled devices to provide immediate mechanical or pharmacological care. Continued stabilization while in transit to a medical facility is a significant challenge.

It is possible that, after being stabilized, the patient could be evacuated in a critical care pod called Life Support for Trauma and Transport (LSTAT). The LSTAT, currently under development, is intended to be a self-contained suite of electrical and mechanical devices that would monitor critically injured patients and render certain life support functions during transport. It will consist of an evacuation platform for life support incorporating a ventilator, suction unit, environmental control system, oxygen generation system, and advanced patient monitoring and closed-loop therapeutic capabilities. Patient and system status can be interrogated and controlled remotely from a base station. The LSTAT will be linked with current medical evacuation vehicles. With the canopy in place, the LSTAT will serve as an environmentally protected, temperature-controlled, preoperative "waiting room" or a postoperative intensive care unit during evacuation. With the canopy off, it will serve as a surgical platform inside a roving surgical van.

Project Promed

ProMed was started in January, 1993, as a cooperative agreement between the Department of Defense and Apple Computer to evaluate the potential use of the Newton Personal Digital Assistant (PDA) device in a medical setting. The project has evolved to encompass other PDA devices, the use of wireless networking in medical settings, and systems that use PDAs as end agents in a wireless, distributed environment. Collectively, these integrated technologies are capable of providing access to remotely based information. Current projects utilizing PDAs and wireless networks include:

- Prototypes that enable access to an intelligent store-and-forward e-mail system
- Access to remotely based referencing information, CD–ROM text, on-line databases, internet-accessible information sources
- Integration of PDA technology into existing legacy systems for access to results reporting and interactive data/order entry (this includes creating API access to the MDIS, the DoD Composite Health Care System [CHCS] and other informatics projects)
- An intuitive method for data collection for nursing and medical corps personnel: the models created in this prototype will enable the efficient, intuitive, and behind-the-scenes collection of data for a wide variety of purposes including staffing needs assessment, outcomes research, and managed care information needs.

Major Paul Zimnik, D.O., of the Medical Advanced Technology Management Office, oversees Project ProMed and others, such as the use of the Internet in DoD for telemedical information access and management, and the development of the Meditag. The Meditag is a nearly indestructible dogtag-like storage device that allows nearly instantaneous access to up to 90 megabytes of secured information. It is envisioned that the

Meditag would allow significantly easier data-capture approaches on the battlefield and would serve as the redundant data archive until the patient reaches the point of definitive care and treatment.

Ongoing Training

Training is an important element of medical readiness. Plans call for medics to participate in more war games and simulations. All new active and reserve component personnel will attend required entry-level military training within 12 months of coming into the military.

Data from past wars reveal that transport times of the wounded can be lengthy, while a combat medic can often reach a casualty often immediately. Since significant numbers of casualties are severe, the survival of a wounded soldier often depends on the actions taken by the medic.

Real Life Gaming—Kernel Blitz '95

One of the largest medical exercises ever conducted by the military originated in Camp Pendleton in California. Kernal Blitz '95 involved 12,000 Navy and Marine sailors. To ensure as realistic an experience as possible for all participants, Walt Disney studios, assisted by field medical school personnel, applied wound moulage to the casualties. The casualties were coached on how to respond to their unique injuries or illness. Casualties were triaged, "treated" and provided after-treatment care, just as in a real contingency.

The objectives of the medical component of the Kernel Blitz '95 exercise were to maximize training of medical personnel, validate medical requirements outlined in the fleet's major regional contingency plans, integrate Navy medical planning and execution requirements with those of the line, and introduce advanced medical information technologies. At the casuality generation level, an extensive test of the Multipurpose Automated Reader Card (MARC) technology was undertaken. MARC is an existing precursor to the Meditag, which was not in production at the time of this exercise. Casuality demographics were written to the MARC at the generation site and the card was then used in the field to record treatment data. The data were subsequently transferred from MARC to the Composite Health Care System (CHCS), the military's legacy health care information system, on the USNS Mercy, USS Peleliu, and Fleet Hospital Operation and Training Command (FHOTC) at Camp Pendleton. This enabled technicians to demonstrate the feasibility of this system in the field for the first time.

Other telemedicine trials included live video-consultations between the USNS Mercy and FHOTC, transfer of digitized x-rays from the surgical company forward to all other echelons, a live broadcast from the exercise to the Mayo Clinic International Telemedicine Symposium, and numerous technical demonstrations for participants' information and interaction.

Among the lessons learned from Kernal Blitz '95 were the importance of information flow and situation awareness for medical commanders, the frequency and bandwidth assignment needed, and equipment compatibility for medical communications. Kernel Blitz '95 showed that the reliability of some communication equipment needs to be reassessed and improved (Andrus, Taylor, and Carey, 1995).

Training of Iowa Reserves

The active duty forces are not the only ones interested in the application of advanced technologies to solve problems. Iowa, the only state to own its own buried fiber optic network, received two large grants (one from ARPA, and another from the National Institute of Health to the University of Iowa) to develop telemedicine through a rural telemedicine network. This fiber optic network connects every National Guard Armory and every county in the state. The National Guard uses the network to train soldiers in low-density specialties and to provide training common to all military units. The network also serves the state in the areas of public education and health services. Iowa telemedicine is to include: (1) the delivery of medical and educational services through a networked support system; (2) the provision of clinical patient services and support for the practitioner on a "just in time" basis to include image transfers from radiology, pathology, dermatology, etc.; and (3) the provision of administrative data processing capability (Edwards, 1994).

Realistic Medical Simulations

The objectives of this program are to achieve virtual representation of the human anatomy for training purposes, including the treatment of combat casualties with operational battlefield requirements. Expanding on the digital images provided by the Visible Human Project of the National Library of Medicine, MusculoGraphics, Inc., of Evanston, Illinois, has developed a high-resolution three-dimensional computer model of the human lower limb to model gunshot wounds. This system will also evaluate injury consequences such as blood loss and loss of muscle function. Using accurate, physically based computer models to simulate gunshot wounds eliminates the need to create such injuries in animals for such testing (MusculoGraphics, Inc., 1995).

IPORT (Individual PORTals)

Simulation is also being used in training soldiers and medics by inserting them into virtual battlefields. One such product, funded by ARPA and under development by Sarcos, Inc., of Salt Lake City, is IPORT—or individual portals to application specific virtual environments. The "warrior link" concept under development has as its goal the insertion of individual combatants into a virtual battlefield for concept definition, training development, mission rehearsal, contingency planning, and analyses and evaluation. "Treadport™" allows users to walk, run, and adopt a full range of postures (kneeling, sitting, prone, etc.) in a virtual battlefield. With its 4-degree-of-freedom torso motion-tracking, it is adaptable to combat medic training. The Individual Soldier Mobility Simulator is being developed as a high-fidelity terrain interactive system, including feedback for mud, rock, stairs, etc. It allows forward, backward and lateral motion, foot and motion tracking, and scenario-based energy expenditure (Sarcos, Inc., 1996).

Conclusion

The United States military endorses the vision of a highly networked health care environment which enables rapid communication across largely horizontal organizations, moving and acting with great speed and precision. Military medicine is reassessing fundamental assumptions about the provision of health care and other services both to patients and the well, in peace and in war. In that reassessment, the senior leadership has embraced the concept of exploiting technology to reengineer missions, organizations, and operating procedures. Experiences gained in the military sector should contribute to a greater understanding of the basic principles of the potentials of technology in the delivery of health services. It is hoped that lessons learned from this sector will be instrumental in developing more effective and efficient delivery of medical care in the civilian sector.

REFERENCES

Altieri, C. K., and R. Ishman. 1995. *Introduction of a fully portable, body-mounted emergency medical information system.* Raleigh, N.C.: InterVision Systems, Inc.

Andrus, K. L., J. K. Taylor, and N. B. Carey. 1995. "Navy Medicine in Kernal Blitz '95," *Navy Medicine,* (November–December): 6–9.

Anonymous. 1995. "Changing Military Medicine: TRICARE, Closings, Europe." *AUSA* News, 17 (5), pp. 3, 10–12.

Bellamy, R. F. 1994. "The Causes of Death in Conventional Land Warfare: Implications for Combat Casualty Care Research," *Military Medicine,* 149: 55–62.

Berry, F. C. 1995. "Army Leads the Telemedicine Revolution," *Army,* (September): 38–46.

Crowther, J. B., and R. Poropatich. 1995. "Telemedicine in the U.S. Army: Case Reports from Somalia and Croatia," *Telemedicine Journal,* 1(1): 73–80.

Department of Defense Draft Telemedicine Test Bed (TTB) Management Plan. 1994. U.S. Medical Advanced Technology Management Office. Unpublished paper, December 16.

Dineen, P., and S. Larson. 1995. "Mental Health Alliance," and "GME Inegration Plan," *TRICARE: Northeast Communique,* 1(1): 5–6.

Dubik, J. M., and G. R. Sullivan. 1994. *War in the Information Age* (Landpower Essay Series no. 94-4). Arlington, VA.: Association of the United States Army, Institute of Land Warfare, May.

Edwards, J. 1994. "Information Paper: Subject: Coordination with Iowa National Guard, Et al., Follow Up." Unpublished memorandum, June 14.

French, Y. 1995. "New Wave: Tofflers Explain Their Theory of Economic Evolution." *Information Bulletin,* 54 (9): 198.

Gunby, P. 1992. "Two Years After Iraqi Invasion, U.S. Military Medicine Studies Desert Shield/Storm While Looking Ahead," *Journal of the American Medical Association,* 268 (5): 577–78.

Gunby, P. 1993. "Satellite Hookup Links Major U.S. Military Hospital with Army Physicians in Mogadishu, Somalia," *Journal of the American Medical Association,* 269 (13): 1606.

Gunby, P. 1994. "Military Medicine Takes Late Gulf War Hits; Aeromedical Evacuation Improves as Result," *Journal of the American Medical Association,* 271 (7): 491.

Haines, J. E. 1994. *Project SeaHawk: Preliminary Requirements Document,* Version 1.1. Roswell, GA: Cmed Corporation.

Harris, E. D. 1995. "Military Medicine Prepares for 21st Century," *Stripe,* 51 (June 9): 8.

Jeffer, E. K. 1994. "Medical Triage in the Post–Cold War Era," *Military Medicine,* 159 (5): 389–91.

Koehler, R. H., R. S. Smith, and T. Bacaner. 1994. "Triage of American Combat Casualties: The Need for Change," *Military Medicine,* 159 (8): 541–47.

Kissick, W. L. 1994. *Medicine's Dilemmas: Infinite Needs versus Finite Resources.* New Haven, CT: Yale University Press.

Medical Advanced Technology Management Office, U.S. Army Medical Research and Material Command. 1994. *Tri-Service Telemedicine Testbed Project: a Plan for Guiding the Testbed Project and Integrating Telemedicine Technologies into the Military Health Services System (MHSS): Working Copy.* Unpublished manuscript.

Medical College of Georgia, Telemedicine Center. 1995. *A Dual-Use Telecommunications System for Delivering Medical Care.* Submitted to Center for Total Access. Unpublished proposal. January.

Mlinaric, J., et al. 1994. "Intensive Care of Severely Wounded Military and Civilian Casualties in Zadar, Croatia." *Military Medicine,* 159 (6): 434–37.

MusculoGraphics, Inc. 1995. *MusculoGraphics' Limb Trauma Simulator.* Evanston, IL: MusculoGraphics, Inc.

Nguyen, D. 1994. "Mass Casualties on the Modern Battlefield: a View from the 1st Armored Division," *Military Medicine,* 159 (11): 683–85.

Oder, J. E. 1994. "Force XXI," *Army,* 37–42 (May).

Pruitt, R. 1993. Assistant Director for Research, Casualty Care Research Center, Department of Emergency and Military Medicine, Uniformed Services University of the Health Sciences. Bethesda, MD: personal communication to Zajtchuck and Sullivan.

Sarcos, Inc. 1996. *IPORT: Application Specific Portals to Virtual Environments.* Salt Lake City, UT: Sarcos, Inc.

Telemedicine Testbed CEO Support Group. 1995. *American Healthline: APN Daily News Briefing,* via the Internet, 4 (45), June 2.

Telepresence: The AKAMAI Project Business Plan. 1995. Unpublished manuscript.

Toffler, A. 1980. *The Third Wave.* New York: Bantam Books.

U.S. General Accounting Office. 1992. *Operation Desert Storm: Full Army Medical Capability Not Achieved.* NSIAD-92-175. Washington, D.C.: GAO, August.

Walters, T. Forthcoming. "Deployment Telemedicine: The Walter Reed Army Medical Center Experience," accepted for publication by *Military Medicine.*

Zajtchuck, R., and G. R. Sullivan. 1995. "Battlefield Trauma Care: Focus on Advanced Technology," *Military Medicine,* January: 1–7.

Zimnik, P. R. 1994. *Bullet Background Paper on Project Promed.* USAFMLO–MCMR–AT. Unpublished paper, December.

Chapter 13

BATTLEFIELD TELEMEDICINE:
THE NEXT GENERATION*

Shaun B. Jones and Richard M. Satava

Introduction

Telemedicine, with its advanced diagnostics and imaging, remote medical and surgical intervention, informatics, virtual prototyping, and simulation, is at the threshold of transforming medical care on the battlefield. Historically, after a battle was fought and the opposing forces retired for the day, surgeons went to the battlefield to care for the wounded. A change occurred in 1792, when Dr. Dominique Larrey, a French surgeon, introduced wagon "ambulances" to accompany the Army of the Rhine's "flying" artillery on the battlefield. These "flying ambulances" evacuated the wounded from the battlefield to the rear echelons where the surgeon could apply his craft more effectively (Barkley, 1978). In World War II and the wars in Korea and Vietnam, the total combat mortality could not be decreased significantly simply by augmenting rear-echelon or hospital-based, combat casualty care. Most deaths still occur on the front lines. Once again, physicians and surgeons must return to the far-forward battlefield to provide rapid, sophisticated medical care. Information-age technologies will provide improved combat-casualty-care training and the ability to project electronically the expertise of the physician and surgeon anywhere there are wounded or sick. With advanced diagnostics, therapeutics, and simulation with sophisticated telecommunications, telemedicine promises to alter the future of battlefield medicine permanently.

During the last three major U.S. wars—World War II, Korea, and Vietnam—the proportion of casualties who died from life-threatening wounds once hospitalized declined substantially from greater than 10

---

*The opinions or assertions herein are the private views of the authors and are not to be construed as official, or as reflecting the views of the Advanced Research Projects Agency, Department of the Navy, Army, or the Department of Defense.

percent to less than 5 percent (Bellamy, 1995). Unfortunately, 88 percent of the combat deaths in the U.S. Army still occurred in the combat zone before the wounded reached hospital-based combat-casualty care. In Vietnam, for instance, two-thirds of those killed in action died within ten minutes of being wounded. Many of the remaining one-third did not arrive for treatment within the "golden hour" for life-saving intervention, and never lived to receive the benefit of the combat medical system (Bellamy, 1995). Telemedicine may make it possible to regain that "golden hour" for the soldier at the front lines. However, this will require not only advanced technology but also correction of centuries-old deficiencies in frontline combat-casualty care.

The Gulf War clearly demonstrated the might of "information warfare" and the advent of the "digital battlefield." Information provided "battlefield awareness," so the Allies knew exactly where the enemy was located. Smart weapons, precisely targeted with minimal collateral damage, and real-time intelligence information provided command and control functions. All these facets of information technology compressed what normally would have been a month- or year-long battle into a mere four days. But ironically, the battlefield medical support during this brief war used decades-old technology and was unable to decrease the mortality rate of the far-forward wounded from that of over 150 years ago.

Three major limitations account for the greatest deficiencies in combat-casualty care today: (1) the inability to determine when, where, and how severely a soldier is wounded; (2) the inability to place surgical expertise at the time and site of wounding; and (3) the limited experience in combat-wound treatment at the front lines. For example, only 5 percent of all Gulf War physicians had experience with combat wounds. Correction of these three deficiencies by the Department of Defense has begun with unique technological approaches, depicted in Figure 1.

To illustrate how these responses would work, we present a realistic scenario already used in military training exercises to simulate the digital battlefield of the future. Every soldier will wear a smart-card electronic "dog tag" and a personnel status monitor (PSM) that provides instantaneous information about his or her precise location and medical status with a multitude of vital signs: whether alive or dead, and if wounded, the severity of the wound. This information will allow the medic who monitors the equipment at the base station to locate the casualty precisely without compromising tactical integrity, to determine the severity of the injury for enroute triage, and to provide immediate

Figure 13-1. Technological Response to Deficiencies in Casualty Care

Deficiency	*Technological Response*
Inability to determine when, where, and how severely a soldier is wounded	Global Positioning Satellite Location
	Identification Friend or Foe
	Remote Biosensing
Inability to place surgical expertise at the time and site of wounding	Telemedical intervention and remote telepresence surgery
Limited experience in combat-wound treatment at the front lines	Distributive, collaborative virtual environment for training

on-site medical assistance. If the soldier can be stabilized on-site, the medic will evacuate him or her in an autonomous micro-miniaturized intensive care unit to the most suitable location for further care. This Life Support for Trauma and Transport (LSTAT), or Trauma Pod, contains a microventilator and oxygen generator for breathing support, automated micro infusion pumps for IV fluids, blood, and medications, suction device for clearing a blocked airway, and multiple vital-sign monitors for continuous assessment of health status. If the soldier cannot be stabilized without surgical intervention, an Armored Remote Telepresence Surgical Unit (ARTSU) in a Medical Forward Area Surgical Telepresence (MEDFAST) vehicle will go to the soldier, and a surgeon in a rear-echelon hospital will be able to perform life-saving remote surgery. Once stabilized, the soldier will be evacuated to the most suitable location for further care.

In this scenario, theater medical information systems will assist with authentication, electronic "notification" of medical status, and real-time triage and transfer routing. Any necessary specialty consultation to the theater of war, from the time of wounding to definitive care, will be available worldwide for both military and civilian hospitals. In addition, training for clinical evaluation, triage, and treatment of these casualties in a predeployment situation is now available through the use of virtual environments embedded with the appropriate technologies. The process of maturation—from proof-of-concept (Ft. Gordon, June, 1993) to deployment in training exercises (Camp Pendelton, April, 1995, and Ft. Gordon, November, 1995)—took about two years.

Modern Medicine on the Battlefield

Modern medicine can be implemented on the battlefield with the advanced technologies already available (i.e., utilizing the new digital communication and information infrastructure).

In trying to assess the importance of information as the key element in the changes that have occurred in medicine, a speaker in an off-the-record estimate given at a 1994 National Science Foundation workshop suggested that 90 percent of the information desired to treat a patient can be acquired electronically (Durlach, 1995). Examples include noninvasive imaging (X-ray, CT, MRI), vital signs, laboratory test results, patient records, treatment protocols, consultations, and database searches. Remote physical examination, simple procedures, and even basic surgery are made possible by sophisticated technologies allowing remote manipulation, palpation, and three-dimensional visualization using computer-based anatomy and the new virtual-reality surgical simulators. The basic clinical information required by the physician to treat a wounded soldier or patient can be acquired in electronic form and subsequently transmitted to a physician at a computer workstation (Satava, 1993a). With the technologies of teleoperation, telemanipulation, and telesurgery, medical expertise can be brought to the place where the soldier is wounded or the patient is sick. In addition, the same computer workstation used for telepresence surgery can provide a surgical simulation of a wounded "virtual" soldier for practicing life-saving procedures on combat injuries. This makes it possible for a virtual combat wound to be created with 3-D graphics and imported into the telepresence surgery system. While waiting for a known casualty, a combat surgeon can "call up" a simulated wound into the telepresence surgery system for practice and training; when the casualty arrives and is placed into the telepresence surgery system, the surgeon simply "flips a switch" and begins doing surgery on the same wound that was practiced on moments before. If the basis of fundamental changes in medicine is the advent of the information age, the cornerstone is the digital physician, who understands the use of the information technologies described above.

Advanced Technologies

Military medicine is responding to this fundamental change of viewing the practice of medicine as a form of information by integrating advanced technology into the modern digital battlefield. Telemedicine is the result, built with the "raw materials" of **advanced diagnostics** and **therapeutics, informatics,** and **virtual reality.** In the following sections, leading examples of remote diagnostics, remote treatment, and informatics are described.

Remote Diagnostics

Remote diagnoses through advanced biomedical sensors provide the initial steps in treating the wounded soldier. The following diagnostic modalities will help the surgeon or medic assess the magnitude of the injury.

Personnel Status Monitor (PSM). The initial PSM is being developed by Sarcos, Inc., with contributions by Bolt Beranek and Newman, Inc., Motorola, Inc., and Sandia National Laboratories (see Figure 13-2). It is a wrist-worn or gauntlet device combining environmental and noninvasive biosensors, global positioning satellite (GPS), and intelligent alerts, all integrated into a sophisticated communications module. The first-generation prototypes' noninvasive biosensors measure heart rate, blood pressure, electrocardiogram (ECG), temperature, and oxygen saturation. The communication module uses modern, secure, wireless technology called Broad-Band Code Division Multiple Access (B–CDMA), a method of sending digital information over communication networks similar to cellular phone systems. Even though initially only soldiers will be wearing PSMs, the same devices should become available in the civilian sector within the next three to five years. Its potential applications range from law enforcement to care for hospitalized or chronic-care patients.

The second-generation sensors, currently still in concept development and design, will noninvasively measure blood electrolytes and metabolites, respiratory rates, blood gases, and levels of consciousness. As these systems are refined, the wrist-worn device will be replaced by flexible, wearable, robust sensors permitting full mobility and body cohesion. In addition, to help reduce the 15–20 percent fratricide rate in armed conflicts (Dupuy, Brinkerhoff, Johnson et al. 1986), the capability to determine whether the soldier is friend or foe (IFF) will be integrated into the unit. The third generation is anticipated to have "smart," multidomain (such as blood pressure, respiration, and temperature) sensors using optoelectronics, infrared, and upconverting phosphors. These sensors will also be able to detect biological and chemical weapons.

Advanced Imagery. The use of advanced imagery on the battlefield today depends primarily on four telemedicine subsystems, including Digital X-ray, Remote Clinical Communications Systems (RCCS), Medi-

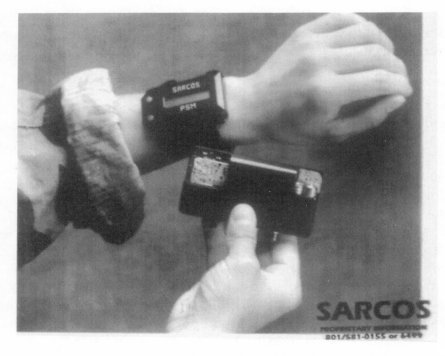

Figure 13-2.

cal Diagnostic Imaging Support (MDIS), and Video Teleconferencing
(VTC) (see Figure 13-3). A solid-state, digital X-ray imaging prototype
has already demonstrated its potential for mobile trauma and battlefield
applications (Ille, 1994). The system consists of a mobile surgical C-arm
modified to provide image acquisition (1024 × 1024 @ 30 fps), a flat
panel detector (8 × 8 cm), and sophisticated image processors and
displays. Larger (20 × 20 cm) solid-state detectors are in the developmental stage. RCCS is a deployable laptop system designed to send diagnostic-
quality medical images using theater communications from remote
locations to specialists and medical centers. The digital images acquired
in the field or at sea can be stored locally on hard disk or optical juke
box. The latter is an integral storage component of the MDIS system
which allows simultaneous access to imaging data by multiple users and
locations. MDIS is a filmless radiology picture archiving and communications system (PACS) developed by the Medical Advanced Technology
Management Office (MATMO) at Ft. Dietrick, Maryland, in cooperation with the Advanced Research Projects Agency (ARPA). In the initial
projects, MDIS linked the hospitals (Project Seahawk) Tripler Army

Medical Center in Hawaii, and military sites in the Pacific Rim (Project Akami). In the future, collaborations already underway will link the battlefield with defense and civilian facilities nationwide and around the world.

Figure 13-3.

The battlefield physician of the future may very well have at his or her disposal tools now only in the prototype phase, such as remote 2-D/3-D hand-held ultrasound, miniaturized portable and robust CT, innovative laser and spread spectrum imaging, and nonometer optical microscopy. Today, however, the technology is based upon digital X-ray, conventional CT and MRI acquisition, data fusion, reconstruction, and image projection. The state of the art, outside of the battlefield, allows digital physicians to augment their cognitive 3-D conceptualizations with real-time, computer-generated, multimodality, 3-D image projections based on the patient's data. Navigational guidance, frameless sterotaxis, multi-level transparency, and haptic interfaces are but a few of the enhancements made possible by advanced technology. However, their implementation on the battlefield is still under conceptual development.

Advanced Laboratory Diagnostics (Pathology). Generally, laboratory analysis is undergoing dramatic change. Off the battlefield, many procedures are now performed at the bedside as a result of the combination of escalating hospital operating costs, improved productivity requiring faster test results, and advanced technological innovation. On the battlefield, medical personnel have long needed a robust, low-cost, hand-held device capable of performing commonly ordered tests that can be integrated into central information systems. Several projects are currently underway around the country; the "i-STAT" system is in the lead in commercially available systems.

Produced by the i-STAT Corporation, this system measures basic electrolytes, glucose, blood gases, and hematocrit using a miniaturized sensor approximately one inch square and requiring less than two drops of blood. Testing is performed on the single-use, disposable sensor cartridges by a hand-held, self-calibrating analyzer in 90 seconds. The analyzer can store up to 50 data records. It is capable of infrared signaling to available information systems and peripheral devices, and uses two 9-volt batteries that provide between 300 (alkaline) and 900 (lithium) tests.

Technologies under development should provide capabilities similar to those described above, but noninvasively (without drawing blood samples or inserting probes into the body) within five to seven years.

Remote Treatment

Trauma Pod. The Mobile Intensive-Care Rescue Facility (MIRF) from the Buchanan Aircraft Corporation of Australia embodies the first-generation trauma pods. These pods are commercially available now, and are suitable for both military and civilian field and hospital use. The MIRF is a 90-kg self-contained intensive-care medical system housed in a rugged composite fiber shell which clips onto a standard North Atlantic Treaty Organization (NATO) stretcher. Self-contained systems use off-the-shelf technologies that include C-size oxygen cylinders, AC/DC converting power sources, and robust medical equipment tolerant of shock and vibration. The medical equipment includes multifunction monitors, defribillators, ventilators, volume and syringe infusers, and suction systems. Similar units were used during the Gulf War on C-141 transport airplanes in medical evacuation flights to Germany and to the continental United States. They are extremely lightweight "pop-up"

NATO stretcher-adapted tents with miniaturized temperature controls, respirators, and monitoring units.

The next generation in trauma treatment is the Life Support Trauma Transport (LSTAT) or "trauma pod" being developed by the Advanced Research Projects Agency (ARPA) in collaboration with Northrop-Grumman and Walter Reed Army Institute of Research (see Figure 13-4). It consists of a miniaturized, autonomous intensive care and trauma unit that can be integrated into a variety of evacuation platforms. Advanced capabilities will be based on sophisticated technologies including lightweight oxygen concentrators and generators, micro-miniaturized defibrillators, power supplies and environmental modulators, autonomous anesthesia-capable ventilators, smart catheters, and multi-function fluid infusers. Onboard microprocessors will also organize the sensor data into

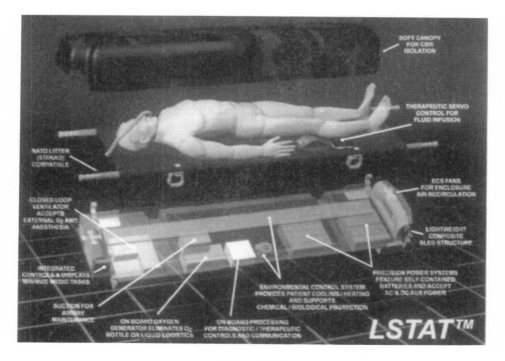

Figure 13-4.

an electronic record compatible with secure theater information systems. The pod itself will be constructed of robust advanced composites responsive to a variety of biological and chemical threats and degradation.

Adaptable configurations for a variety of medical evacuation platforms (tracked vehicles, helicopters, surface and subsurface vessels) are now under investigation. These advanced technologies are being prototyped, including virtual environments using the Virtual and Distributed Prototyping Tool of the Future (VDPTF). This tool was developed as part of the "Operating Environments of the Future" project, a collaborative effort between ARPA and an academic and industrial consortium at the Massachusetts Institute of Technology (Kaplan, 1995). Future generations of trauma pods, optimized in the virtual environment, may have capabilities for medical imaging, telemedicine, and telepresence surgery.

Telepresence Surgery. The successful return of the surgeon to the battlefield will require novel telemedical approaches to combat-casualty care. To meet the unique demands of battlefield medical care, SRI International, Inc. is developing a Telepresence Surgery System that enables a trauma surgeon, located away from hostile fire, to operate on a wounded soldier in a combat zone. The system consists of a surgeon's workstation and a remote surgical unit. The surgeon's workstation integrates a full-color stereoscopic, real-time view of the operative field with two servo-controlled surgical instrument handles and stereophonic audio to create a fully immersive yet transparent interface with the remote site. The remote surgical unit incorporates two high-resolution color CCD cameras with electronic zoom lenses and two servo-controlled manipulators with interchangeable surgical instruments.

The first generation, the "Green Telepresence Surgery System," developed by SRI International, Inc., consisted of a single-handed manipulator providing four degrees of freedom, opening and closing surgical end-effectors, and force reflection (Hill and Green, 1995). During use, movements of the surgeon's hands at the workstation are precisely and instantaneously translated to instruments at the ends of the manipulator arms of the remote unit. The surgeon perceives the instruments and the operative field to be directly in front of him or her as in conventional surgery. Further, the force-reflective manipulators reproduce a haptic interface normally experienced during conventional surgery. Although this is telepresence rather than virtual reality, the surgeon operates on a virtual three-dimensional image.

The current generation system is a two-handed, five-degree-of-freedom system with paired CCD cameras to improve three-dimensional vision. A prototype of the system has been successfully demonstrated in field demonstrations as well as in animal studies. Surgical procedures ranging from femoral artery repair to repair of stomach and liver lacerations have been performed successfully in porcine models (Bowersox, 1995). This system promises to change future battlefield medical scenarios dramatically. Trained, experienced trauma surgeons will perform surgical procedures on critically wounded soldiers to enhance necessary medical and surgical life-saving interventions. Related systems, intended for smaller-scale ophthalmologic procedures rather than battlefield care, are being developed separately by the Georgia Institute of Technology, Microdexterity Systems, and the Massachusetts Institute of Technology (Hunter, Doukoglou, LaFontaine et al., 1993).

Future generations will have greater degrees of freedom and may incorporate emerging technologies such as advanced holographic displays, sensory substitution, olfaction, voice interaction, remote manipulation based upon microelectromechanical systems (MEMS) of "muscles" and actuators, and dexterity enhancement using specialized instruments (mechatronics, lasers), tactile detectors, and displays.

Medical Information Architecture and the Global Grid

The requirements of modern warfare have created new levels of command and control which exceed the existing capabilities of computing, networking, and information services. Legacy systems have been operating poorly at these levels because of the difficulty in updating and inserting new technology, prolonged developmental lag times, and a lack of interoperability. Accordingly, the defense information infrastructure was developed to ensure overall military superiority using secure, scaleable information technologies to connect global yet disparate fixed station and mobile systems. It is founded on adaptive and distributive information systems embedded in tactical and operational environments. Its implementation requires a high-speed, broadband, global infrastructure merging modeling and simulation, Command, Control, Communication, Computers, and Intelligence (C4I), resource registration, discovery and situation sensitive applications, networks, and secure transactions. This same information infrastructure will be used for telemedicine to rapidly integrate advanced technologies. The main purpose is to enhance capability within diagnostic, triage, treatment, and information management functions.

Information management systems on the battlefield will be used to coordinate mass casualty trauma care in hostile environments. This is uniquely challenging whether in traditional military or civilian settings. Too often in the past, critical time was lost locating, triaging, and evacuating the wounded. Inconsistent treatment protocols and haphazard patient record-keeping and reporting have created chaos on the battlefield, at all levels of command and control. Now, trauma-care information management systems are being designed to improve the decision-making ability of medical personnel on the battlefield. These systems will also facilitate information flow and patient-processing in various civilian trauma-care situations. Combat-casualty care will become largely automated and distributive (mobile), using such enabling technologies as personal monitors, electronic "dog tags," intelligent alerts and agents, hand-held digital assistants, medical anchor desks, and computerized medical records. High-quality, timely care throughout the theater of operations will be provided with a minimum of impact on the tactical integrity and security of war operations, enhancing situational awareness and operational readiness.

An example of such enhancement was a naval exercise (Project Challenge Athena II, 1994), involving 105 patient evaluations aboard the aircraft carrier USS George Washington (CVN-73) during a Mediterranean deployment. In this exercise, it was shown that teleradiology and teleconsultation, as part of a trauma-care management system, eliminated 31 unnecessary shipboard evacuations. This is equivalent to 31 instances in which the tactical and operational integrity of the fleet was not degraded in support of an evacuation, saving over $100,000 in operations, as well as improving conservation of the fighting force. Improved utilization of resources was not limited to medical applications. The shipboard material managers reduced inventory and transshipment requirements by using the teleradiology systems during off-hours to "diagnose and treat" damaged machine parts in collaboration with their colleagues back in the United States.

The global grid is still largely a concept rather than an actuality. But when it is implemented, it could integrate in real time sea, air, and land assets in a virtual battlespace projectable to any location where missions might be conducted, forces deployed or trained, and soldiers wounded and sick. The global grid is intended for distribution unfettered, requiring little location preparation. It will be able to integrate advanced medical capabilities, active and call-up forces, modeling and simulation, intelligence resources, smart processing with strategic and tactical communications, and interoperability with public telecommunications. Although often referenced in discussions of global telemedicine, its use for casualty care in real-world missions and in military and civilian telemedicine is just now under consideration by ARPA and the Department of Defense.

Virtual Battlefield

Virtual environments (VE) are emerging as a powerful milieu in which to optimize telemedical training and operational planning. Combat-casualty care providers practice virtual on-site first aid, triage, evacuation, and medical and surgical procedures in both battlefield and civilian-disaster simulations. Scenario-dependent casualty or mortality rates for operational planning can be estimated, and doctrine, policy, and procedure given a virtual "reality check," before developing further plans. The latter is especially important because many of the remaining short-comings on the battlefield involve situational awareness in non-acute casualty-care domains (command, control, and communication (C3), medical triage, resourcing and contingency planning, regulation, registration, and record-keeping). An additional challenge is the escalating complexity of operational planning as the "rear echelons" participate in the added dimensions of joint (tri-service), multinational peacekeeping and Operations Other Than War (OOTW).

Training of Medical Forces

Virtual environments embedded in the enabling advanced technologies can now be used to train personnel for the evaluation, triage, and treatment of casualties. Medical trainers and medical forces planners are beginning to use such environments to insert medical personnel into virtual simulated battlefields alongside individual soldiers. Individual soldiers are represented by autonomous agents or avatars, or avatars under human control, allowing training of medics and physicians in the treatment of battle wounds (e.g., control of hemorrhage, wound debridement, and even surgical procedures). Thus, as virtual soldiers are wounded, medical personnel will provide casualty location, enroute triage, advanced trauma life saving, and evacuation.

Individual portals into the virtual environment, called "I–Ports," are being developed to provide soldiers and medical personnel with access to a virtual battlefield over the defense simulation network (SIMNET/ DIS). "I–Ports" consist of an upper-body conforming exo-skeleton that tracks hand, arm, and head motion, and a motion platform upon which the soldier is placed, registering body movements and locomotion in the virtual battlefield, with variable resistances appropriate for changes in terrain features. Body movements register almost identically on the virtual soldier, the avatar, as the real soldier performs actions. What the real soldier sees is a cartoon-level figure of himself and other members of the platoon, in addition to any other battlefield participants. Avatars (MEDIVAC JACK/Dexter) have been developed that exhibit realistic anatomy when wounded (Visible Human/WDMET). Personnel practice combat-casualty evaluations, emergency treatment, triage, evacuation, and other essential skills in realistic task-oriented scenarios in collaborative and cooperative interactive virtual battlefields. The degradation of casualty-care training concomitant with the recent elimination of animal-based combat surgical training can also be minimized.

Virtual reality's promise to simulate both realistic and imaginary worlds faithfully cannot be totally fulfilled until olfactory stimuli are also included. In many engineering problems, odors are not regarded as important. However, olfactory components are significant both in surgical training and the in training of dismounted medical and combat personnel. Artificial Reality Corporation and with Monell Chemical Senses Laboratory are collaborating on a three-year ARPA-sponsored project to add olfactory stimuli to these applications. For surgical training, delivery systems and medically relevant odors are being developed. For dismounted personnel, portable delivery systems and a wide-area wireless tracking system are being engineered that will allow individuals to "follow their noses" in virtual reality. The effectiveness of olfactory cues in improving training, performance, and other uses of the sense of smell in medical applications will also be explored. Finally, the development of biosensors to recognize medically salient odors and their coupling with the delivery systems is being contemplated to enhance telepresence environments. The next generation may incorporate emerging technologies (miniature gas chromatographs and mass spectrometers) based on chemical sensors the size of a postage stamp. The latter projects are in the prototyping phase.

VR Surgical Simulators

Surgical education and training on the virtual wounded require simulators incorporating interactive, life-like virtual anatomy. The most important factors to be considered are visual realism or fidelity, object interactivity, object physical and physiologic properties, and sensory input. To date, simulators incorporate the first three factors; the latter two are in rudimentary development. The considerable computing power each of these components requires is a barrier to progress. Current computing platforms are not sufficiently powerful to perform the necessary calculations for all five components simultaneously. Consequently, there are trade-offs. For example, there may be extremely realistic-appearing organs which do not deform or change position in real time. Continuing increases in available computing power and sophisticated algorithms, however, will soon provide the power needed to achieve more realism.

The first generation of virtual environment surgical simulators recreated geometric anatomy, with the ability to interact with the organs simply by moving them around like solid objects (Satava, 1993b). The second generation, currently available, has organs with physical properties, such as viscoelastic deformation, collision detection, and higher fidelity of visual realism using texture mapping—the first to incorporate the National Library of Medicine's Visible Human Project dataset. The most advanced have full three-dimensional, highly realistic muscles, bones, joints, and neurovascular anatomy combined with elaborate underlying physical and kinematic modeling of the dynamics of the thigh and knee (Delp, 1995). Other maturing biomechanical simulators include human muscle (Chen and Zeltzer, 1992) and complex kinematics (McKenna, 1994). The utilization of ballistic wounding databases from Vietnam, the Wound Data and Munitions Effectiveness Team (WDMET) database, is beginning to provide realistic rendering of battlefield leg wounds for trauma training and education. In a related effort to add realistic wounding to the surgical simulation, MusculoGraphics, Inc. is developing an interactive, high-resolution, three-dimensional, computer-based, lower limb trauma simulator to help train medical personnel for the battlefield. The lower-limb simulator, used to predict muscle-tendon transplant outcomes in hip and knee reconstructive surgery (Delp, Loan, Hoy et al., 1990), includes a library of battlefield injuries and anatomy based upon the Visible Human data set. For battlefield surgical training in virtual reality the Green Remote Telepresence Surgery system is the most successful. Although not intended for simulated battlefield training, three major commercially available systems have individual surgical simulators—a colon simulator (Hon, 1994), a gall bladder simulator (McGovern, 1994), and leg simulators (Rosen, 1992).

Third-generation simulators, which will depict physiologic properties such as bleeding or leaking of various fluids, exist only as early prototypes. Future generations will require higher-resolution data sets and innovative image acquisition, processing, and reconstruction. They are expected to include microscopic anatomy such as microglandular or organocellular structures (generation 4), complex biochemical system modeling such as immunologic, neuroendocrine or pathologic states of shock (generation 5), and sub-cellular and molecular systems and pharmaceutical interactions (generation 6). We can look forward to the day when "clinical trials" will be first performed using virtual humans, virtual pathological states, and virtual therapeutics while providing virtual outcomes.

Another use of virtual reality, indirectly related to casualty care, is rehabilitation after wounding. Regrettably, even with sophisticated medical care, most survivors of lethal battlefield injuries have long-term physical deficits requiring rehabilitation. As a rehabilitation tool, virtual reality can create a challenging environment in which the disabled can practice necessary skills before going into the real world. Using special devices (e.g., wheelchairs), instrumented in a virtual environment, disabled persons can practice navigating through shopping malls, using public and private transportation, going to medical and dental facilities, and performing other daily activities. Using unique adaptive interfaces, virtual interfaces can be designed to enhance the ability to interact with the real world. Eye trackers, for example, allow those completely paralyzed to move a cursor on a video monitor or an object in a virtual world and use a computer to communicate with and regain the ability to function in the world around them.

Virtual Distributive Simulation

Virtual environments allow collaboration among diverse teams, such as disaster relief, humanitarian assistance, law enforcement, and fire fighting, without the need to expend sizable resources. Telemedical assets, advanced technologies, and realistic teaming scenarios can be incorporated into the simulation environment enabling powerful testing, training, and education. The state of the art in distributive virtual environments robustly represents human and real-time interactions in a hierarchical network of variable resolution dynamic environments (NPSNET). The participant's individual viewing position and orientation are computed on the basis of joint articulations as well as body posture. They are updated intuitively while the head is tracked with a head-mounted display (HMD) sensor. An interface has been developed to allow a user with arm magnetic sensors to control the avatar's arms in real time without using an exoskeleton or sensor suit. Using such technology, real-time upper-body motions can be tracked and used with efficient multicast networks for virtual interactions independent of limited scripted animations. Input devices are also being tested to tire medical personnel training on the virtual battlefield (O'Keefe, 1992).

In distributed simulations, a simple representation can be used to identify the location on the battlefield where a virtual soldier is wounded. In the lower-resolution battlefield environment, multiple ballistic wounds can be displayed, adequately indicating the location of the injuries and the need for medical attention. View switching between the low-resolution battlefield and the higher-resolution medical environment is now possible, and it is driven by the task and scenario. In NPSNET networked, autonomous medical evacuation units can be dispatched to the location of the wounded soldier. Upon reaching the victim, a two-man avatar stretcher team disembarks from the rear of the ambulance, walks to and retrieves the casualty, and returns to the ambulance. The ambulance then returns to the location from which it was dispatched. Improved collision-detection techniques have enabled this networked representation of dynamic objects that can be picked up, passed between entities, or left behind in the virtual environment. Further development of hierarchical networking, rendering, and collision-detection schemes will support the simultaneous, rather than switched, display of high-resolution medical and lower-resolution battlefield environments, allowing more detailed wound representations and physiological modeling.

In the future, the distributive virtual environment will network each wounded soldier's sensor-suit data, providing training in location, evaluation, treatment, triage, and information management. A more complex environment, including biological and chemical threats, is also under consideration and would model the potential effects of weather, vectors, and epidemics on operational planning and logistics.

Virtual Prototyping

To ensure compatibility or interoperability between all the components of the virtual battlefield, advanced technologies will be fielded in the Operating Environment of the Future (OEF). These are under development by ARPA and the Massachusetts Institute of Technology (Hon, 1994). Virtual prototyping will be achieved using the Virtual and Distributed Prototyping Tool of the Future (VDPTF), which will link designing, costing, and manufacturing, and has several advantages over traditional and simulation-based design tools. The VDPTF anticipates design and system integration requirements and allows for the development of modular devices that can be used across a wide range of platforms. Distinct from computer-aided design (CAD) renderings and animations, the VDPTF simulates these devices and systems in a multisensory, unencumbered, immersive environment revealing subtle design tolerances to both designer and user prior to the high cost of construction.

The Operating Environment of the Future will provide a distributive, networked environment linking teams of designers, manufacturers, and users from government, academia, and industry in a first-of-its-kind teleprototyping facility. The first phase of this effort will create a virtual surgical room, based upon the technologies of wireless networking, smart materials, image fusion, telementoring, and teleoperation, embedded noninvasive sensors, and autoregulation. Once the environment is created, battlefield instruments and equipment, such as the personnel status monitor (PSM), next-generation intelligent stretcher (trauma pod), telepresence surgical system, and telemedical vehicle, will be modeled in order to optimize battlefield compatibility. Soldiers, combat medics, physicians, nurses, hospital commanders, administrators, and material managers can test and use the virtual-battlefield medical equipment before it is actually built. In this way, the virtual prototype can be used synchronously to optimize its design while also using it for education and training. Through the OEF project, health care providers and researchers in different geographic locations can collaborate in real and virtual worlds with these tools.

Maturing the Technology

Virtual environments have demonstrated significant potential for training and education. Although they offer unique opportunities for sensing and analyzing the inhabitant's activities, they lack consistent criteria for both human and machine performance. Acceptable methods are just now being developed for validation, standardization, scalability, and flexibility. Special efforts will be made to determine their cost-effectiveness and their amenability to infrastructure integration. Critical technologies that require the most immediate refinement include virtual-human interfaces, multisensory acquisition and registration, and integration into dynamic single or multiuser protocols.

Before advanced technologies for training and education in virtual environments can be evaluated, four critical issues must be resolved: (1) the expectations of the users within VR environments; (2) the tools the users will find most intuitive for authoring and stylizing virtual expression;

(3) the determination of the tasks and scenarios for which multimedia and VR are effective and for which they are not; and (4) how to design virtual environments that will probe the broadest spectrum of human sensory, motor, and cognitive capabilities (Bellman, 1994). Although in the formative stage of development, the Computer-Aided Education and Technology Initiative (CAETI) at ARPA is addressing all four of these concerns.

Conclusion

The goal of military telemedicine is to conserve and enhance the fighting force. The cornerstone of the system is the digital physician, employing advanced diagnostics and imaging, remote medical and surgical intervention, informatics, virtual prototyping, and battlefield simulation. That physician will be supported by robust networks (indifferent to interruptions, interceptions, alterations, and partial physical destruction), whose open architecture will accommodate both emerging technologies and legacy systems. The integration of the digital physician, the hardware foundation, and the information architecture accommodating multimedia acquisition, intelligent processing, recordkeeping, remote archive retrieval, high-speed broad-band communications, and the Global Grid will result in an ubiquitous global presence for telemedicine.

In the grand scheme of global force projection, however, military telemedicine is just one of many powerful tools in the armamentarium of the deployed force commander. Today, that commander can employ a substantial array of sophisticated communications and information management tools in demanding combat and noncombatant environments. Telemedicine's role in such environments will depend upon the same sophisticated tools. For example, the communications supporting battlefield telemedicine constitute subsets of the overall battlefield information architecture. They require robust, multilevel, mobile, secure information flows in a hostile environment. Finally, telemedicine's seamless interface with tactical, operational, and strategic sustainment bases is a prerequisite for success in joint (tri-service) and multinational deployments under Global Command and Control Systems (GCCS). Although a formidable proposition even with advanced technology, the implementation of global military telemedicine has a high likelihood of success because it is less constrained than civilian telemedicine by industrial, economic, sociocultural, and geopolitical obligations or medicolegal and regulatory barriers.

REFERENCES

Barkley, K. T. 1978. The Ambulance: *The Story of Emergency Transportation of Sick and Wounded through the Centuries.* Hicksville, NY: Exposition Press.

Bellamy, R. F. 1995. ATACC Workshop, May.

Bellman, K. L. 1994. "Playing in the MUD: Broadening Our Concepts of Virtual Reality." Keynote address to the NATO DRG Panel 8 Workshop: Advanced Technologies for Training Design. October.

Bowersox, J. C. 1995. Chairman, Department of Vascular Surgery, Madigan Army Medical Center. Personal communication.

Chen, D. T., and D. Zeltzer. 1992. "Pump It Up: Computer Animation of a Biomechanically Based Model of Muscle Using the Finite Element Method." *Computer Graphics* 26: 89–98.

Delp, S. 1995. Personal communication.

Delp, S., P. Loan, M. Hoy, F. Zajac, E. Topp, and J. Rosen. 1990. "An Interactive Graphics-Based Model of the Lower Extremity to Study Orthopaedic Surgical Procedures." *IEE Transactions on Biomedical Engineering* 37 (August): 8.

Dupuy, T. N., J. R. Brinkerhoff, C. C. Johnson et al. 1986. *Handbook on Ground Forces Attrition in Modern Warfare.* Fairfax, VA: Data Memory Systems.

Durlach, N. 1995. Personal communication.

Hill, J. H., and P. S. Green. 1995. Personal communication.

Hon, D. A. 1994. "Laparoscopic Surgical Skills Simulator." Proceedings, Medicine Meets Virtual Reality II. San Diego, January 27–30.

Hunter, I. W., T. D. Doukoglou, S. R. LaFontain et al. 1993. "A Teleoperated Microsurgical Robot and Associated Virtual Environment for Eye Surgery." *Presence* 4: 265–80.

Ille, K. 1994. AUSA demonstration by GE Corporate Research and Development, GE Medical Systems and EC&G Inc. Optoelectronics Group, September.

Kaplan, K. A. 1995. "Virtual Environment for a Surgical Room of the Future." In Morgan, K., R. Satava, H. Siegurg et al. *Interactive Technology and the New Paradigm for Healthcare.* Washington, DC: IOS Press, pp. 161–67.

McGovern, K. 1994. "The Virtual Clinic: A Virtual Reality Surgical Simulator." Proceedings, Medicine Meets Virtual Reality II. San Diego: January 27–30.

McKenna, M. A. 1994. "A Physically Based Human Figure Model with a Complex Foot and Low-Level Behavior Control." Ph.D. Dissertation. Massachusetts Institute of Technology, June.

O'Keefe, J. 1992. "Army Science and Technology Master Plan." Headquarters, Department of the Army, Washington, DC, Volume 2 (November): A-219–20.

Rosen, J. 1992. "From Computer-Aided Design to Computer-Aided Surgery." Proceedings, Medicine Meets Virtual Reality. San Diego, June 1–2.

Satava, R. M. 1993a. "Surgery 2001: A Technologic Framework for the Future." *Surgical Endoscopy* 7: 111–13.

Satava, R. M. 1993b. "Virtual Reality Surgical Simulator: The First Steps." *Surgical Endoscopy* 7: 203–5.

Chapter 14

TELEMEDICINE IN CORRECTIONAL SYSTEMS

James H. Hipkins

Introduction

Most telemedicine systems being developed are designed to address the needs of rural populations living in remote areas determined to be medically underserved. There is another population, however, for which the potential of telemedicine to increase accessibility to specialty care, to improve the quality of care, and to reduce its costs significantly is perhaps even greater: our prison population. This chapter examines the special health and health care problems of this population, discusses the potential of telemedicine to ameliorate some of these problems, and describes the development and implementation of a telemedicine system within one state's correctional system.

The Prison Population

Prisoners in correctional systems comprise a large and growing population whose complex medical and social needs seriously challenge the health care profession. The Bureau of Justice Statistics reported that, in the summer of 1994, for the first time in U.S. history, over one million men and women were in state and federal prisons (Holmes, 1994). This figure does not include the half million persons in county and city jails. The incarceration rate in the United States increased from 139 per 100,000 residents in 1980 to 373 per 100,000 in 1994. Furthermore, the incarceration rate is expected to accelerate even further as tougher crime legislation is passed, harsher sentences are imposed, the war on drugs continues, and children of the baby boom reach their more violence-prone years.

The Health of the Prison Population

This increasingly large population poses a series of special health and health care problems. To start with, prison populations are generally less healthy than the general population. Some differences may be due to lifestyles preceding incarceration; others may result from conditions of the prison environment itself (Krupp et al., 1987). In the first instance, most prisoners are from groups generally considered most at risk for poor health. The reasons relate to poverty, unemployment, poor housing and sanitation, inadequate nutrition, and limited educational achievement (Wilson, 1993). Additionally, few prisoners have regularly received much in the way of preventative health care services. Many have a history of intravenous drug use and most have had little if any previous contact with any professional health care system. As a result, their health status upon arriving at prison is very low. This information is not new and was noted as early as 1971 (Anonymous, 1971).

Today, there is a high prevalence and incidence of infectious diseases among prisoners, including tuberculosis, gonorrhea, viral hepatitis, AIDS, and other sexually transmitted diseases (Hutton et al., 1993). For example, in New York, according to the State Department of Correctional Services, an estimated 8,000 inmates of a total population just over 65,000 are believed to be HIV positive. Of these, some 4,000 have AIDS. In most systems, if inmates test positive for HIV, they are housed with the general inmate population. This "mainstreaming" of HIV-positive inmates may increase the risk of spread of the infection. When inmates develop symptoms of AIDS, they may be removed from the general prison population, usually for medical reasons (Edwards, 1989; Greenberg, 1988).

Prison systems are also being overwhelmed by mentally ill inmates. Since no national studies of the mentally ill in prisons have been conducted, accurate estimates are difficult to obtain. Nevertheless, based on limited statewide estimates, it is believed that from 7 to 15 percent of inmates have mental disorders and one-third of these diagnoses include severe forms of schizophrenia, manic-depression, or depression (Foderaro, 1994; National Alliance for the Mentally Ill and Public Citizens Health Research Group, 1992). In one study, indirect evidence suggestive of mental illness among inmates was observed in their excessively high rates of hospitalization for insertion of foreign bodies into the gastrointestinal tract (Krupp et al., 1987).

The increase in numbers of mentally ill inmates can be traced, in part, to the deinstitutionalization begun in the early 1960s and the subsequent failure of the community mental health system to provide adequate housing for the more seriously mentally ill. One observer has coined a new term for the plight of the mentally ill—"transinstitutionalization" —moving from one bad institution to another (Foderaro, 1994). Another contributing factor is the drug abuse among the mentally ill and the crimes and convictions associated with it. Finally, there appears to be increasing concern about the insanity defense and an increased resistance to it. For example, questions are being raised pertaining to standards used to determine insanity (Weinstein et al., 1991); the credibility of neuropsychological evidence (Faust, 1991); and contradictory psychiatric testimony (Peters, 1990). The ultimate question of abolishing the insanity defense is also being raised (Ciccone and Clements, 1987). As this debate leads to fewer acquittals or sentencing to mental facilities, inevitably more mentally ill criminals will be sentenced to prison. Without increased attention to this problem and the potential contribution of telemedicine in the form of telepsychiatry, it is likely that prison mental health services may revert to the inmate model that characterized correctional mental health services from 1900 to about 1975 (Ferrara and Ferrara, 1991).

Evidence suggesting that the prison environment may be an important contributor to the health problems among inmates is increasing. Overcrowded conditions, lack of adequate ventilation, insufficient exercise, physical and verbal abuse, and inadequate health care have been implicated (Walker, 1991). Tuberculosis, for example, because it is spread by airborne transmission, can move easily through poorly ventilated areas crowded with inmates (Greifinger et al., 1993). Other diseases have been associated with poor ventilation as well (Hoge et al., 1994). Drug use and homosexual intercourse enhance the spread of infectious diseases such as AIDS and hepatitis. And those who are mentally ill cannot be expected to improve in the harsh conditions imposed by prison life unless access to appropriate psychiatric services and rehabilitative programs are provided (Bednarowski, 1993).

The problems of access to care, adequacy of services, availability of qualified health personnel, and escalating health care costs affect law enforcement and corrections agencies. These problems may be even more pronounced for correctional health systems owing to the severity of need among their patients and constraints on gaining access to services imposed by the need to maintain security and the safety of the general population.

Prison Medical Care

Health care is increasingly becoming a priority for correctional systems. Burgeoning prison and jail populations have brought more and sicker inmates. Large numbers of intravenous drug users increased the rates of communicable disease associated with needle-sharing. Longer sentences mean aging populations with more degenerative diseases and disabilities. Health problems preceding incarceration worsen from neglect. Thus, prisons have become a major component of the nation's public health system, and criminal justice agencies are asked increasingly to address public health concerns and to provide health care to large segments of the population. Most often, these demands are made with little regard to the availability of and accessibility to resources (Anno, 1993).

Medical care in prison is, therefore, very different from that available to the general population (Thorburn, 1992). Two major characteristics influence health care in U.S. prisons. First, health care occurs in an environment separated from mainstream medicine by difficulties in recruiting qualified providers, restricted access to referral and diagnostic facilities, and a profoundly altered patient-physician relationship. Second, health care services within prison walls can become more a product of the correctional system than of the other half of their split identity—a health care facility (Novick, 1987).

For many professionals and nonprofessionals alike, the concept of "prison health care" is perplexing. Today, physicians, politicians, ethicists, and the public debate the meaning. To understand where we are today in correctional health care, some understanding of the recent evolutionary process is necessary.

Evidence for inadequate health care in prisons began to emerge in the early 1970s from various sources. For example, in a paper prepared for presentation at the 1970 meeting of the American Public Health Association, James Hoffa noted prisons are "unbelievably bad for those who enter their gates either with incipient physical or mental health problems, or even for those who have no more than average resistance to physical or mental health problems, or even for those who have no more than average resistance to physical or mental stress" (cited in Goldsmith, 1975).

Twenty years ago, it was not uncommon for the warden to screen inmates for illness or for medical services to be provided by fellow inmates. One could easily predict the problems with such a system: discrimination, victimization, missed diagnoses, and in some cases, even willful injury. In 1972, Zalman wrote that prisoners were considered to be "slaves of the state and entitled only to the rights granted them by the basic humanity of whims of their jailers" (Zalman, 1972). One of the primary grievances in a particularly destructive riot at a prison in Attica, New York, pertained to medical care (Goldsmith, 1975).

Efforts at intervention by professional medical and inmate advocacy groups as well as rulings by the Supreme Court have radically changed correctional health care in the United States. The American Medical Association developed standards and a voluntary accreditation program for jails; the first were accredited in 1977. In the early 1980s, the program evolved into the National Commission on Correctional Health Care, which now includes representatives from some three dozen professional associations.

Despite these good intentions, reform in correctional health care would not have occurred without the courts. The federal courts, including the Supreme Court, heard numerous cases (e.g., *Ramsey* v. *Ciccone*) and concluded that depriving inmates of necessary medical care was cruel and unusual punishment and illegal under the Eighth Amendment of the Constitution. This was most firmly established in 1976 Supreme Court case *Estelle* v. *Gamble*. The court concluded that the infliction of unnecessary suffering is inconsistent with contemporary standards of decency, and held:

> We therefore conclude that deliberate indifference to serious medical needs of prisoners constitutes the "unnecessary and wanton infliction of pain" proscribed by the Eighth Amendment. This is true whether the indifference is manifested by prison doctors in their response to the prisoner's needs or by prison guards in intentionally denying or delaying access to medical care or intentionally interfering with the treatment once prescribed. Regardless of how evidenced, deliberate indifference to a prisoner's serious illness or injury states a cause of action . . . (*Estelle* v. *Gamble* 1976).

The Court declared that the provision of medical care for inmates was an obligation of government, and added that omission of medical care might produce "torture of lingering death" (Kemmler, 1980). Or, in less serious cases, "denial of medical care may result in pain and suffering which no one suggests would serve any penological purpose." But the finality of the Court's decisions and directives is perhaps more apparent than real as litigation continues over the rights of inmates to expensive treatments such as organ transplants, heart by-pass surgery, or experimental intervention (Kolata, 1994). Difficult questions remain, such as "Should the correctional system provide expensive care and scarce organs to convicted felons?"

Perhaps equally as difficult is implementing the directives within the prison environment. Philosophically, the courts are clear and direct regarding inmates' rights to medical care. But the implementation of the directives is complex and difficult. Some of the more pertinent issues are addressed in subsequent sections.

Organization of Correctional Health Care

In 1980, the elements of a sound State correctional health care system were spelled out by the courts in *Lightfoot* v. *Walker* (1980). These include:

(a) selection of a chief of medical services with responsibilities for budget, health-care planning, and the supervision of all medical staff;

(b) a prompt medical history;

(c) an appropriate number of staff with training specific to performing required medical services and dispensing of medication;

(d) a sick-call procedure with written protocols supervised by individuals trained in physical diagnosis and triage;

(e) regularly scheduled clinic visits;

(f) a system for monitoring inpatients;

(g) adequate laboratory and x-ray services;

(h) emergency medical care, 24 hours a day, 7 days per week.

Particularly important to the potential role of telemedicine in the delivery of health care to inmates were other court decisions. For example, in 1982, the State's obligations to provide medical care to inmates were summarized in *Capps* v. *Atiyeh* (1982).

> The State's obligation is three-fold. First, prisoners must be able to make their medical problems known. Second, the medical staff must be competent to examine inmates and diagnose their illness. Third, staff must be able to treat the inmate's medical problems or to refer the inmates to outside medical sources who can.

The introduction of telemedicine in this legal environment may help correctional systems to live up to both the letter and the spirit of the laws.

In the following sections, the Georgia State Correctional System and the organization of medical care within it are described briefly. Subsequently, special problems of medical care delivery in the system and the potential for telemedicine to ameliorate them are discussed.

Telemedicine in the Georgia State Correctional System

Approximately 35,000 inmates are in the Georgia State Correctional System. The number is projected to increase to more than 50,000 by the year 2000. Today, they are housed in about 100 institutions, 39 of which are large state institutions. The remainder is made up of other types of facilities including detention centers, county correctional centers, transitional centers, and boot camps. The largest state facility, with about 2,000 inmates, is located south of Atlanta in the small town of Jackson. Some facilities hold as few as ten State inmates; the others would be county inmates. In addition to the range in size, there is also a range in the levels of security, from maximum to minimal.

The medical care of the inmates throughout the state system is the responsibility of the Medical Director and the Health Services Section of the Georgia Department of Corrections. Their responsibilities include developing standards of practice and policies pertaining to the delivery of health care. Access to health care is a major concern as it pertains to management of emergencies and providing medications within a formulary setting. Of considerable importance are problems relating to access to services outside the system, including specialty consultation, surgeries, and invasive procedures. The services may range from seeking a diagnosis for a skin rash to therapeutic bone-marrow transplants for complex malignancies. The Medical Director is responsible for assuring that medical care is delivered and monitored regularly.

Insofar as possible, medical care is provided within the facilities of the Correctional System. Components of the medical care delivery system reflect the range of correctional institutions within the state. Thus, medical care may consist of a single physician who works two or three hours per week in a small facility, or, at the other end of the spectrum, of five or six full-time physicians in the largest facilities.

The Augusta Correctional Medical Institution (ACMI), opened in 1983, is the system's hospital. It was originally designed as a centralized referral center to provide acute and specialized medical services for male and female inmates, primarily as transients or out-patients. However, it does not have an intensive care unit or respiratory support. It has two surgical suites, and many uncomplicated, same-day surgeries, including hernias, hysterectomies, endoscopies, and excisional biopsies, are conducted routinely. Unfortunately, due to the large and continually growing number of inmates in the system and the increased incidence of chronic illnesses as well as conditions associated with infectious agents such as hepatitis and the human immunodeficiency viruses, demand for treatment and beds at ACMI significantly exceeds the capacity.

In addition, accidents or acute illness cannot be handled within the inmate's resident institution or within any of the Correctional System's facilities. In such instances, inmates are usually transferred to the Augusta Regional Medical Center, a private community hospital. In an emergency, inmates may be treated in local hospitals around the state. This design requires the transport of inmates out of their resident prison to another location where required services are available. Transport of inmates always poses a security risk, and requiring additional resources and raising costs.

Not surprisingly, many community physicians and hospitals are unenthusiastic about shackled inmates in their waiting rooms. There have also been cases of community exposure, specifically to drug-resistant tuberculosis, when inmates were transported to unsuspecting community hospitals where respiratory isolation was not available (Center for Disease Control, 1992).

Insofar as many of the larger correctional facilities are located in small towns in rural areas, inmates needing to use "local" medical care physicians and hospitals, or requiring the consultation of specialists, are subject to the same constraints facing the general population.

Special Concerns of Correctional Medicine

Certain other important constraints on the delivery of medical care within a correctional system are not relevant to the general population, however. These concerns pertain to security and associated costs, something called "secondary gain," liability and litigation, and the dynamics of the patient-physician relationship.

Security and Cost

To insure public safety, when an inmate travels to a local provider, he or she must be accompanied by security personnel. This increases the cost of medical care significantly. In some cases, two or three security personnel must accompany a high-security inmate. The cost of care escalates rapidly for these inmates and for those requiring convalescence in special facilities. Security personnel must remain with the inmate continuously while he is recuperating from treatment in a nonsystem facility. Further, it is estimated that, on any given day, up to 50 inmates may be traveling across the state for medical care or at ACMI or local hospitals. In the corrections system, eliminating or reducing transport of inmates is critical to reduction of costs and increasing public safety. Insofar as telemedicine eliminates or reduces travel, it lowers the costs of medical care and decreases threats to public safety.

Secondary Gain

The ability of corrections' physicians to interact directly with specialists via telemedicine helps to solve another problem, namely, secondary gain. Inmates sometimes attempt to gain secondary advantages using a medical care encounter, making "simple" requests of visiting specialists outside the system. For example, an inmate with a back problem may request a double-thick mattress. If recommended by the specialist, the mattress will be provided. While this may seem trivial, even the simplest things may give an inmate a sense of status or influence and lessen the tension within an institution. With physicians consulting directly with one another via telemedicine, frequently in the presence of the inmate, this source of potential problems is diminished.

Telemedicine consultation may also reduce malingering on the part of inmates. X-rays may be taken at the prison of an inmate who has fallen or otherwise claims to suffer from a back problem, and an orthopedic specialist or neurologist may be consulted directly and promptly for an examination. The need for and type of follow-up treatment may be determined promptly. This reduces the need to transport the patient and lessens the chance that the inmate is trying to evade work.

Liability and Litigation

Access to care and the promise of telemedicine to improve it are also important in correctional medicine for another reason. If access to medical care for an inmate is delayed or perceived as being denied, litigation and liability claims may result (Kolata, 1994; Brushwood, 1992). Excessive delays in delivering medical care within a correctional system may contribute to the deterioration of a health condition and, subsequently, cause expensive emergency department visits and hospitalizations (Keller et al., 1993). An inmate's refusal to travel or the lack of transport results in such delays, but in other instances, the delay may be due to the periodic excess of demand over the capacity of the medical facilities. Should an inmate feel that a delay in medical care or that "inadequate" medical care contributed to a significant deterioration of a health problem, litigation is likely. One prison physician reported: "Prisoners always tell me, 'don't take this personally, but I'm going to sue you,'" (Edwards, 1989). Telemedicine has the potential to reduce these delays in obtaining consultations from specialists. In addition, a video-taped record of the encounter provides an almost infallible electronic medical record of care, and the patient has less recourse to challenge the medical care received.

Physician Education and Respect

Physicians within the correctional system will benefit from telemedicine consultation in the area of education. The telemedicine consult not only contributes to the care of the patient but also to the education of the physician, increasing the physician's competence and helping to gain the inmates' respect. Inmates test physicians constantly, challenging their medical knowledge (Edwards, 1989). Lack of respect for primary physicians can be another source of contention. In its absence, inmates may continue to request visits to subspecialists of various types. Direct consultation via telemedicine between corrections system physicians and medical center specialists reduces this possibility and, indirectly, enhances the credibility of the corrections system physician. Additionally, telemedicine consultations have the potential to bring the system physicians into the mainstream of medical care.

Development of Telemedicine

Enthusiasm for the potential benefits of telemedicine to medical care in the Georgia Correctional System derived in large part from the development and demonstration of telemedicine at the Medical College of Georgia Telemedicine Center. Telemedicine was an obvious answer to the specific problems of health care delivery in the correctional setting discussed earlier.

A telemedicine consultation room was identified at the ACMI, and equipment was installed. Three high-resolution cameras can be controlled from remote or local sites. Included in the equipment are an electronic stethoscope, otoscope, and a universal connection for endoscopic equipment. One camera is used for projection and transmission of documents, x-rays, and electrocardiograms. All electronic images are recorded and accessible on line, and can be referenced later to evaluate any changes in condition of the inmate. All interactions are controllable from either consult site, so the operator at the distant site can control the interaction. Using this original telemedicine linkage, physicians at the ACMI are consulted by physicians at the Men's Correctional Institution at Hardwick. In turn, ACMI patients can be presented to specialists at the Medical College of Georgia for further consultation.

Prior to the establishment of telemedicine, inmate patients typically were transferred to the ACMI for appointments with consultants or hospitalization. The remote telemedicine sites have eliminated thousands of inmate-miles of transportation by obtaining diagnostic and therapeutic services on line via telemedicine.

Perhaps the best example of the impact of telemedicine is the Dermatology Clinic at ACMI. On average, it provides consultation for as many as 20 patients in a half-day clinic. It is estimated that 18 of every 20 patients can be treated successfully and completely via telemedicine, leaving only two or three requiring transport to the ACMI. Substantial savings accrue to the Department of Corrections in costs of travel, security personnel, and public safety. Outcome studies have demonstrated that telemedicine performs well in these encounters and does not compromise the quality of care. Additionally, many inmates have expressed a preference for the "camera consult" versus the long and uncomfortable bus ride to the ACMI.

The major scheduling center for all subspecialty services and outside referrals is located at ACMI. Requests for consultations are generated at each of the remote institutions and forwarded to ACMI for scheduling. A major effort is underway to identify those consultations that could be completed via telemedicine. When requests for consultations in a particular subspecialty suffice to fill a complete clinic, it is then scheduled as a telemedicine clinic. The main computer program generating the clinic schedules is now available to track consultation requests that providers have identified as eligible for telemedicine, and will "generate" a telemedicine clinic when requests are sufficient. Arrangements are then made with the subspecialist and the clinic is conducted.

Data are then collected on all aspects of the telemedicine clinics including number of patients, length of each visit, materials and documents exchanged, and recommendations of the subspecialist. The consultant writes his completed diagnosis and recommendations and transmits the information to the remote institution via facsimile machine. The transmitted copy in the patient's medical file is replaced by the original when it is received.

The mobile unit increases the flexibility and range of the telemedicine program. Eliminating the need for dedicated clinic space within an institution, it can be moved quickly from one location to another as required. While the initial outlay of funds is substantial, the telemedicine unit is not nearly as costly as it would be to establish permanent consultant sites in the facilities to be served. Thus, the mobile unit is an attractive alternative in periods of decreased funding for correctional health care.

It is envisioned that in the future, permanent telemedicine consult sites or access to mobile telemedicine vans will be available to all major institutions on the Corrections System, a total of about 40 institutions. It is hoped that, when the telemedicine system is completed, minimal need to utilize facilities outside the system will be minimal except for scheduled surgeries or emergency hospitalization.

Finally, to handle the increasing numbers of chronically ill inmates in the system, a chronic-care facility has been proposed. It will be centrally located in the state and house up to 700 chronically ill inmates. Most of the inmates requiring on-going maintenance will be located here. A telemedicine unit in this facility will permit remote out-patient clinic consultation.

Conclusion

Correctional facilities may be the optimal environment for managed care and telemedicine. Telemedicine should significantly reduce prison referrals outside as well as within the medical facilities of the correctional system. Telemedicine will reduce inmate transfer and the associated costs and security risks of off-site medical care. It will also facilitate prompt and appropriate medical care of the inmates. The system's physicians' continuing medical education will improve, as will their credibility with inmates. Telemedicine consultation should also reduce inappropriate attempts at secondary gains among inmates.

Telemedicine has been available in Georgia since 1989. Two major correctional institutions are now on line and telemedicine services are planned for an additional six facilities. Early, largely anecdotal, evidence points to the success of telemedicine in ameliorating some of the more important concerns of health care delivery in the Georgia Correctional System. The correctional systems of a number of other states are also incorporating telemedicine. The initial experience in Georgia suggests that the benefits of telemedicine are substantial. Given the continuing increases in the incarceration rates, without the intervention of telemedicine, Georgia would be forced to build extra tertiary care centers or to use local facilities and providers at exorbitant costs to the state. Telemedicine provides a viable alternative to this scenario, while at the same time delivering care more promptly and of higher quality.

REFERENCES

Anderson, G. 1990. "Sick Behind Bars: Health Care in Prison," *America*, 162: 124–26, 134.

Anno, B. J. 1993. "Health Care for Prisoners," *Journal of the American Medical Association*, 269: 633–34.

Anno, B. J. 1991. *Prison Health Care: Guidelines for the Management of an Adequate Delivery System.* Washington, DC: National Institute of Corrections.

Anonymous. 1971. "Medicine Behind Bars: Hostility, Horror, and the Hippocratic Oath," *Medical World News*, 23: 26–29.

Bednarowski, J. 1993. "Creating a Treatment Culture for Special Needs Inmates," *Corrections Today*, 55: 100–04.

Brushwood, D. 1992. "Prison Pharmacist may be Liable for Refusing to Dispense Anticonvulsants to an Inmate," *American Journal of Hospital Pharmacy*, 49: 636–39.

Camp, G. M., and C. G. Camp. 1993. *The Corrections Yearbook.* South Salem, NY: Criminal Justice Institute.

Center for Disease Control (CDC). 1992. "Transmission of Multidrug-Resistant Tuberculosis Among Immunocompromised Persons in a Correctional System — New York 1991," *Morbidity and Mortality Weekly Report*, 41: 507–09.

Ciccone, J. R., and C. Clements. 1987. "The Insanity Defense: Asking and Answering the Ultimate Question," *Bulletin of the American Academy of Psychiatry Law*, 15: 329–38.

Edwards, K. 1989. "Medicine Behind Bars," *Ohio Medicine*, (June): 441–47.

Faust, D. 1991. "Forensic Neuropsychology: The Art of Practicing a Science that Does Not Yet Exist," *Neuropsychological Review*, 2: 205–31.

Ferrara, M., and S. Ferrara. 1991. "The Evolution of Prison Mental Health Services," *Corrections Today,* 53: 198–204.

Foderaro, L. 1994. "For Some Mental Ill Inmates in New York State, Punishment is the Treatment," *The New York Times,* 144: 17.

Greenberg, M. 1988. "Prison Medicine," *American Family Practitioner,* 38: 167–70.

Goldsmith, S. 1975. *Prison Health.* New York: Prodist.

Greifinger, R. B., N. J. Heywood, and J. B. Glaser. 1993. "Tuberculosis in Prison: Balancing Justice and Public Health," *Journal of Law, Medicine, & Ethics,* 21: 332–41.

Hoge, C. W., M. R. Reichler, E. A. Dominguez, et al. 1994. "An Epidemic of Pneumococcal Disease in Overcrowded, Inadequately Ventilated Jail," *The New England Journal of Medicine,* 331: 13–19.

Holmes, S. 1994. "Ranks of Inmates Reach One Million," *The New York Times,* 144: A1, A25.

Hutton, M., G. Cauthen, and A. Bloch. 1993. "Results of a 29-State Survey of Tuberculosis in Nursing Homes and Correctional Facilities," *Public Health Reports,* 108: 305–14.

Keller, A., R. Link, N. Bickell, M. Charap, A. Kalet, and M. Schwartz. 1993. "Diabetic Ketoacidosis in Prisoners Without Access to Insulin," *Journal of the American Medical Association,* 269: 619–21.

Kolata, G. 1994. "U.S. Refuses to Finance Prison Heart Transplant," The New York *Times,* 143: 6.

Krupp, L. B., E. Gelberg, and G. Wormser. 1987. "Prisoners are Medical Patients," *American Journal of Public Health,* 77: 859–60.

National Alliance for the Mentally Ill. 1992. Joint report with the Public Citizen's Health Research Group, "Criminalizing the Seriously Mentally Ill: The Abuse of Jails as Mental Hospitals." Arlington, VA: National Alliance.

Novick, L. F. 1987. "Health Services in Prisons," *Journal of Community Health,* 12 (1): 1–3.

Peters, U. H. 1990. "The Hinckley Case and Some Sequelae for Psychiatry," *Fortschritt für Neurologie und Psychiatrie,* 58: 339–42.

Sanders, J. H., and F. J. Tedesco. 1993. "Telemedicine: Bringing Medical Care to Isolated Communities," *Journal of the Medical Association of Georgia,* 82: 237–41.

Thorburn, K. 1992. "Health Care Moves to the Forefront," *Corrections Today,* 54: 8.

Walker, B., Jr. 1991. "Prisons Population Pressures," *Journal of Environmental Health,* 54: 18–22.

Weinstein, R. M., E. P. Mulvey, and R. Rogers. 1991. "A Prospective Comparison of Four Insanity Defense Standards," *American Journal of Psychiatry,* 148: 21–27.

Wilson, J. 1993. "Childbearing within the Prison System," *Nursing Standard,* 7: 25–28.

Zalman, M. 1972. "Prisoners' Rights to Medical Care," *Journal of Criminal Law, Criminology, and Police Science,* 2: 185–89.

COURT CASES

Capps v. Atiyeh, 559 F. Supp. 894 D. Ore. 1982.

Estelle v. Gamble, 429 U.S. 98, 97 S.Ct. 285 (1976).

In re Kemmler, 136 U.S. 436, 447, 10 S.Ct. 930, 933 (1980).

Lightfoot v. Walker, 486 F. Supp. 504 S.D. Ill. 1980.

Ramsey v. Ciccone, 310 F. Supp. 600 (W.D. Mo. 1970).

SECTION V
TELEMEDICINE OF THE FUTURE

CHAPTER 15

TELEMEDICINE OF THE FUTURE:
A PRAGMATIC SPECULATION

RICHARD M. SATAVA

Introduction

The purpose of this chapter is to speculate about the telemedicine of the future and, as a direct corollary, the future of telemedicine. This type of speculation can be carried out at two levels. One level, and that which is most pragmatic, forecasts the development and use of future technology only as an extension of the present. In this context, we can only forecast the likely outcomes of current design innovations not yet on the market, but whose design features, characteristics, and capabilities are known with a considerable degree of certainty. This type of forecast is limited in its temporal and conceptual horizons and, therefore, represents only short-term future developments. In telemedicine, this type of forecast assumes a particular fixed technological form. But an additional purpose here is to stimulate discussion of the vast potential of the information age to affect health care delivery. Therefore, we will try to "push the envelope" and to "think outside the box" to permit a greater understanding of how information age technologies can enhance our abilities to provide a higher quality of patient care and promote and improve health status among the general population. Therefore, at this more speculative level, the assumption of a fixed technological form can be relaxed and an integrated, concept-based forecast is possible that is not bound by a fixed technology or extrapolation from current design innovations.

The future development of any technology obviously does not occur in a vacuum, making accurate forecasts and predictions even more difficult. The social and cultural milieu may alter, limit, or even prevent development of new ideas and technological development as well as the acceptance and/or implementation of both. We will try here to strike some

393

balance between the two levels of prediction. The forecasts presented are certainly more limited and practical than those of the sixteenth century astrologer Michel de Notredame; nevertheless, they build on the recognition that current telemedicine technologies represent only one phase in its development and only one possible technological form.

Telemedicine

To provide the highest quality of patient care, telemedicine must address the three cornerstones of medical practice, namely, **diagnostics, therapeutics,** and **education** of both the patient and physician. Neglecting or, for that matter, placing undue emphasis on any one of these elements at the expense of the others necessarily detracts from comprehensive progress. Through **diagnosis,** the physician learns about the patient's problem, and available **therapeutics** provide the method for maintenance or cure. **Education** has a twofold purpose: to educate the patient about his or her disease, an essential activity for future prevention and maintenance; and to provide continuing education for physicians so that they remain informed about current standards of medical practice. Telemedicine has the potential to contribute to each of these dimensions of medical practice by interconnecting physicians with each other as well as with patients, both with data bases and educational resources. If successful, and when fully implemented, telemedicine can enhance the efficiency and effectiveness of the health care system, improving the health status of the population.

Telemedicine, in application, has many interrelated meanings, but the lack of a precise definition is not necessarily deleterious. In fact, a precise definition might limit speculation about the future. Therefore, a very broad perspective of telemedicine is taken here, providing a practical clinical approach that opens the prospects for several current and potential applications. As discussed in other chapters of this book, telemedicine in clinical and other applications ranges from radiology, pathology, and psychiatry to community health information networks. Telemedicine is a manifestation of the information age. It consists, therefore, not only of the transmission of information, but also of that information's necessary decomposition and subsequent restructuring to permit its manipulation and reconstitution in a recognizable form. By taking physical entities such as x-rays, laboratory test reports, medical records, and electrocardiograms and reducing them to a common form including digital images, it is possible to disseminate information much more rapidly over much greater distances than was possible under the constraints of actual "holistic" physical transfer of the information.

Such technological transmission of information is now spreading rapidly to include medicine. Relatively commonplace now are largely separate efforts such as transmission of electronic medical records, use of simple artificial intelligence programs for clinical decision support, and transmission of radiological images over telephone lines as well as video images of endoscopic and laproscopic procedures. The common requirement in these activities is the reduction of information about the patient in some manner to an electronic signal. Subsequently, the signal is processed and transmitted to a physician, or other provider, who usually receives the information at a remote site via a coder/decoder on a video monitor with accompanying audio.

The trend in biochemical and vital sign diagnostics is toward smaller, portable, integrated devices that acquire data in direct digital format. Biosensors will become noninvasive, such as the pulse oximeter, which measures oxygen concentration in the blood by shining infrared light through a finger or earlobe and measuring the transmitted wave length of light. In the future, there will be no need to draw blood samples. Display of data on hand-held or strategically-placed video monitors at the point-of-care will be almost immediate at the bedside, out-patient clinic, or emergency room. Some sensors will be hand-held by physicians and nurses, while others will be worn like a Sony Walkman® on the body of the individual or patient. The military, for example, has a prototype device worn on the wrists of soldiers called a personnel status monitor (PSM). The PSM provides a soldier's geographical location via a geo-position satellite locator. It also has a vital sign monitor including pulse rate, oxygen saturation, and EKG via a single lead. Should a soldier become wounded or otherwise suffer a health problem, an alert, including the location of the soldier and his vital signs, is transmitted to the nearest medic. Such advances in technology and communication are eminently applicable to victims of auto accidents, postoperative patients, high-risk homebound or nursing home patients, as well as to the population at large.

Currently, telemedicine therapeutics are progressing much more slowly than the telemedicine diagnostics. As discussed in chapters dealing with clinical applications of telemedicine, clinicians find themselves constrained by current technology insufficiently developed to permit remote surgical and other necessary "hands on" diagnostic and therapeutic procedures. Some experimental technologies, such as "telepresence surgery," now exist that will remove this barrier in many cases. In this technology, the surgeon's (or other practitioner's) hand motion is converted into electronic signals. The signals are transmitted to a remote site where they move the tip of the surgical instrument in response to the hand motions of the surgeon. Thus, the physician can not only obtain diagnostic information from sound and visual sources, but can also conduct physical procedures such as biopsies.

Audio and visual information obtained from the patient may actually be so much enhanced by electronic surveillance that a revision of clinical diagnostic protocol and standards may be required. Once made available, telepresence surgery also may require new training for practitioners to accommodate greater or lesser sensitivity to surgical movements resulting from electronic transfer of information through time and over space. Potentially, therefore, telemedicine has the capability not only to match but even to enhance in-person clinical evaluation and intervention.

While no telemanipulators exist for use in palpation or minor surgical procedures such as biopsies, the technology is mature and both techniques have been performed under strict laboratory demonstration conditions. The use of remote control laparoscopic cameras during laparoscopic surgery (for telementoring) and teleoperation of microscopes for telepathology diagnostics are both in early clinical trials. Also under institutional review are sterotactic neurosurgery systems that permit a surgeon to control the surgical instruments from a workstation adjacent to the patient (rather than holding the instruments in the surgeon's own hands). This is the first step toward telepresence surgery; the physician can now move into the realm of surgical intervention and therapy that can be conducted electronically from a remote site. Teleneurological procedures that can be used between a remote neurosurgical subspecialist and a general neurosurgeon in a distant city are also currently under institutional review. In addition, exploratory efforts are underway to build micro-robots that would replace endoscopes, to provide remote access to parts of the body where today's endoscopes cannot travel. Due to such various constraints as legal liability, restrictions on licensing, and credentialing issues, as well as the substantial cost of telecommunications, the extensive implementation of these and other technological advances is not likely in the near future. But the reduction of these barriers would be associated with a significant proliferation of both technology and procedures.

The value of this perspective of telemedicine is that the various physical modalities can be integrated via image fusion, distributed via networks, and modified via digital signal processing of images. They can be enhanced via microscopic scale of motion or vision, and once they have been reduced to their digital equivalents, they can be implemented simultaneously via interactive collaboration. These functions are not possible with the physical derivative of the original activity. For example, during a neurosurgical procedure, it is not physically possible (nor desirable, for that matter) to wrap an x-ray film around the brain to determine the exact location of a tumor. However, it is possible to use the video image of the brain during surgery and electronically fuse or superimpose it over the digital image of the patient's CT or MRI scan. Thus, the precise internal anatomy and tumor location can be viewed, identified, and reconciled for the invasive procedure. Given the potential of telepresence surgery, it may not be necessary for the surgeon and the patient to be in the same room even when more complicated procedures are required.

There are numerous similar analogies using networking, digital signal processing, and other functional activities. What they all have in common is that the technology may be used to enhance the evaluative and therapeutic functions of the physician beyond that currently possible in order to provide a higher quality of care for an individual patient, at least in a technical sense. The development of the technology and the design of the systems must focus on the well-being of the patient.

Current Status

The current status of the clinical and community applications of telemedicine is well documented in this volume, which also offers an extended discussion of current telemedicine technology. Nevertheless, it is useful to review briefly the current status of telemedicine prior to discussing trends in application and technology.

A wealth of developing and mature technologies is available today in the practice of medicine. And supporting architecture is in place in three broad categories: (1) information infrastructure software (such as massive heterogeneous databases, artificial intelligence, and decision support systems); (2) high-performance computing hardware (such as massively parallel processors and super computers); and (3) networking (such as the information superhighway).

While none of these technologies was designed specifically for medical practice, they do constitute "the enabling technologies" that will permit and accommodate medical applications. The investment in the information infrastructure has been made over the past three decades. The health care sector does not need to reinvent, redevelop, or reinvest the large sums of money necessary to set up a separate health care infrastructure. Rather, it must take advantage of these existing technologies. The rapidity with which medicine (and health care generally) accepts telemedicine will be determined by a number of factors including the level of technology available, its utility, and acceptability, as well as by a host of factors unrelated to technology. The change will constitute a radical shift from current practice, but the adoption process may proceed incrementally. Laparoscopic surgery, however, was a sudden change and had rapid introduction. Each development must proceed from an established base. From this perspective, the total impact of information technology embodied in the development and diffusion of telemedicine on the practice of medicine remains difficult to predict.

That impact may also depend upon a potential shift in the conceptualization of information production and processing, and the integration of the two. The concept of information as a method of thinking must include not only the actual information to be transferred as actual data and the infrastructure described above, but also the use and integration of information age manufacturing. Information age manufacturing pertains to such things as the production of millions of computer chips in silicon, the use of robotics rather than hand wiring, and "flexible manufacturing" that turns out custom articles as cheaply as those mass-produced. This also includes the implementation of "just in time" inventories, eliminating the need for large warehouses, to respond quickly to demand. Collaborative and interactive business practices will also be involved. Examples might be multiple corporations working on a specific project as a "virtual corporation" and dissolving back into the parent companies when the project is complete; or temporary multilevel consortia of government agencies, academic institutions, and large corporations in geographically diverse areas of the country, networked together as a single entity. Another example might be distributed product bases, in which large corporations own many smaller subsidiaries that produce all the components assembled by the parent corporations into a single system (e.g., aircraft, automobiles, or computers).

Thus, information can indeed provide higher-quality products and simultaneously lower their cost. It is only the older industrial-manu-facturing paradigm that suggests that higher quality requires a higher production cost. If medicine is to adapt to the rapidity of the changes brought about by the information highway, it will be necessary to embrace the requisite perspectives as well as technology. Of course, the expected efficiencies, quality improvements, and cost reductions and/or contain-ments that would result from the application of information in the medical sector must be demonstrated.

Infrastructure

To understand these fundamental changes better and to extrapolate into the future, it is necessary to elaborate on the supporting architecture. The technologies of the future must be derived from this infrastructure. The software portions of the information infrastructure include: artifi-cial intelligence, decision support, and genetic programming. In addition, internet protocols and intelligent agents are important software tools.

The health information infrastructure includes such things as the electronic medical record, "clinical associate" programs, large heteroge-neous databases, and as previously mentioned, real-time distributive, collaborative, and interactive networking. For high-performance com-puting, examples include the various powerful silicon chips (RISC, ASIC, VLSIC, etc.), multichip modules, super computers and "massively parallel" processors, image generators, rendering engines, high-density, and holographic devices. The combination of hardware and software will permit total interconnectivity via the information superhighway on a global scale. This network will be facilitated by the worldwide internet, fiber-optic cables, satellites, and wireless and cellular communications.

This architecture and these enabling technologies can support the current generation of telemedicine, and can provide the means to the future. But certain barriers could delay or even prevent implementation of possible applications. Several such barriers are discussed briefly below.

Medical records, electronic or not, are presently held in numerous isolated, noncompatible databases distributed across the nation (i.e., in many different databases that cannot share the information contained in each). These databases are not accessible, shared, or even transmissible outside local hospitals and institutions. In addition, access to information in retrievable form frequently requires excessive time and tedium, discouraging those desiring access to the database. The proposed solution to this problem in the health information infrastructure is to create an intuitive protocol, integrating the nation's massive, heterogeneous databases and providing "intelligent agents" or "knowbots" that can rapidly search and retrieve information in response to inquiries expressed in everyday language. To provide access to the databases to a wide range of physicians, the Health Information Infrastructure should provide knowledge, not only data. Intelligent interfaces will be able to learn an individual physician's needs and customize the data and information to provide answers to questions in an easily interpretable format. The physician is not replaced; he or she is made more effective by having "relevant" information. This type of infrastructure could provide the physician with information about the patient at the "point of service," wherever the data might be retrieved or stored.

For high-performance computing, the barriers include difficulties in acquiring the information, especially the massive amounts of data included in medical images; processing and storing the information; and then retrieving and displaying the information in a form easily recognized and understood. Super-computers strategically placed around the world could retrieve, almost instantly, complex images from large-scale storage devices such as optical discs or holographic cubes. These are new, advanced information storage methods based on laser writing and reading of data, which are replacing standard floppy disks and hard drives that read magnetically. The images could then be transmitted to a physician on a video monitor with hologram capabilities or via a device such as "virtual reality glasses." Thus, the information could be provided "right now" to the physician's office or surgery or at a patient's bedside. In addition, the physician would choose the form of data presentation, for example, as a three-dimensional image or a table of laboratory data.

The problem of physical connectivity, however, must be solved before the information can be shared in real time via the information superhighway (ISH). This requires a very broad bandwidth over secured networks and sophisticated switching for widespread distribution. Very high-bandwidth fiber-optic and satellite networks that use terrestrial cables or wireless communication pathways are required. These pathways could be switched almost instantaneously via asynchronous transfer mode (ATM) and opto-electrical connectors, and relayed across the world. This highly integrated network would allow patients and physicians to communicate and interact simultaneously as if each were present in the same place at the same time, reducing the isolation now experienced in remote communities and institutions by physicians and patients alike.

Into the Future of Telemedicine

Telemedicine continues to evolve. Some predictions of the future will be discounted and, perhaps, discarded as impractical, ineffective, or too expensive. Others, however, will become standardized procedures and technology in the delivery of health care. In spite of this uncertainty, a number of broad and general trends can be reasonably projected. They can serve to focus telemedicine's general direction, and to envision the limits of its future. It is generally agreed that the health care system, for better or worse, is moving toward larger, managed care organizations, many of which will encompass large geographic areas. Within this context, an additional level of care can be considered with the development of telemedicine. Currently, the health care system is largely composed of primary physicians, secondary services available at community and regional hospitals, and a tertiary level of care associated with academic and other major medical centers. It is not difficult, due to telemedicine's fiberoptic and satellite networks, to speak of developing an integrated "quaternary care" or virtual medical "facility." Physically, virtually, and organizationally, the quaternary-care facility incorporates the three subsumed levels of care into a single virtual regional entity. If developed and managed correctly, the quaternary-care facility based on telemedicine will be much more integrated and, therefore, more synergistic than the current health care system. Clearly, the quaternary facility provides a much higher level of integration and (potential) cooperation among the components than presently possible. Improved access by remote areas and providers, increased quality of care, continuity of care, and, one hopes, reduced costs through greater efficiency of integration of care, all mean that, in the quaternary facility, the value provided is much more than the sum of all the individual components.

Beyond this, there is also a suggestion of organizing and developing a quinternary level of care. This reflects the emerging (and the possibility of emphasizing) national or international "centers of excellence" that could provide "super-specialty" care in selected areas such as multimodal brain imaging or breast disease. The quinternary-care facility would

aggregate resources clearly currently too expensive for any single tertiary medical center. If the real-world problems of organization and scheduling can be resolved, the information and expertise in such centers of
excellence could be almost instantaneously available on a national or
international basis.

With rare exceptions, telemedicine today is practiced as a point-to-
point communication from physician to physician or between physician
and patient. Usually, this provides access to a medical center from a
remote clinic or hospital. Diagnostic sensors and imagers are employed,
therapeutic modalities instituted, and educational resources are exchanged
or other transactions occur. The most frequent result is the projection of
specialty expertise to the remote site.

The diagnostic component of telemedicine consists largely of sensors
and imagers. Current remote sensors include the adaptation to specific
clinical purposes of large, "off-the-shelf" equipment placed in a physician's
office, clinic, or hospital consultation room. The name of the telemedicine
specialty, such as teleophthalmology or teleoncology, derives largely
from the specialist using the equipment from time to time. In any event,
there is a process of local (remote) acquisition via teleophthalmoscopes,
otoscopes, stethoscopes, and the like, of the sensed or imaged information
and conversion of the information into a form easily transmitted to a
central site. Images are acquired using local equipment such as video
cameras, microscopes, x-rays, and ultrasound. These images are then
"decoded" and "digitized" and transmitted. Some data, such as those
from EKGs, ultrasound, and video-conferencing, can provide real-time
information from the remote to the central sites. When this occurs,
acquisition of information is interactive, as opposed to passive. Currently, laboratory data require drawing of blood, testing samples, typing
of reports, and faxing the results to the medical center.

The transmission of the video image of the specialty physician to the
patient and from the patient to the remote physician is a *sine qua non* of
telemedicine in its truest form. This is the bridge crossed by the physician
in a distant medical center to the patient and vice versa. This provides, so
far as is possible with the intervening technology, the human experience
of empathy and sympathy that comes across through body language and
facial expression. It is this process that preserves the *art* and the science of
medicine. The importance of this interactive video communication cannot be overemphasized. The picture is important even when it is transmitted at a low bandwidth, resulting in a fuzzy image or a jerky motion

at a few frames per second. It is "seeing" the doctor that enhances, ensures, and enables the continuation of the traditional doctor-patient relationship. Improved technology in the form of higher-fidelity images, "wrap-around" sound, tactile sensors, and other inputs will improve the degree of scientific and medical accuracy. Perhaps even more importantly, patients will feel an increased sense of their doctor's "presence."

As the sense of presence increases, so too does the opportunity for more effective interactive education. Through improved communication, patients will gain a greater understanding of their health problems and, one hopes, assume greater control of their own health and health care. Increased quality of interaction and communication via telemedicine also enhances the likelihood of more effective continuing medical education for remote health care providers. Current access to information sources such as MEDLINE and other medical databases via CD–ROM will be supplemented by a high level of face-to-face interactive accessibility to the specialist through telemedicine consultation. This should diminish the isolation felt frequently by remote providers. Improved and increased provider education should translate, in turn, to a higher quality of patient care.

The value of education with regard to telemedicine in other dimensions cannot be overlooked. For decades, education has been the leader in the use of telecommunications and information technology. Examples include computer-aided instruction and training, and access to national resources such as MEDLINE or the National Library of Medicine. In addition, "distance learning" has incorporated televideo communication in conferences. Improved technology can further enhance and "individualize" the educational experience. For example, the new science of "virtual reality" provides a method of accessing and interacting with information in graphic form. The "Visible Human" project of the National Library of Medicine has provided a three-dimensional human form now being explored as an educational tool and as a potential subject for medical and surgical simulation. While the current laboratory investigations are limited to large high-performance computers, the future holds the potential for access to these virtual cadavers all along the information highway.

At another level, because the computer network never becomes impatient or bored, it is possible for the next generation of telemedicine-learning computers to customize an individual student's learning experience to his or her pace and level. Moreover, this educational revolution will most likely take place through the desktop computer, which already has access to massive amounts of information, both locally and through-out the nation and world on the World Wide Web. The emergence of powerful "search engines" with transparent and friendly user interfaces and intelligent browsing software, such as Gopher, Mosaic, or Netscape, permit even computer and network neophytes to search efficiently and effectively.

As discussed elsewhere in this volume, teleconferencing has brought specialty education to areas of the country not previously reached. The diffusion of telemedicine brings about entirely new possibilities for life-long learning. In addition to formal courses, telemedicine can pro-vide continuous professional "on the job" training by telementoring during teleconsultations. Anecdotally, this experience has decreased the professional isolation of many rural physicians. The remote specialist who educates both the local physician and/or provider and the patient during one-on-one interactions can emphasize and promote preventive medicine as never before. Education can be brought to the patient and physician at the point of care and at the moment of patient-physician encounter; no separate effort to search for the educational opportunity is required.

Conclusion

However great the opportunities for telemedicine to improve diag-nostics, therapeutics, and education, great challenges remain. They are both pragmatic and environmental. One very pragmatic issue is the problem of "the last mile," or bringing suitably high bandwidth commu-nication capabilities to clinics, offices, and in particular, to individual homes. If we are successful in the latter, we can bring health services and the point of care to the home. Experiments are being conducted with "electronic house calls" that place a small automatic health station in the homes of chronically ill persons, which then connects them through a television and network with a clinic. Many simple telediagnostic instru-ments such as a stethoscope, otoscope, spirometer, and EKG can then be used or self-administered by the patient, if necessary, under the supervi-sion of a remote nurse or physician. This will provide timely diagnostic capability to patients before a condition becomes sufficiently acute to warrant a trip to the clinic, emergency room, or admission to a hospital. If high bandwidth to the home becomes available, we could then place unobtrusive, noninvasive sensors in strategic locations throughout the home. Sensors in the "millennium toilet," for example, can detect colon cancer and hematuria; sensors for protein or creatinine can detect chronic renal failure; metabolites of specific medications such as digitalis can provide early detection of occult disease or monitor a chronic disease state. Glucose microsensors in the handle of a telephone could detect imbalances important to chronic cardiac or diabetic patients. Ultrasound sensors in the shower could be used to detect mammographic lesions. These types of sensors could quietly monitor an individual's daily health and alert him or her when they detect an abnormality.

The potential for telemedicine to improve the delivery of medical care is probably limited only by the boundaries of our intellect and ability to think futuristically. However, it is essential to justify each stage in the progression of telemedicine by its contributions to the health status of the general population, and by its affordability, however we agree to measure that. Ultimately, the value of any new health technology must be measured against the benefit that it brings to the patient.

GLOSSARY OF TERMS AND ACRONYMS

Terms that appear in boldface print have their own entries.

ACR	American College of Radiology
Analog signal	A continuously varying electrical signal in the shape of waves, whose size and number change as the information source varies. Fluctuations in the signal produce the differences in loudness, voice, or pitch heard by the user.
ANSI	American National Standards Institute
ARPA	Advanced Research Projects Agency. An agency of the Department of Defense involved in tele-medicine use, development, and research. ARPA established the network that subsequently became the Internet.
ARPANET	Advanced Research Projects Agency Network
ATLS	Advanced Trauma Life Support. Refers to basic trauma resuscitative skills used by emergency care personnel.
ATM:	Asynchronous transmission mode A start/stop mode of transmission in which each character is preceded by a start signal and followed by one or more stop signals. There may be a delay or time interval between characters.
B–CDMA	Broadband code division multiple access. A secure, wireless technology used to send **digital** information over communication **networks** similar to cellular phone systems.
Bandwidth	A measure of the information carrying capacity of a communications channel. Higher bandwidths can carry more information. Bandwidth is measured in **megahertz** (cycles per second) in **analog** transmission, and in megabits/second (**Mbps**) for **digital** transmission.
Baud	A unit of **digital** transmission indicating the speed of information flow. Expressed as bits per second (**bps**), the rate represents the number of events processed in one second. The **baud rate** is the standard unit of measure for data transmission capability. (Typical rates are 1200, 2400, 9600, and 14,400 baud.)
Bit	Binary digit.

	The smallest piece of **digital** information that a computing device handles, represented by "on" or "off" (0 or 1). All characters, numbers, symbols, etc. are processed as electronic strings of bits.
Bps	**Bytes** per second.
	The number of characters transmitted per second.
bps	Bits per second.
	The number of **binary digits** transmitted per second.
Brightness	**Luminance** as perceived by the human eye. The terms luminance and **brightness**, though often used interchangeably, are not equivalent.
Broadband	A telecommunications medium capable of carrying a wide range of frequencies and with bandwidth in excess of that required for high-quality voice transmission. Examples include television, **microwave**, and **satellite** transmission.
Byte	A string or grouping of 8 **bits** used to represent a character or value.
CAD	Computer-aided design.
CCD	Charge-coupled device.
	A photoelectric device that converts light information into electronic information by means of sensors that collect and store light as a buildup of electrical charge. The resulting electrical signal can be converted into computer code and reconstructed to form an image. CCDs are commonly used in television cameras and image scanners.
CCITT	Consultative Committee on International Telephone and Telegraph.
	Now the International Telecommunications Union Consultative Committee for Telecommunications (See **ITU–T**). An international agency involved in developing standards for telecommunications, as well as FAX and video coder-decoder (**CODEC**) devices.
CD–ROM	Compact-disk, read-only memory.
	A storage medium that can hold large amounts of data, comparable to about 220,000 pages of text.
CDC	Centers for Disease Control
CDS	Clinical decision support.
	Information about a patient, the patient's health problem, and alternative tests/treatments used to help a clinician diagnose or treat the patient. Also referred to as clinical decision support systems (CDSS).
CEN	European Technical Committee for Normalization.
	An international agency involved in setting standards in health care **informatics.**
CHIN	Community health information network.

A common communications system designed for common use by health professionals, patients, and the community. CHINs integrate hospital information systems (see **HIS**) with medical databases, community health information, and on-line computer services.

CME	Continuing medical education
CMHCs	Community Mental Health Centers
Co-processor	A device in a computer to which specialized processing operations are delegated, such as mathematical computation or video display. The advantage of a co-processor is that it significantly increases processing speed.
Coaxial cable	One or two transmission wires covered by an insulating layer, a shielding layer, and an outer jacket. Given its high **bandwidth,** coaxial cable can be a **broadband** carrier and can transmit data, voice and video.
CODEC	Coder-Decoder.
	A device that coverts a **digital** signal to an **analog** signal at one end of transmission, and back again to a digital signal at the other end.
Compression ratio	The ratio of the number of **bits** in an original image to that in a compressed version of that image. For example, a compression ratio of 2:1 would correspond to a compressed image with one-half the number of bits of the original.
Computed radiography (CR)	A method of producing **digital** radiographic images using a storage phosphor plate (rather than film) in a cassette. After the plate is exposed, a laser beam scans it to produce the digital data which are then converted to an image.
CPRI	Computer-based Patient Record Institute, Inc.
	An independent institute involved in developing and recommending standards for computerized patient records.
CPU	Central processing unit.
	The device in a computer that executes instructions in software programs and performs calculations and other operations on data.
CRT	Cathode ray tube.
	Refers to the monitor or display device in a computer system.
CT	Computed tomography.
Data compression	Methods to reduce data volume by encoding it in a more efficient manner, thus reducing image processing, transmission times, and storage space requirements.
DDS	Digital data system.
	A system for carrying telephone traffic in **digital** format between major switching hubs. DDS permits digital transmission of voice and data as part of the **analog** telephone system (**POTS**).

Dedicated line	A permanent telephone line owned by, or reserved exclusively for one customer, available 24 hours a day. Dedicated lines generally offer better quality than standard dial-up telephone lines, but may not significantly increase the performance of data communications. Also called "leased" or "private" line.
DICOM	Digital imaging and communications in medicine. A vendor-independent standard for data formats and transfer developed by the American College of Radiology and the National Electronic Manufacturers Association. The standard emphasizes point-to-point connection of **digital** medical imaging devices. DICOM 3.0 is the current version.
Digital camera	An image-forming lens system composed of one or more light-sensitive integrated circuits, a matrix of light sensitive elements, and circuits for timing, nonlinear amplification, and encoding color.
Digital image	An image composed of discrete **pixels,** each of which is characterized by a digitally represented **luminance** level. For example, a common screen size for digital images is a 1024 by 1024 matrix of pixels × 8 bits, representing 256 luminance levels.
Digital signal	An electrical signal in the form of discrete voltage pulses. Digital signals transmit audio, video, and data as **bits,** which are either on or off, in contrast to continuous variable analog signals. Communications signals can be compressed using digital technology, permitting efficient and reliable transmission rotes.
Digitize	The process by which **analog** (continuous) information is converted into **digital** (discrete) information. This process is essential in computer imaging applications because visual information is inherently in analog format and most computers use only digital information.
Direct capture	A process by which image data are acquired directly from a source, such as an image file or a video display console, permitting high quality image reproduction. For example, images produced from image files are identical to the originals, regardless of the device used to produce them (e.g. CT, MRI). In direct video capture, the video signal is digitized from the display, a process that is more efficient and produces better quality images than acquisition through scanning.
Dpi (dots per inch)	Film resolution for digitized images, commonly expressed as dots (**pixels**) per inch. In conventional radiography resolution is given in line pairs per millimeter (lp/mm).
DS0, DS1, DS3	**Digital** telecommunications channels, capable of carrying high volume voice, data or compressed video signals. DS1 and DS3 are also known as **T1** and **T3** carriers.

Transmission rates are:

DS0:	64 Kbps
DS1 (T1):	1.544 Mbps
DS3 (T3):	45 Mbps

Dynamic range The ability of a communications or imaging system to transmit or reproduce a spectrum of information or **brightness** values.

EDI Electronic data interchange.

The transmission of data without paper or human intervention between two devices or applications, using a standard data format.

Ethernet A communications protocol that runs on different types of cable at a rate of **10 Mbps.**

Fiber optic cable Insulated, flexible glass-core cable that uses light pulses instead of electricity to transmit audio, video, and data signals. Fiber optics allows high capacity transmission, at very high speeds, e.g. billions of **bits** per second, with low error rates.

Film alternator A motor driven device that displays multiple films for interpretation and moves them under the control of an operator. An alternator may be thought of as multiple banks of moving view boxes.

Filmless radiology The practice of radiology without film, using devices that acquire **digital** images and related patient information and transmit, store, retrieve, and display them electronically.

Gbps Gigabits per second.

A measure of **bandwidth** and rate of information flow in **digital** transmission.

Gigabyte (Gbyte) A measure of computer memory and storage capacity. One gigabyte equals 1.074 billion bytes or 1,000 Mbytes.

Gray scale Refers to the number of different shades or levels of gray that can be stored and displayed by a computer system. The number of gray levels, or gray scale, is directly related to the number of **bits** used in each **pixel:**

6 bits = 64 gray levels	10 bits = 1024 gray levels
7 bits = 128 gray levels	12 bits = 4096 gray levels
8 bits = 256 gray levels	

Gray-scale monitor A black to white display with varying shades of gray, ranging from several levels to thousands, and thus suitable for use in imaging. Also called a **monochrome monitor.**

HDTV High-definition television.

A television system with 1125 lines of horizontal resolution, capable of producing high quality video images.

HIS Hospital information system.

An integrated computer-based system to store and retrieve patient information. May include or be linked to laboratory and radiology information systems (**LIS and RIS**).

HISPP   Health Care Information Standards Planning Panel.
                     A body established by **ANSI** to coordinate the development
                     of standards by standard-setting organizations in health care.

HL7   Health Level 7.
                     A communications protocol at the application level specify-
                     ing standards for transmission of health-related data. Typi-
                     cally used within a single institution, HL7 permits integration
                     of multiple applications, such as bedside terminals, radio-
                     logical imaging stations, and patient accounting, into one
                     system.

HMO   Health maintenance organization

HPCC   High Performance Computing and Communications. A fed-
                     erally coordinated, interagency program of research and
                     development, established to expedite the introduction and
                     use of the next generation of high performance computer
                     systems.

IEEE   Institute of Electrical and Electronic Engineers

IITF   Information Infrastructure Task Force.
                     Established by the Clinton Administration, the IITF is com-
                     posed of the Federal agencies active in information and tele-
                     communications technology development and application.
                     The mandate of the task force is to articulate and implement
                     a vision for the **National Information Infrastructure (NII)**.

Imaging   Any device or technology used to produce a medical image,
modality   such as **computed tomography**, magnetic resonance, or con-
                     ventional x-ray.

Informatics   The application of computer science and information tech-
                     nologies to the management and processing of data, informa-
                     tion and knowledge.

ISDN   Integrated Services Digital Network.
                     A **digital** telecommunications channel and protocol that per-
                     mits the integrated transmission of voice, video, and data at
                     high speed, and provides connections to a universal **network**.

ISO   International Organization for Standardization.
                     A non-treaty organization, composed of national standard-
                     setting bodies, that is involved in setting norms for all com-
                     munications fields except electrotechnical.

ISO/OSI   International Organization for Standardization/Open Sys-
                     tems Interconnections. The standard reference model for local
                     area **network** (LAN) architecture. The model consists of seven
                     hierarchical layers (physical, data link, network, transport,
                     session, presentation, and application) that address LAN
                     design, from the specification of the physical transmission
                     medium to the capabilities of user interaction with LAN
                     services.

ITU	International Telecommunications Union. A treaty organization composed of government telecommunications agencies involved in setting standards for radio, telegraph, telephone, and television.
ITU-T	International Telecommunications Union Consultative Committee for Telecommunications. Formerly the Consultative Committee on International Telephone and Telegraph. See **CCITT**.
JCAHO	Joint Commission on Accreditation of Healthcare Organizations.
JPEG	Joint Photographic Experts Group. An algorithm and the standard for the **digital** compression of still images.
Kbps	Kilobits per second. A measure of **bandwidth** and rate of information flow in **digital** transmission. One Kbps is 1,024 kilobits per second.
Kilobyte (Kbyte)	A measure of computer storage and memory capacity. One Kbyte equals 1024 **bytes.**
LAN	Local Area Network. A small private **network** of computers with limited reach, such as for a building or campus, linked to allow access and sharing of information and computer resources by users. See also **MAN** and **WAN**.
LIS	Laboratory Information System. An integrated system for processing laboratory data and an essential part of clinical and hospital information systems. See also **HIS** and **RIS**.
Lossless	A form of data compression, usually of an order of less than 2:1, that involves no loss of original digital image information upon reconstruction.
Lossy	A method of data compression at a relatively high ratio, that leads to some permanent loss of information when the image is reproduced.
LSTAT	Life Support for Trauma and Transport.
Luminance	The amount of light given off by an object, expressed in footlamberts. Perceived luminance is denoted as brightness. The terms **luminance** and **brightness**, though often used interchangeably, are not equivalent.
MAN	Metropolitan Area Network. **Network** services within a metropolitan area. MANs can be used to link telemedicine applications at a data rate comparable to **DS1**. In some MANs, cable companies offer links to off-network services such as the internet, airline reservation systems, and commercial information services, in addition to data exchange capabilities. See also **LAN** and **WAN**.

MATMO	Medical Advanced Technology Management Office. An imaging system combining **PACS** and teleradiology networks, developed and implemented by the Department of Defense. See also **MDIS**.
Mbps	Megabits per second. A measure of **bandwidth** and rate of information flow in **digital** transmission. One Mbps equals one million bits per second.
MCU	Multipoint control unit. A centrally located service offered by **switched network** providers to which three or more users can be connected, permitting audio and video teleconferencing.
MDIS	Medical Diagnostic Imaging Support. The precursor to **MATMO**.
MEDIX	Medical Data Interchange Standard. A data communication protocol at the application level developed by the **IEEE**.
Megabyte (Mbyte)	A measure of computer storage and memory capacity. One Mbyte is equal to 1.024 million bytes, 1,024 thousand bytes or 1,024 kbytes.
Mhz	Megahertz. A measure of **bandwidth** and rate of information flow for **analog** transmission. One Mhz equals 10^6 cycles per second.
Microwave link	A communications system using high frequency radio signals (above 800 **megahertz**) for audio, video, and data transmission. Microwave links require line of sight connection between transmission antennas.
Modem	Modulator/De-modulator. A device that converts **digital** signals to pulse tone (**analog**) signals for transmission over telephone lines and reconverts them to digital form at the point of reception.
Monochrome monitor	Another term for a **gray-scale monitor**. A computer display that presents images as different shades of gray from black to white.
MRI	Magnetic resonance imaging. A form of radiographic imaging for the analysis of soft tissue.
Multiplexer	Equipment that permits transmission of multiple lines of audio, video or data information in one high-capacity communications channel, by combining and interweaving low-capacity channels in discrete time or frequency slices.
Narrowband	A telecommunications medium that uses low frequency signals, generally up to 1.544 **Mbps**.
NCHSR	National Center for Health Services Research. Former name of the Agency for Health Care Policy and Research (**AHCPR**).
NEMA	National Electrical Manufacturers Association

Network	Interconnected telecommunications equipment used for data and information exchange. Three common network types are Local Area Networks (**LAN**), Metropolitan Area Networks (**MAN**), and Wide Area Networks (**WAN**).
NHSC	National Health Service Corps program
NII	National Information Infrastructure. A U.S. government policy initiated by the Clinton Administration. See **NTIA**.
NIMH	National Institute of Mental Health
NLM	National Library of Medicine
NTIA	National Telecommunications and Information Administration. The federal agency in the Department of Commerce responsible for the National Information Infrastructure initiative. See **NII**.
NTSC	National Television System Committee. An independent body that sets standards for broadcast television in the U.S. The NTSC standard is also known as "composite video," because all video information is combined into one **analog** signal.
Optical disk	A computer storage disk used primarily for large amounts (**Gbytes**) of data.
OSI	Open Systems Interconnect. The standard reference model for local area network (**LAN**) architecture. See **ISO/OSI**.
OTA	Office of Technology Assessment. Agency of the U.S. Congress; terminated in 1996.
Packet	A basic message unit for communications in networks, that includes data, call control signals and error control information.
Packet network	A network that parcels out data bits in packets.
Packet switching network	Also called packet switched network (PSN), the term refers to the transmission of **digital** information using addressed packets that travel along different routes in a network. Packet switching is more efficient than **modem** transmission, because the channel is occupied only during **packet** transmission, rather than throughout the transmission. See also **asynchronous transmission mode (ATM)**.
PACS	Picture Archiving and Communication System: A system that acquires, transmits, stores (archives), retrieves (from storage), and displays **digital** images and related patient information from a variety of **imaging modalities**. Other capabilities include image processing, linkage to radiology and hospital information systems (see **RIS, HIS**), and use of alternative methods of information input/output, such as speech recognition systems.

PBX   Private Branch Exchange.
                        A computerized private telephone exchange, or switchboard, with an expanded range of data and voice services. A PBX typically serves one organization and is connected to the public telephone **network.**

PCM   Pulse code modulation.
                        A method to encode audio signals.

Peripheral   A device that is externally attached to a computer. Examples include scanners, mouse pointers, printers, keyboards and monitors.

PET   Positron emission tomography

Phosphor   The coating on the inside of a cathode ray tube (**CRT**) or monitor that produces light when struck by an electron beam.

Pixel   Stands for picture element, the smallest piece of information that can be displayed on a **CRT.** Represented by a numerical code in the computer, pixels appear on the monitor as dots of specific color or intensity. Images are composed of many, many pixels.

POTS   Plain Old Telephone Service

PPO   Preferred provider organizations

PPRC   Physician Payment Review Commission

RAM   Random Access Memory.
                        A computer's temporary memory space where programs are run, images are processed, and information is stored. The amount of required RAM varies according to the application. Information stored in the RAM is lost when the power is shut off.

Real time   The capture, processing, and presentation of data, audio, and/or video signals at the time that data are originated on one end and received on the other end. When frames containing such data are transmitted at rates of 30 per second, real time is achieved.

RIS   Radiology Information System.
                        An integrated system for the electronic processing, storage and transmission of radiographic images. RIS permit the remote interpretation of radiographic images— teleradiology— and can be connected to Hospital Information Systems (**HIS**) and Laboratory Information Systems (**LIS**).

RMDS   Navy Remote Medical Diagnosis System

ROC   Receiver operating characteristic. A statistical method to assess the ability of a diagnostic tool to discriminate between healthy and diseased persons. ROC curves are used frequently in observer performance evaluations of the feasibility and performance of diagnostic imaging systems.

ROM	Read Only Memory.
	The permanent memory space of a computer. Programs and information stored in ROM are not lost when the power is shutdown.
RSNA	Radiological Society of North America.
	A professional organization involved in setting practice and telecommunications standards in health care.
SAF	Store and forward.
	A type of telemedicine interaction that produces a multimedia electronic medical record. Data and images are captured and stored for later transmission, consultation or downloading.
Satellite connections	A communications system that uses radio signals sent to and from a satellite orbiting the Earth. This mode of transmission permits connections between points at great distance from each other on the Earth's surface, between which direct transmission is difficult, as well as to remote areas lacking cables for telephone lines.
SCSI	Small Computer Systems Interface.
	An interface protocol used to link **peripherals** such as disk drives, scanners, and tape back-up units, to computers. Note: SCSI is known as "scuzzy".
SDM	Shared decision making.
	A participatory style of decision-making in health care that involves patients in treatment decisions, by offering them greater control over the choice of treatment, and thereby giving them a greater sense of responsibility for their care and health. Also known as shared decision programs (SDP).
Slow-scan	Refers to the speed of still video image transmission, which is usually over narrow communications channels such as standard telephone lines.
SONET	Synchronous optical networks.
	A **broadband,** wide area communications service capable of transmitting very high capacity data, such as interactive video, at very high speeds of 150 **Mbps** to 10 **Gbps**. SONET services are useful for **real-time digital** telemedicine applications.
Spatial resolution	Property of distinguishing two equal sized adjacent objects in the same place. Refers to the number of **pixels** in a specified area of a matrix.
SPECT	Single-photon emission computed tomography.
Switched network	A telecommunications system where each user has a unique address and any two points can be connected directly, using any combination of available routes in the **network**.
Switched service	A telecommunications service, usually based on telephone technology, that changes (switches) circuits to connect two or more points.

T-carrier   A time division, **multiplexed digital** transmission facility, usually provided by a phone company and operating at an aggregate rate of 1.544 **Mbps** and higher.

T1, T3   A **digital** transmission system for high volume voice, data, or compressed video traffic. T1 and T3 transmission rates are 1.544 **Mbps** and 44.736 **Mbps**, respectively. Also known as **DS1, DS3**.

Tbps   Terabits per second.
A measure of **bandwidth** and rate of information flow in **digital** transmission. One **Tbps** equals one trillion bits per second.

TCP/IP   Transmission Control Protocol/Internet Protocol.
The underlying communications protocols that permit computers to interact with each other on the Internet.

TDM   Time division multiplexing. Transmission of multiple lines of information in one high-capacity communications channel using time as the means to separate channels.

Telehealth   The electronic transfer of health information from one location to another for purposes of preventive medicine, health promotion, diagnosis, consultation, education, and/or therapy. Although telehealth is sometimes considered broader in scope than telemedicine, there is no clear-cut distinction between the two.

Telemetry   The science and technology of automatic measurement and transmission of data by wire, radio, or other means from remote sources to receiving stations for recording and analysis.

Terabyte (Tbyte)   A measure of computer memory and storage capacity. One terabyte equals one trillion bytes, or ten to the twelfth power **bytes**. See also **gigabyte, kilobyte,** and **megabyte**.

Throughput   The amount of data actually transmitted in a given period of time, expressed in **bits** per second. Throughput rates are related to **baud** rates, but are usually a little lower because transmission conditions are never ideal. Typically, higher baud rates will allow higher throughput.

UHF   Ultrahigh frequency.
A radio frequency in the second highest range of the radio spectrum, from 300 to 3,000 **MHz**.

VGA   Videographics array.
A measure of image size, indicating the capacity to display 640 × 480 lines, such as on viewing monitors for personal computers.

VHF   Very high frequency.
A radio frequency in the very high range of the radio spectrum, from 30 to 300 **MHz**.

Voxel	Volume element.
	A voxel is a three-dimensional version of a **pixel** and is generated by computer-based imaging systems, such as **CT** or **MRI**.
VRAM	Video random access memory
VTC	Video teleconferencing.
WAN	Wide Area Network.
	A data communication network that links together distinct networks and their computers over large geographical areas. WANs are typically larger than metropolitan area networks. See also **MAN** and **LAN**.
Window width	The range of the **gray scale** of the image appearing on a screen. The middle value is the "window level."
WRAIR	Walter Reed Army Institute for Research
WWW	World Wide Web (the "Web"). An internet information resource through which information-producing sites offer hyperlinked multimedia information to users.

Sources: This glossary was developed from multiple sources and includes primarily terms and acronyms used in this volume. Readers interested in a more comprehensive glossary are referred to the on-line **Telemedicine Glossary**, compiled by Jonathan Smith and published by the Health Science Center-Syracuse, State University of New York, 1995. The World Wide Web site address is:

http://www.hscsyr.edu/~wwwserv/telemedicine/glossary.html.

NAME INDEX

421

SUBJECT INDEX